T0229167

Electrical Machines and Drive Systems

Third Edition

John Hindmarsh BSc (Eng), CEng, MIEE
Formerly Senior Lecturer, University of Manchester Institute of Science and Technology

Alasdair Renfrew BSc, PhD, CEng, MIEE
Senior Lecturer, University of Manchester Institute of Science and Technology

Newnes
OXFORD AMSTERDAM BOSTON LONDON NEW YORK PARIS
SAN DIEGO SAN FRANCISCO SINGAPORE SYDNEY TOKYO

Ad Maiorem Dei Gloriam

Newnes
An imprint of Elsevier Science
Linacre House, Jordan Hill, Oxford OX2 8DP
225 Wildwood Avenue, Woburn MA 01801-2041

First published 1982
Second edition 1985
Third edition 1996
Reprinted 1999
Transferred to digital printing 2002

British Library Cataloguing in Publication Data
A catalogue record for this book is available from the British Library

Library of Congress Cataloguing in Publication Data
A catalogue record for this book is available from the Library of Congress

ISBN 0 7506 2724 7

Printed and bound in Great Britain by Antony Rowe Ltd, Eastbourne

Contents

Authors' preface to the Third Edition

This book commenced its life following a suggestion that there was a need for practical assistance to be given in the subject through the medium of illustrated examples, to enable students to get to grips with machine theory. This still remains a major purpose of the book and the comprehensive coverage of both worked and tutorial examples, from fairly basic to quite advanced level, should be of assistance in giving deeper understanding, quite apart from meeting the needs of the two parties on opposite sides of the examination system and providing a handy reference volume on output calculations for recent graduates in industry and postgraduate studies.

From the practical viewpoint, too, as in many engineering disciplines, machines are more and more regarded as only one of many elements in a much larger system which, in so many cases, demands knowledge of the techniques needed to control this system and how the machine interacts with these. A major part of this system is the incorporation of fast, heavy-duty electronic switches – the power-electronic drive is now almost considered as a system in itself. The second edition placed more emphasis on this and included more detail on the simulation of such 'elemental systems' for greater understanding.

However, events have moved on further in that the micro-electronic revolution has permitted previously unachievable aims; for example, to make the cheap, rugged cage-rotor induction motor act, in the transient state at least, like the more expensive and more simply controlled d.c. machine and thereby replace the latter's previous position as the fast-response, heavy-overload-duty variable-speed drive.

The third edition of this book has been occasioned by this increasing pace of change in electrical drives as a result of the foregoing developments. Announcement of 3.3 kV IGBTs has brought compact, fast-switching electronic devices into the fields of heavy-rail traction and industrial drives in the MW range and the ability of the modern drive to incorporate microprocessor-based modelling has produced substantial development in control theory for electrical drives.

The revisions to the text include enhanced coverage of power-electronic systems and new material on closed-loop control systems as part of the chapter on Transient Response. A substantial section of the following chapter attempts a simplified explanation of the basis of Field-oriented (vector) control of cage-rotor induction motors, giving equations from which torque and flux can be estimated and thereby, through a fast current-control system, manipulated to copy the torque-efficient mechanism of the d.c. machine. It includes a description of how these increasingly common drives operate, from both control and electromagnetic viewpoints. The authors hope that this section will be particularly useful to undergraduate and postgraduate students and academic staff seeking to understand 'vector-control' using only the bases of simple algebra and electrical circuits.

New examples have been added to accompany the additions and revisions to the text and there is increasing emphasis on the electrical machine as part of a system which includes a motor, sensors, control equipment and power-electronic converters. The book retains its original objectives; it is still possible, using the worked examples, for the engineer or engineering student to analyse the electromechanical performance of an electrical drive system with reasonable accuracy, using only a pocket calculator.

J. H.
A. C. R.
Sale, Cheshire

Author's preface to the Second Edition

The response to the first edition included a suggestion that the final section on simulation was somewhat limited in scope. This topic really requires a separate text to deal in depth with such a wide-ranging subject. Nevertheless, an attempt has been made in the present edition to augment the introductory treatment so that with sufficient interest, extensive simulations could be undertaken on the basis of the material given. Although this is really tending towards project or even long-term postgraduate work, requiring substantial computer-program development, its immediate importance is in the facility it provides to display computed performance, especially of power-electronic/machine circuits, which should give a better understanding of their special features. The author gratefully acknowledges permission to use the computational facilities at UMIST for this purpose.

There are other changes in the text to include reference for example to unbalanced operation, permanent-magnet machines and the universal motor. Also, the additional worked examples and tutorial examples should ensure that the rather wide topic of Electrical Drives is given sufficiently comprehensive coverage. Overall, the intention is to support the suggestion of teaching the subject by means of worked examples, after due preparation on the basic equations and with discussion of the problems and solutions.

J. H.
Sale, Cheshire
March 1985

Author's preface to the First Edition

Discussion of the summary and objective of this book is deferred to the first section of Chapter 1 lest this preface, like many others, goes unread. However, there are some matters which need to be covered here, especially those which concern the author's debt to others. Every author likes to believe that he has created something original or shed new light on an old topic. But, inevitably, much of the work must have been derived from his experiences as a student, from his own teachers, from books read, from his own students and from his colleagues. This last is particularly true in the present case. A major influence has been the author's long association with Dr. N. N. Hancock, whose deep perception of the subject and ever-sympathetic ear when difficulties of understanding arose have been a source of constant sustenance. To Dr. A. C. Williamson, with his remarkable facility for cutting clean through the theoretical fog to grasp the essential nature of tricky machines problems, especially those associated with power-electronic circuits, I am deeply indebted. Much time was spent by him in kindly checking and correcting formative ideas for the material on constant-current and variable-frequency drives. Sections 4.3, 5.5, 7.3 and 7.4 are heavily reliant on his contributions. Dr. B. J. Chalmers' experience, with saturated a.c. machines particularly, was very important to the clarification of this section of the work. I am grateful, too, to Dr. M. Lockwood for his suggestions on simplifying simulation problems.

With regard to the examples themselves, an attempt has been made to cover thoroughly the basic machine types, but the subject is very wide. Very small and special machines have been omitted, quite apart from study of the economic and environmental factors which influence the decision in choosing an electrical drive, though some brief comments are made as appropriate. This selective treatment seemed to be the best way of meeting the many requests in response to which this book was written. Many of the examples are taken, or modified, from examination papers set at the University of Manchester Institute of Science and Technology (UMIST), and permission to publish these is gratefully acknowledged. Other examples, especially those in the Appendix, are drawn from a variety of sources. The author cannot deny that some of them may have originated in form from other books read over the years since first meeting the topic as a student, and from discussions and contact with present and past colleagues.

Finally, the author would like to record his thanks to the Consulting Editor, Professor Percy Hammond, for his encouragement, for reading the text and making his usual perceptive comments and suggestions to get the balance right. To the Managing Editor, Mr. Jim Gilgunn-Jones, and his colleagues at Pergamon Press, who have been so patient in spite of delays and last-minute changes, I tender my grateful appreciation.

J. H.
Sale, Cheshire
August 1981

Symbols

The following list comprises those symbols which are used fairly frequently throughout the text. Other symbols which are confined to certain sections of the book and those which are in general use are not included, e.g. the circuit symbols like R for resistance and the use of A, B and C for 3-phase quantities. Some symbols are used for more than one quantity as indicated in the list. With few exceptions, the symbols conform to those recommended by the British Standards Institution BS 1991.

Instantaneous values are given small letters, e.g. e, i, for e.m.f. and current respectively.

R.M.S. and steady d.c. values are given capital letters, e.g. E, I.

Maximum values are written thus: $\hat{E}$, $\hat{I}$.

Bold face type is used for phasor and vector quantities and for matrices, e.g. **E**, **I**. In general, the symbol E (e) is used for induced e.m.f.s due to mutual flux and the symbol V (v) is used for terminal voltages.

At — Ampere turns.

B — Flux density, in teslas (T) (webers/metre^2).

d — Symbol for direct-axis quantities.

d — Armature diameter, in metres.

e — Base of natural logarithms.

E_f — Induced e.m.f. due to field m.m.f. F_f.

f — Frequency, in hertz (Hz) (cycles per second).

F — Magnetomotive force (m.m.f.) in ampere turns. Peak m.m.f. per pole per phase.

F'_a — Effective d.c. armature-winding magnetising m.m.f. per pole.

F_a — Peak armature-winding m.m.f. per pole.

F_f — Peak field-winding m.m.f. per pole.

(Note that the suffices a and f are also used with the symbols for currents, fluxes and resistances of armature and field respectively.)

F_r Peak resultant m.m.f. per pole.
$I_{f.l.}$ Full-load current.
I_0 Current in magnetising branch.
I_m Reactive or magnetising component of I_0.
I_p Power component of I_0.
J Polar moment of inertia (rotational inertia), in kgm^2.
k Coefficient. A constant.
k_f Generated volts per field ampere or per unit of m.m.f.
k_{fs} Saturated value of k_f.
k_ϕ Flux factor = generated volts per radian/sec or torque per ampere.
l Conductor length. Magnetic path length.
l (or l_1, l_2, etc.). Leakage inductance.
L General inductance symbol; e.g. L_{11} = self-inductance of coil 1; L_{12}, L_{13}, etc., for mutual inductances.
m Number of phases.
M Alternative mutual-inductance symbol.
n Rev/sec.
n_s Rev/sec synchronous = f/p.
N Number of turns. Rev/min.
N_s Rev/min synchronous = $60f/p$.
p Operator d/dt.
p Number of pole pairs.
p.u. Suffix for *per-unit* quantities.
P Power.
$P_{control}$ Power at control terminals.
P_{elec} Power (total) at electrical terminals (P_e = per phase).
P_{gap} Air-gap power (total) (P_g = per phase).
P_m Mechanical power converted (per phase) ($m \cdot P_m = \omega_m \cdot T_e$).
P_{mech} Power at mechanical terminals ($\omega_m \cdot T_{coupling} = P_{coupling}$).
q Symbol for quadrature-axis quantities.
R_m Magnetising resistance, representing iron losses.
s Fractional slip = $(n_s - n)/n_s$.
S *Per-unit* relative motion n/n_s (= $1 - s$).
$T_{coupling}$ Torque at mechanical shaft coupling.
T_{loss} Sum of all machine internal loss torques.
T_e Torque developed electromagnetically, in newton metres.
T_m Torque arising mechanically = T_e in steady state.
v Velocity, in metres per second.
V Voltage measured at the terminals of a circuit or machine.
x (or x_1, x_2, x_{al}, etc.). Leakage reactance.
X General reactance symbol.
X_m Magnetising reactance.
X_{ms} Saturated value of $X_m \cdot X_{mu}$. Unsaturated value of X_m.

X_s	Synchronous reactance = $X_m + x_{al}$.
z_s	Number of series-connected conductors per phase or per parallel path of a winding.
Z_s	Synchronous impedance.
α	General angle. Slot angle. Impedance angle $\tan^{-1} R/X$. Firing-delay angle. Abbreviation for 'proportional to'.
δ	Load angle. Chopper duty-cycle ratio or modulation factor.
δ_{fa}	(or δ_T) Torque angle.
η	Efficiency.
Λ	Magnetic permeance, webers/ampere-turn.
μ_0	Magnetic constant = $4\pi/10^7$.
μ_r	Relative permeability.
μ	Absolute permeability = $B/H = \mu_0\mu_r$.
φ	Power-factor angle. N.B. This must be distinguished from the symbol for flux ϕ below.
ϕ	Instantaneous value of flux. Flux per pole, in webers.
ϕ_m	Mutual flux, in webers, due to resultant m.m.f.
Φ	Flux time-phasor.
θ	Shaft angular position in electrical radians (θ_m mechanical angle). Temperature rise. General variable.
$\dot{\theta}$	Shaft speed in electrical rad/s.
τ	Time constant.
ω	Angular velocity of rotating time-phasors = $2\pi f$ radians/sec.
ω_m	Mechanical angular rotational velocity = $2\pi n$ radians/sec.
ω_s	Synchronous angular velocity = $2\pi n_s = 2\pi f/p$ radians/sec.

Note: SI units (Systeme International d'Unites) are used in the text unless specifically stated otherwise, as in the case of *per-unit* quantities and the so-called Engineers' units; i.e. 1 hp = 746 watts; 1 lbf ft = (746/550) Nm.

1 Introduction and review of basic theory

1.1 Aim of the book

On entering the world of electrical machines, the student meets many conceptual difficulties not experienced for example in the early studies of digital systems, with their simple and precise 2-state operation. More assistance is required to permit the new-comer to gain confidence in dealing with non-linear, 3-dimensional, rotating electromagnetic devices. The purpose of this book is to provide this aid to understanding by showing how, with a limited number of equations derived from basic considerations of power flow and elementary circuit and electromagnetic theory, the electromechanical performance can be explained and predicted with reasonable accuracy.

Such an aim, which will permit the calculation of power-input/output characteristics almost close enough in engineering terms to those of the device itself, can be achieved by representing the machine as a simple electrical circuit – the equivalent-circuit model. This concept is explained in many books, for example in the author's companion volume *Electrical Machines and Their Applications*. Though more detailed theoretical treatment is given there, substantial portions of the present text may be regarded as suitable revision material. This expanded 3rd edition can, as a whole, be considered as a textbook with particular, but not exclusive, emphasis on Electrical Drives, taught through worked examples, for a reader having some familiarity with basic machine theory.

Perhaps it is appropriate to point out that complete and exact analysis of machine performance is so complex as to be virtually impossible. The additional accuracy achieved by attempts to approach such methods is primarily of interest to the specialist designer who must ensure that his product will meet the user's needs without breakdown and he must judge when the analytical complication is justified. For the user, and for the engineering student who is not yet a specialist, the simpler methods are adequate for general understanding and provide a lead-in if necessary for later specialisation.

There are many features of all machine types which are common, the obvious example being the mechanical shaft equations. But apart from these and the fundamental electromagnetic laws, the input/output relationships and modes of operation have many similarities. These are brought together where possible and also in this first chapter, some elementary mechanical, magnetic and circuit theory is discussed briefly, as a reminder of the basic knowledge required. Students should beware of underestimating the vital importance of this material, since experience shows that it is these very points, improperly understood, which hold back progress in coming to feel at ease with machines problems.

However familiar one may become with theory, as a student, the true test of an engineer is his ability to make things work. First steps to this goal of confidence are reached when a student is prepared to commit himself to selecting equations and inserting values in the algebraic expressions, producing answers to a specific problem. Hence the importance of practice with numerical examples. Understanding grows in proportion to one's ability to realise that the equations developed really can be used in a systematic fashion to solve such problems, since they describe the physical behaviour in mathematical terms. Appreciation of this last statement is the key to successful problem-solving.

The chapters are planned to sequence the examples at increasing levels of difficulty. Much theoretical support is given, in that the equations are discussed either at the beginning of each chapter, or as the need arises. Solution programmes indicate the kind of problems which can be formulated for the three basic types of rotating machine: d.c., induction, and synchronous. Readers are encouraged to adopt an ordered approach to the solution; for example it is a good idea to incorporate the question data on a diagram. One of the difficulties of machines problems often lies in the amount of data given. By putting the values on a simple diagram, assimilation is easier and it helps to avoid mistakes of interpretation, especially when working with 3-phase circuits. In following this recommended pattern, it is hoped that the text will help to remove the mystery with which some students feel the machines area is shrouded.

The emphasis is on machine terminal-characteristics, rather than on the internal electromagnetic design. In other words, the electrical-drives aspect is uppermost since this is the area in which most engineering students need to have some good knowledge. It is worth noting that about 60–70% of all electrical power is consumed by motors driving mechanical shafts and virtually all this power is produced by generators driven through mechanical shafts, so that the subject is of considerable importance to engineers. The problems and solutions are discussed where appropriate, to draw out the engineering implications. Electromechanical transients, stability and control are not neglected and opportunity is also taken to consider the effects introduced by the impact of power-electronic circuits, so often intimately associated with machine control. In general, the usual methods of analysis are still reasonably effective in predicting machine performance. Full account of the influence of this important environment, in which harmonics proliferate, is a somewhat specialised topic but some indication is given of the means used to deal with the machines problems which arise. Detailed study of machine/semiconductor systems requires the use of mathematical and computer simulation procedures, which have tended to become the province of those who market commercial computer software packages. However, Chapter 8 considers this topic in sufficient depth to provide a better understanding of such investigations. Finally, in Appendix D, some tutorial examples are given along with the answers. Some of the worked examples in the text have been taken from Appendix E of *Electrical Machines and Their Applications*, but many of these remain, as further exercises for the determined student.

1.2 Foundation theory

Excitation calculations

Virtually all machines have iron in the magnetic circuit to enhance the flux value and/or to reduce the excitation requirements. The price to pay for these advantages is reflected in iron loss and non-linearity. Figure 1.1a shows a typical iron magnetisation-characteristic. The economic operating point is beyond the linear region and well clear of full saturation, at about $B = 1$ tesla, though certain short parts of the magnetic circuit, like armature teeth, may exceed this by 50% or more. Under transient conditions too, this limit can be exceeded. The equation governing the excitation requirements follows from:

$$B = \mu_0\mu_r H = \mu H = \mu IN/l.$$

Multiplying by area A:

$$B \times A = \mu \frac{IN}{l} \times A = IN \times \frac{\mu A}{l}.$$

In words:

Flux	=	Magnetomotive force	×	Permeance (or 1/Reluctance)
$\phi\ (= BA)$	=	$F(= IN)$	×	$\Lambda(= \mu A/l)$

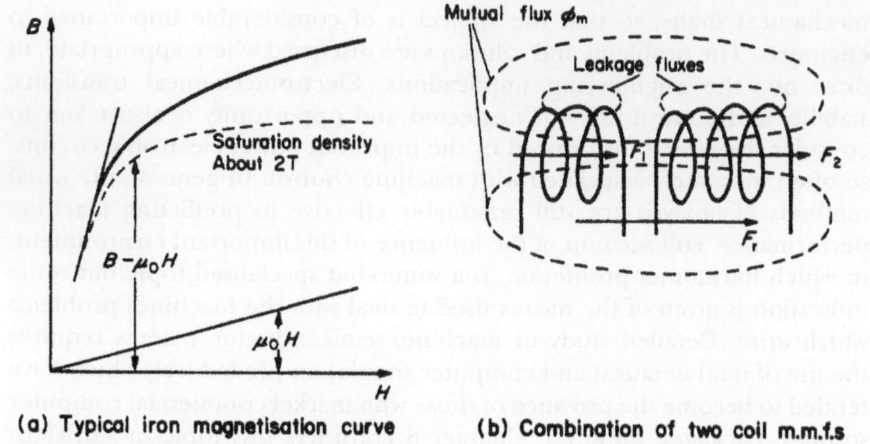

Figure 1.1 *Magnetic excitation.*

The m.m.f. is shown in ampere turns (At) (turns N being dimensionless) but is effectively the current enclosing the magnetic circuit.

The calculation of excitation m.m.f. (F) is often required for a given flux and magnetic geometry, to determine the design of the coils. Frequently there are two (or more) such coils so that the resultant excitation F_r is the combination of F_1 and F_2 which produces ϕ_m, see Figure 1.1b. The two m.m.f.s may be produced on opposite sides of a machine air gap; F_1 say, due to several stator coils, while F_2 similarly may be due to several rotor coils. Often, sinusoidal distribution of m.m.f. is assumed and the coils can be designed to approach this closely. 'Vector' techniques can then be used to combine these two 'sinusoidal quantities' giving $\mathbf{F}_1 + \mathbf{F}_2 = \mathbf{F}_r$ and ϕ_m, the mutual flux = function (F_r). However, m.m.f. is not a vector but a scalar, so a different term, space phasor, is becoming accepted as an appropriate designation for such representations of sinusoidal space variations. It is sometimes convenient to take the positive magnetising senses of $\mathbf{F}_1$ and $\mathbf{F}_2$ to be in the same direction, though in practice, the one is usually magnetising in the opposite sense to the other and would then be negative with respect to this.

Electromagnetic theory

The most important equations for present purposes are:

$$e = N\,\mathrm{d}\phi/\mathrm{d}t; \quad e = Blv; \quad \text{and} \quad Force = Bli;$$

most practical machines having the directions of B, v and i at right angles to one another.

For a fixed magnetic geometry:

$$e = N\frac{d\phi}{di}\frac{di}{dt} = L\frac{di}{dt}$$

where:

$$L = N\frac{d\phi}{di} = N\frac{iN\Lambda}{i} = N^2\Lambda$$

and will fall with the onset of saturation, so the inductance L is flux/current dependent. For a sinusoidally time-varying current:

$$i = \hat{I}\sin 2\pi ft = \hat{I}\sin\omega t,$$

then:

$$e = L \times \omega\hat{I}\sin(\omega t + 90°)$$

and in r.m.s. and complex-number expressions:

$$\mathbf{E} = j\omega L\mathbf{I} = jX\mathbf{I} = \mathbf{V},$$

and **I** lags **V** by 90°. These quantities are scalars but their sinusoidal variation can be represented by time phasors, see Figure 1.2. The word phasor alone will often be used in the text as an abbreviation for time

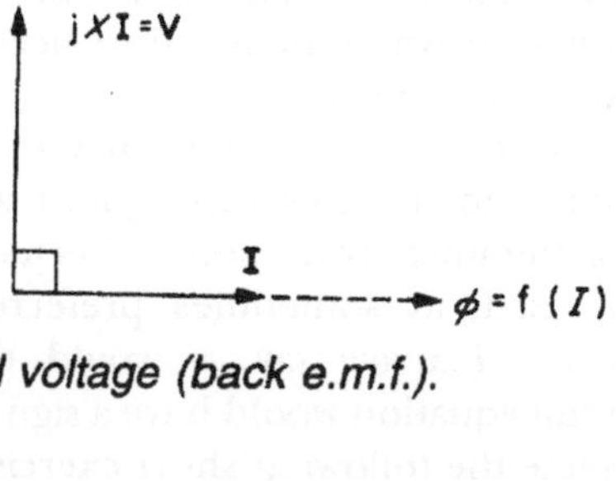

Figure 1.2 *Induced voltage (back e.m.f.).*

phasor. The use of the back e.m.f. expression ($+L\,di/dt$) instead of the forward e.m.f. expression ($-L\,di/dt$) is seen to be preferable, since the current **I** comes out directly as lagging **V** by 90° for the inductive circuit, instead of having to deal with the concept of two identical but phase-opposed voltages.

For the general case with varying geometry, $e = d(Li)/dt = L\,di/dt$ (transformer voltage) + $i\,dL/dt$ (motional voltage).

Circuit-theory conventions

Figure 1.3a shows a representation of a machine with its instantaneous e.m.f. and resistive and inductive voltage-drops. The voltage arrowheads

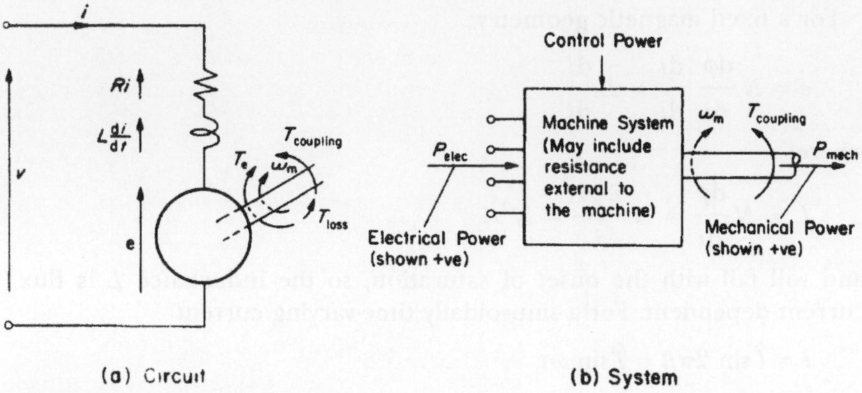

Figure 1.3 *Motor conventions.*

are the assumed +ve ends. The directions of the arrows for the instantaneous terminal voltage v and for e may be assigned arbitrarily but Ri and $L\,di/dt$ must oppose i, since the voltage arrowheads must be positive for +ve i and +ve di/dt respectively. The direction of i may also be assigned arbitrarily but the decision has consequences when related to the v and/or e arrows. As shown, and with all quantities assumed to be +ve, then the machine is a power sink; i.e. in a MOTORING mode; the vi and ei products are both positive. For GENERATING, when the machine becomes a power source, ei will then be negative; e or i reversed.

The above is called the MOTORING convention and it is often convenient in electrical-drives studies to use this throughout and let a negative ei product indicate a generating condition. Alternatively, a GENERATING convention could be used, as sometimes preferred in power-systems studies. By reversing the i arrow say, ei would then be positive for generating and the circuit equation would have a sign reversed. It would be a good check to complete the following short exercise to see if the above statements are properly understood.

Write down the MOTOR equation;	with MOTOR conventions:	$V = E \quad RI$
Write down the GENERATOR equation;	with MOTOR conventions:	$V = E \quad RI$
Write down the GENERATOR equation;	with GENERATOR conventions:	$V = E \quad RI$
Write down the MOTOR equation;	with GENERATOR conventions:	$V = E \quad RI$

The mechanical equation can be expressed as a simple extension of the above. The motor (as a mechanical power source) produces (generates) an electromagnetic torque T_e, and in equilibrium at steady speed, this is balanced by the total mechanical torque T_m, part of which is due to the internal mechanical resistance T_{loss} and the remainder is the load torque at the coupling 'terminals', $T_{coupling}$.

So: $T_e = T_m = T_{coupling} + T_{loss}$ (cf. electrical source, $E = V + RI$).

This is also a MOTORING convention. For a generator, with rotation unchanged, both T_e and $T_{coupling}$ would be negative using this convention.

To illustrate how these conventions affect the machine considered as a system, with electrical-power terminals and mechanical-power 'terminals' – excluding for the moment the control-power terminals – consider Figure 1.3b. In general, either or both terminal powers can be negative and here, a motoring convention is being considered. The three practicable conditions are:

Electrical power positive;	Mechanical power positive;	MOTORING	(A)
Electrical power; negative;	Mechanical power negative;	GENERATING	(B)
Electrical power positive;	Mechanical power negative;	BRAKING	(C)

In the last mode, it will be noticed that both mechanical and electrical 'terminals' are accepting power into the machine system. All the power is in fact being dissipated within this system, which may include resistance external to the machine itself. The mechanical power is usually coming from energy stored in the moving parts, and since this cannot be released without a fall of speed, the action is one of braking. The machine is generating; not feeding power into the electrical supply, but assisting this to provide the power dissipated; see Section 3.5.

To understand how the mechanical 'terminals' respond to these three modes, assume that T_{loss} is 1 unit and T_e is 10 units. Let the speed be positive and remembering that power is (torque × speed), use the mechanical balance equation to find:

$$T_{coupling} = T_e - T_{loss}$$

Mode A; Motoring $T_{coupling} = 10 - 1 = +9.$ $\omega_m T_{coupling}$ +ve.

Mode B; Generating $T_{coupling} = -10 - 1 = -11$. $\omega_m T_{coupling}$ –ve.
[T_e will be –ve for +ve ω_m]

Mode C; Braking (i) $T_{coupling} = -10 - 1 = -11$. $\omega_m T_{coupling}$ –ve.
[ω_m +ve. ∴ $T_{loss} = +1$
T_e will be –ve]

Mode C; Braking (ii) $T_{coupling} = +10 - (-1) = +11$. $\omega_m T_{coupling}$ –ve.
[ω_m –ve. ∴ $T_{loss} = -1$
T_e will be +ve]

Note that if rotation reverses, T_{loss} will reverse because it always opposes rotation. In mode C, the sign of T_e is opposite to that of ω_m because the machine itself is generating, so for either rotation, the mechanical 'terminal' power, $\omega_m T_{coupling}$, is negative.

Sinusoidal a.c. theory

Most a.c. sources are of nominally constant r.m.s. voltage so the voltage phasor is taken as the reference phasor. It need not be horizontal and can be drawn in any angular position. A lagging power factor cos φ means *current* lagging the voltage as shown on Figure 1.4a. The instantaneous power vi, which pulsates at double frequency, is also shown and has a mean value of VI cos φ. If φ were to be greater than 90°, the power flow would have reversed since I cos φ would be negative as seen on the phasor diagram for a current I'. Note that the phasor diagrams have been drawn at a time $\omega t = \pi/2$ for a voltage expressed as $v = \hat{V} \sin \omega t$.

For the reverse power-flow condition, if the opposite convention had been chosen (with v or i reversed), then VI' cos φ would have been positive. This is shown on Figure 1.4b where it will be noted that the current is at a leading power factor. Taking Figure 1.4a as a motoring condition, it shows electrical power being *absorbed* at lagging power factor

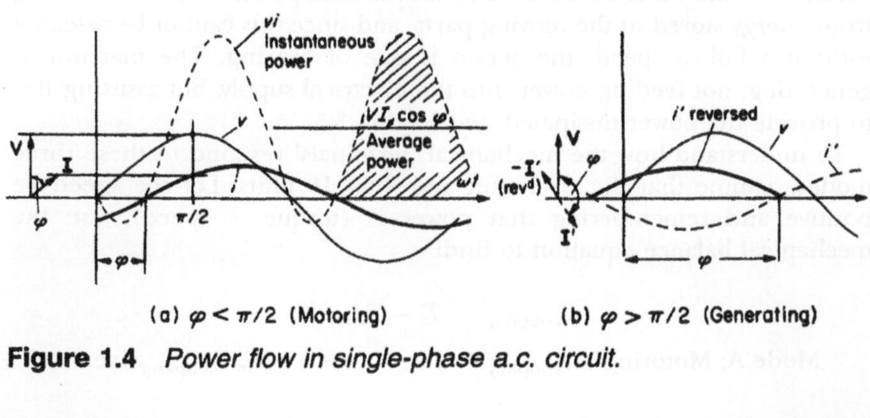

Figure 1.4 *Power flow in single-phase a.c. circuit.*

whereas Figure 1.4b shows electrical power being *delivered* at a leading power factor.

Phasor diagram including machine e.m.f.; motoring condition

The equation, allowing for inductive impedance, is:

$$\mathbf{V} = \mathbf{E} + R\mathbf{I} + \mathrm{j}X\mathbf{I},$$

and is shown as a phasor diagram on Figure 1.5 for two different values of *E*. Note that on a.c., the e.m.f. may be greater than the terminal voltage *V* and yet the machine may still operate as a motor. The power factor is

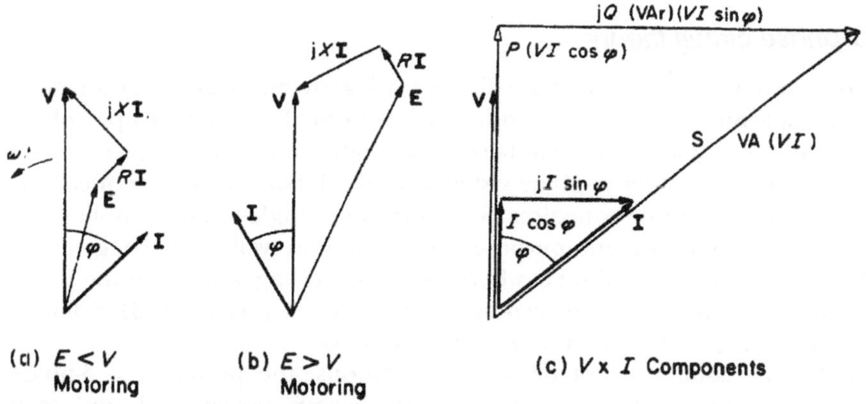

Figure 1.5 *Phasor diagrams.*

affected but the power flow is determined by the *phase* of **E** with respect to **V**. Frequently, the current is the unknown and this is found by rearranging the equation as:

$$\mathbf{I} = \frac{\mathbf{V} - \mathbf{E}}{R + \mathrm{j}X} = \frac{(\mathbf{V} - \mathbf{E})\ (R - \mathrm{j}X)}{R^2 + X^2} = \frac{|\mathbf{V} - \mathbf{E}|}{Z}\left[\frac{R}{Z} - \frac{\mathrm{j}X}{Z}\right]$$

$$= I(\cos\varphi + \mathrm{j}\sin\varphi)$$

N.B. φ will normally be taken as –ve for lagging power factor.

The appropriate exercise to check that these phasor diagrams are understood is to draw the corresponding diagrams for a generator using (a) motor conventions and (b) generator conventions.

Meaning of V × I components

Multiplying the I, $I\cos\varphi$ and $I\sin\varphi$ current phasors by V gives:

VI (S voltamperes, VA)

$VI\cos\varphi$ (P watts) and

$VI\sin\varphi$ (Q voltamperes reactive, VAr)

and a 'power phasor diagram' can be drawn as shown on Figure 1.5c.

Power devices are frequently very large and the units kVA, kW and kVAr, ($\times 10^3$), and MVA, MW and MVAr, ($\times 10^6$), are in common use. The largest single-unit steam-turbine generators for power stations are now over 1000 MW = 1 GW, (10^9 W).

3-phase circuit theory

For many reasons, including efficiency of generation and transmission, quite apart from the ease of producing a rotating field as in any polyphase system, the 3-phase system has become virtually universal though there are occasions when other m-phase systems are used. For low powers of course, as in the domestic situation for example, single-phase supplies are satisfactory. For present purposes consideration will only be given to balanced 3-phase circuits, i.e. where the phase voltages and also the phase currents are mutually displaced by 120 electrical degrees ($2\pi/3$) radians). Electrical angles are given by $\omega_t = 2\pi ft$ radians.

On the assumption of balanced conditions, the power in a 3-phase system can be considered as available in three equal power 'packages', each handling 1/3 of the total power, i.e.

$$\frac{\text{Total power}}{3} = V_{\text{phase}}\, I_{\text{phase}} \cos\varphi$$

where φ is the same for each phase. The pulsating components of power cancel, giving steady power flow.

There are two symmetrical ways of connecting the three phases as shown on Figure 1.6:

in STAR (or wye);
for which it is obvious that the current through the line terminals is the same as the current in the phase itself, or:

in DELTA (or mesh);
for which it is obvious that the voltage across the line terminals is the same as the voltage across the phase itself.

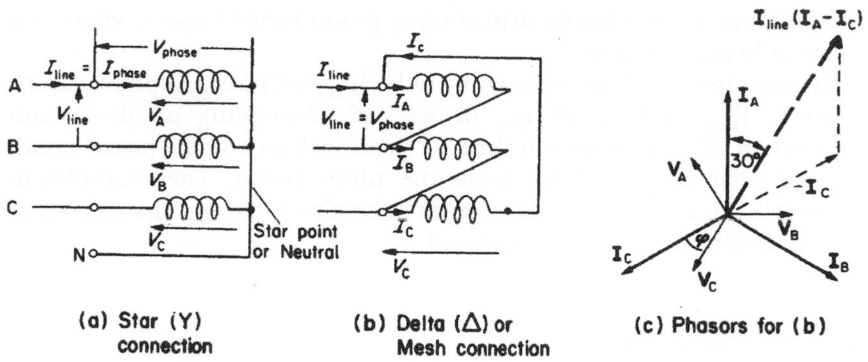

Figure 1.6 *3-phase circuits.*

For the delta case shown:

$$\mathbf{I}_{\text{line}} = \mathbf{I}_\text{A} - \mathbf{I}_\text{C} \text{ and } \mathbf{I}_\text{B} - \mathbf{I}_\text{A} \text{ and } \mathbf{I}_\text{C} - \mathbf{I}_\text{B}.$$

For the star case shown:

$$\mathbf{V}_{\text{line}} = \mathbf{V}_\text{A} - \mathbf{V}_\text{B} \quad \text{and} \quad \mathbf{V}_\text{B} - \mathbf{V}_\text{C} \quad \text{and} \quad \mathbf{V}_\text{C} - \mathbf{V}_\text{A};$$

assumed positive senses of phase currents and voltages being indicated.

The 120° displacement means that the magnitude of the line quantities in these two cases is equal to $\sqrt{3}$ times the magnitude of the phase quantities and there is a ±30° displacement between line and phase phasors; ± depending on which phasors are differenced. Only one line value involves a $\sqrt{3}$ factor, hence for both star and delta circuits:

$$\text{Total power} = 3 \times \frac{V_{\text{line}} I_{\text{line}}}{\sqrt{3}} \times \cos \varphi$$

$$= \sqrt{3} \times V_{\text{line}} I_{\text{line}} \times \cos \varphi$$

$$(\text{or Power} = \text{Voltamperes} \times \cos \varphi)$$

which can be solved for any one unknown. Frequently this is the current, from the known power and voltage ratings. Sometimes, for parameter measurements, φ is required for dividing currents, voltages or impedances into resistive and reactive components. The total voltamperes for a 3-phase system are thus given by $\sqrt{3}VI$ where V and I are here, line values, or alternatively, three times the phase VI product. A.C. devices are rated on a VA (kVA or MVA) basis, since they must be big enough magnetically to deal with full voltage, whatever the current, and big enough in terms of the electrically-sensitive parameters to deal with the current-carrying capacity specified in the rating, whatever the voltage. This means for example, that at zero power-factor, at full voltage and current, the temperature rise will be as high as or even higher than at unity power factor (u.p.f.), where real power is being converted.

3-phase circuits will be analysed by reducing everything to phase values: *Power*/3, V_{line} or $V_{line}/\sqrt{3}$, I_{line} or $I_{line}/\sqrt{3}$, depending on the circuit connection. The problem can then be dealt with as a single-phase circuit but the total power is three times the phase power. The examples in Chapter 2 are especially valuable for revising this topic and the analysis of unbalanced circuits is also introduced.

Torque components

T_e, the electromagnetic torque, will have different expressions for different machine types. It is basically due to the sum of all the tangential electromagnetic forces between the currents in rotor and stator conductors. If the windings are so distributed that the m.m.f. space waves are sinusoidal in shape and of magnitudes F_{stator} and F_{rotor}, the axes being displaced by the torque angle δ_T, then T_e is proportional to the products of these m.m.f.s and the sine of the torque angle. Because of the sinusoidal distribution, the two m.m.f. space waves can be combined vectorially to give a resultant m.m.f. wave of magnitude F_r, which produces the resultant mutual (air-gap) flux ϕ_m. An alternative expression, invoking the sine rule, gives T_e as proportional to the product of F_r with either F_{stator} or F_{rotor} and the sine of the different angle between them. This alternative is used for a.c. machines (cf. Figure 5.2) and since the angle is a function of load, it is called the load angle δ. For d.c. machines, the torque angle is fixed by the brush position (usually at maximum angle 90°) and so T_e can be expressed as $K \times \phi \times I_a$, where $K \times \phi$ can be combined as one coefficient k_ϕ which will be shown in Chapter 3 to be directly proportional to flux. The expression shows that k_ϕ is equal to the torque in newton metres, per ampere, (T_e/I_a). It is also equal to the generated e.m.f. per radian/second, (E/ω_m).

T_{loss} is due to internal machine friction, windage and iron-loss torques.

$T_{coupling}$ is the terminal torque, supplying the load in the case of a motor. The load torque may have an active component due to gravity or stored energy in the load system. This may oppose or assist the rotation. The passive components of the load torque, like friction, can only oppose rotation and will therefore reverse with rotation. The loss torques should be small and since they are mostly similar in nature to the passive load torques, it is convenient to combine them. At the balance point, where the speed is steady therefore, we have:

$$T_e = T_{loss} + T_{coupling} = T_m.$$

T_m is not a simple function of speed but is sometimes expressed in the form:

$$T_m = f(\omega_m) = k_1 + k_2\omega_m + k_3\omega_m{}^2$$

where k_1 is the idealised 'Coulomb' friction.

$k_2\omega_m$ is the viscous friction, proportional to speed and corresponding to 'streamline' flow. It occurs when the torque is due to eddy currents.

$k_3\omega_m{}^2$ is the torque due to 'turbulent' flow; as an approximation. It occurs, above the 'streamline'-flow speed, with fan- and propeller-type loads, e.g. windage losses.

Regulation; speed/torque curves; 4-quadrant diagram

The important characteristic of a power device is the way it reacts to the application of load. For a generator, the natural tendency is for the terminal voltage to fall as load current is taken. This fall is called the regulation and can be controlled by various means. The corresponding characteristic for a motor is the way in which speed changes as load torque is applied. With the d.c. motor as an example, the speed is nearly proportional to terminal voltage and the torque is proportional to current I_a, so the speed/torque axes follow the voltage/current axes for the generator. Figure 1.7 shows the natural characteristics for the various machine types. The d.c. machine (which could include 'brushless' machines) can easily be given a variety of curve shapes and two distinct forms are shown, for shunt and series motors. The synchronous machine runs at constant speed and as load increases the speed does not fall; the load angle increases to a maximum whereupon the speed will collapse, see Figure 8.5. The induction machine, like the d.c. machine, reacts to torque by a decrease of speed until it too reaches a maximum and stalls. The d.c. machine has a much higher maximum though it rarely reaches it without damage.

Also shown on Figure 1.7 is a typical load characteristic $T_m = f(\omega_m)$. Where this intersects the motor $\omega_m = f(T_e)$ characteristic, we have the balancing (steady-state) speed. There is not a universal practice in the assignment of axes and sometimes the torque axis is drawn vertically following the mechanical characteristic where T_m is the dependent variable. The usual practice for d.c. machines will be retained for all machine types in this text.

It will be noted that the axes have been continued into the negative regions, giving a 4-quadrant diagram with all combinations of positive and negative speeds and torques. Electromagnetic machines operate in all four regions as will be illustrated in later chapters.

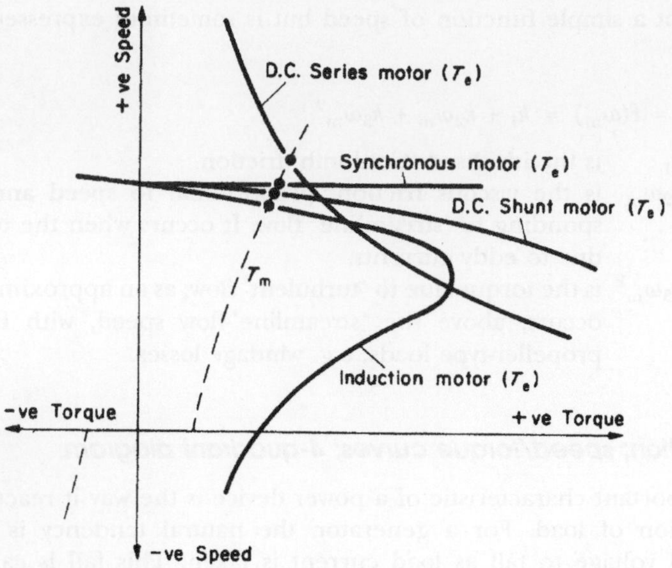

Figure 1.7 *Speed/torque characteristics.*

1.3 Equivalent circuits

These circuits represent a physical system by an electrical circuit. The simplest example is that for a battery, since over a limited range, the terminal voltage falls in proportion to the current taken. The battery behaves as if it consisted of a constant e.m.f. E, behind a resistance k equal to the slope of the 'regulation' curve of V against I, see Figure 1.8a.

The d.c. machine can be represented by the same equivalent circuit with the modification that the e.m.f. is controllable, being a function of speed and flux ($E = k_\phi \omega_m$); Figure 1.8b. An a.c. machine can also be represented this way, with the further modification that inductance must be included. Normally, the inductive reactance is appreciably larger than the resistance. The reactance may be considered in components corresponding to the leakage fluxes (which are relatively small and proportional to current because of the relatively large air-path reluctance), and the mutual flux (non-linear with excitation and confined largely to a path having its reluctance sensitive to iron saturation), see Figure 1.1b.

The equivalent-circuit parameters are often measured by conducting open-circuit and short-circuit tests. On open circuit for example, the current I is zero and the measured terminal voltage V is then equal to E. On short circuit, if this is possible without damage, i.e. if E is controllable, then the e.m.f. in the circuit is equal to the impedance drop since $V = 0$, so the impedance is obtained on dividing E by the current.

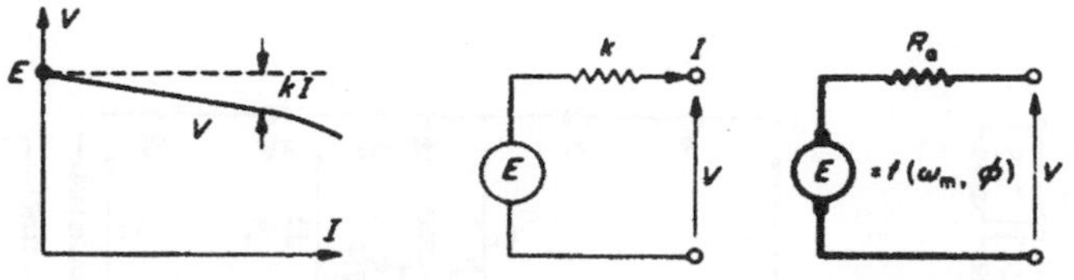

Figure 1.8 *Equivalent circuits.*

A full consideration of the induced voltages in the machine windings leads to circuit equations which can apply either to the machine or to another circuit which has the same equations. Starting with the 2-coil transformer, Figure 1.9a, we arrive at the circuit shown in which R_1 and x_1 are the primary resistance and leakage reactance respectively. X_m is the magnetising reactance and represents the effect of the mutual flux common to both primary and secondary windings. $R_2' + jx_2'$ is the 'referred' secondary leakage impedance, which is the actual value multiplied by the (turns ratio, $N_1/N_2)^2$. This is equivalent to replacing the secondary having N_2 turns with another secondary having the same number of turns as the primary, N_1. This would increase the voltage by N_1/N_2 and reduce the current by N_2/N_1. Hence the impedance (voltage/current) would be increased by $(N_1/N_2)^2$. Since it is not possible to tell from measurements on the primary side how many turns there are on the secondary, this replacement by a 1/1 ratio is convenient, expressing secondary voltages and currents in primary terms. The ideal transformer at the end of the circuit converts these referred values back to actual values. Note that the positive directions of $\mathbf{I}_1$ and $\mathbf{I}_2$ have been taken in the same sense magnetically because this is convenient when developing the equations. Generally, however, the positive sense of $\mathbf{I}_2$ is taken in the opposite direction to $\mathbf{I}_1$ and the magnetising branch in the middle carries $\mathbf{I}_1 - \mathbf{I}_2' = \mathbf{I}_0$, as usually designated.

For rotating machines, the above treatment can be adapted and extended by considering the stator m.m.f. as being produced by one specially distributed coil. Similarly the rotor m.m.f. is treated as due to one coil. The difference from the transformer is that the rotor coils move with respect to the stator coils, though their m.m.f.s are always in synchronism for the steady-state condition. The fluxes follow the same general pattern in that there is a common mutual flux, crossing the air gap and linking both stator and rotor windings, and leakage fluxes associated with each winding individually.

The way in which the equivalent circuit is modified from the transformer depends on the machine type. For a.c. machines, the m.m.f. of the stator

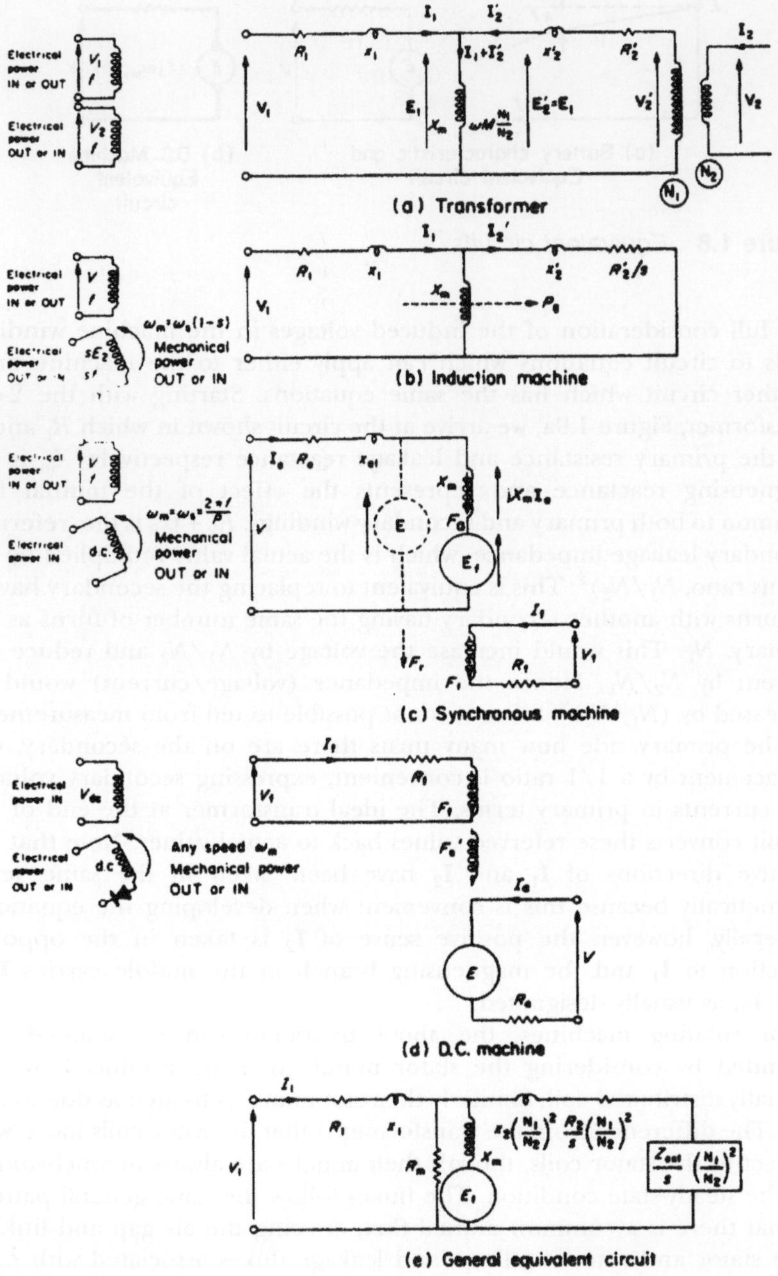

Figure 1.9 *Machine equivalent circuits.*

(usually) produces a rotating field at synchronous speed $\omega_s = 2\pi n_s$ rad/s, where $n_s = f/p$ rev/s. The induction machine runs at a speed $\omega_m = \omega_s(1 - s)$ where the slip s expresses the relative movement of the rotor with respect to the synchronous-speed m.m.f. wave. The rotor e.m.f. is reduced to $s\mathbf{E}_2$, where $\mathbf{E}_2$ is the e.m.f. at standstill with stator and rotor windings stationary as in a transformer. The value of x_2 is also reduced so the rotor current $\mathbf{I}_2 = s\mathbf{E}_2/(R_2 + \mathrm{j}sx_2) = \mathbf{E}_2/(R_2/s + \mathrm{j}x_2)$, so that the only difference from the transformer equivalent circuit is the replacement of R_2' by R_2'/s (Figure 1.9b) and all parameters are *per-phase* values.

For the synchronous machine, $s = 0$ since $\omega_m = \omega_s$ and the right-hand side of the equivalent circuit carries no induced current on steady state since $R_2'/s = \infty$. A d.c. current has to be provided in the 'secondary' winding which now becomes the field; Figure 1.9c. The effect on the 'primary' winding is now expressed as $\mathbf{E}_f = f(\omega_s, I_f)$ instead of through $\mathbf{I}_m \cdot X_m$ as in the transformer and induction machines. The resultant m.m.f. F_r produces the mutual (air-gap) flux ϕ_m and e.m.f. E.

For the d.c. machine, both terminal currents are d.c. so the reactive elements may be omitted. The effect of the armature m.m.f. F_a on the field m.m.f. is more complex than for the other machines and is represented by its net magnetising action F_a', which is usually negative. Figure 1.9d shows the equivalent circuit.

This rather rapid review of equivalent-circuit development is obviously deficient in many details but is dealt with fully in Reference 1. Figure 1.9e shows a general equivalent circuit which is applicable to all machine types discussed, with appropriate modifications. For the transformer, E_f is omitted and the value of s is unity. For the induction machine, s takes on any value. For the synchronous machine, $s = 0$ and the right-hand side of the equivalent circuit is omitted and E_f is inserted. For the d.c. machine, the reactances are omitted for steady-state operation. It will also be noted that there is an additional element, R_m. The power dissipated here (E^2/R_m) represents the iron loss (per phase). When the circuit represents an a.c.-excited device like a transformer or an induction motor, this power is provided by the electrical supply. The value of R_m is relatively high and does not normally affect the calculations of the remaining currents very significantly. For Figures 1.9(a) and (b), X_m (and R_m) may be shown directly across V_1 to give the 'approximate' circuit (see Example 2.8) with leakage impedance $R_1 + R_2' + \mathrm{j}(x_1 + x_2')$.

1.4 Power-flow diagram

Figure 1.10 is an extension of Figure 1.3b showing more details of the power distribution within the machine. The expressions for the various power components sometimes differ as between the different machine

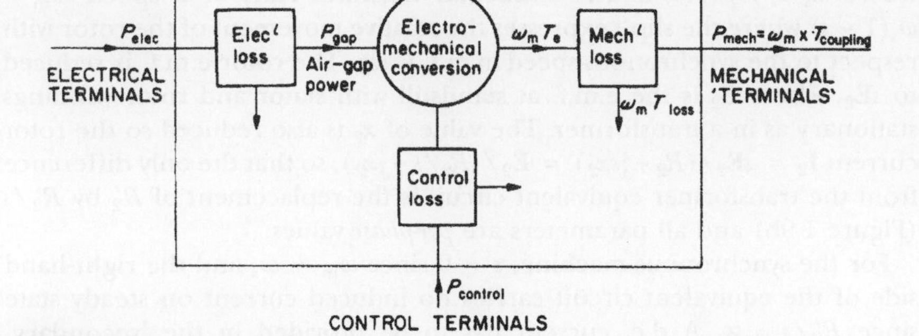

Figure 1.10 *Power-flow distribution between machine terminals (motoring convention).*

types but the general pattern is the same. The power flow for motoring operation is from left to right, the electrical terminal power P_{elec} and the mechanical terminal power P_{mech} being both positive. For generating operation, these are both negative, power flowing from right to left. For braking, P_{mech} is negative and for reverse-current braking, power flow is inwards from both ends. Electromechanical power conversion takes place through the air gap and the power P_{gap} is less than P_{elec} for motoring, by the amount 'Electrical loss'. The power converted to mechanical power is

Table 1.1

	D.C. motor	*3-ph induction motor*	*3-ph synchronous motor*
P_{elec}	VI_a	$\sqrt{3}V_L I_L \cos\varphi_1 = 3V_1 I_1 \cos\varphi_1$	$\sqrt{3}V_L I_L \cos\varphi = 3VI_a \cos\varphi$
'Electrical loss'	$I_a{}^2R_a$ + brush loss	$3I_1{}^2R_1$ + Fe loss	$3I_a{}^2R_a$
Air-gap power P_{gap}	$EI_a = \omega_m T_e$	$3E_1 I'_2 \cos\varphi_2 = 3I_2{}^2R_2/s = \omega_s T_e$	$3\ E_f I_a \cos(\varphi - \delta) = \omega_s T_e \simeq 3VI_a \cos\varphi$
Control loss	$I_f{}^2R_F$	$3I_2{}^2R_2 = sP_{gap}$ if s.c.	$I_f{}^2R_F$
$P_{control}$	$V_f I_f = I_f{}^2R_F$	$3V_3 I'_2 \cos\varphi_3$ = 0 if secondary s.c.	$V_f I_f = I_f{}^2R_F$
P_{mech}	$\omega_m T_{coupling}$	$\omega_m T_{coupling}$	$\omega_s T_{coupling}$

Note. The Fe loss mostly manifests itself as a torque loss, part of T_{loss}, though it is usual on induction machines to show it as part of the electrical loss.

always (speed × electromagnetic torque) = $\omega_m T_e = P_{gap}$, where $\omega_m = \omega_s$ for the synchronous machine. For the induction machine the speed is $\omega_m = \omega_s (1 - s)$ so that there is a power sP_{gap} converted not to mechanical power, but to electrical power in the secondary circuits. Continuing along the motoring power-flow path, the remaining elements are the same for all machine types. $\omega_m T_e$ is reduced by the mechanical loss $\omega_m T_{loss}$ to $\omega_m T_{coupling}$ which is the output motoring power at the mechanical terminals. There is a further set of terminals for the control power, $P_{control}$. For the d.c. and synchronous machines, this is absorbed entirely by the field Cu loss $I_f^2 R_F$. For the induction machine, the control loss is the secondary Cu loss, $3I_2{}^2R_2$ and if the secondary terminals are short circuited, this is provided by the 'transformer-converted' power, sP_{gap} and $P_{control}$ is zero. Sometimes however, for slip-power recovery schemes, control power is exerted from an external voltage source V_3 and this power may be inwards (for super-synchronous speeds) or outwards from the machine (for sub-synchronous speeds). Figure 1.10 is also useful for explaining the concepts of efficiency, control and a method of measuring the losses, by setting the power to zero at two-sets of terminals and measuring the input at the other set.

In Table 1.1 the power components for the various machine types are listed, using the symbols adopted in the later text where the equations will be explained.

2 Transformers

Although the transformer is not an electromechanical converter like the other devices to be covered in this text, it forms the basis from which all the other equivalent circuits are derived. The theory is relatively simple, built up from Faraday's Law of Electromagnetic Induction. Only steady-state operation will be dealt with here but it is very important to the understanding of the remaining chapters; for example, in coming to grips with the referring process by means of which two coupled windings of different voltage and current ratings can be replaced by a simple series/parallel circuit. There is also much vital practice to be obtained in the analysis and manipulation of 3-phase circuits. Multiple windings and the combination of winding m.m.f.s, together with basic work in a.c. circuit theory and use of complex numbers, are all illustrated by the various examples. Finally, an unbalanced transformer load condition is analysed, using Symmetrical Components.

2.1 Solution of equations

In this chapter, since some of the groundwork has already been covered in Chapter 1, equations will mostly be discussed as the need arises. The first example requires the important e.m.f. equation, which applies, with some slight modifications, to all a.c. machines. It relates the r.m.s.-induced e.m.f. E to the maximum flux $\hat{\phi}$ in webers, frequency f and number of turns N:

From the law of induction, the instantaneous e.m.f. = $N\mathrm{d}\phi/\mathrm{d}t$ and for a sinusoidal variation of flux, expressed as $\phi = \hat{\phi} \sin 2\pi ft$, it is readily shown by differentiation and substitution that:

$$E = 4.44 \times \hat{\phi} \times f \times N \text{ volts} \qquad (2.1)$$

Note especially that E is r.m.s. e.m.f. and flux is the maximum value.

Example 2.1

A transformer core has a square cross-section of 20 mm side. The primary winding is to be designed for 230 V, the secondary winding for 110 V and a further centre-tapped, 6/0/6-V winding is to be provided. If the flux density $\hat{B}$ is not to exceed 1 tesla, find a suitable number of turns for each winding, for a frequency of 50 Hz. Neglect all transformer imperfections.

The low-voltage winding is always designed first because voltage ratios can rarely be obtained exactly and since there are the fewest turns on the low-voltage winding and the actual number of turns must always be integral, the adjustment is also the coarsest.

The maximum flux must not exceed $\hat{\phi} = \hat{B} \times A = 1 \times (20 \times 10^{-3})^2$.

From the e.m.f. equation: $6 = 4.44 \times 400 \times 10^{-6} \times 50 \times N_3$ from which $N_3 = 67.57$ and the nearest integral number is 68, to avoid the specified maximum flux density being exceeded. It will now be slightly lower than $\hat{B} = 1$.

For the secondary winding: $\dfrac{N_2}{N_3} = \dfrac{110}{6}$, so $N_2 = \dfrac{110 \times 68}{6} = 1246.7$ say 1247 turns.

For the primary winding: $\dfrac{N_1}{N_3} = \dfrac{230}{6}$, so $N_1 = \dfrac{230 \times 68}{6} = 2606.7$ say 2607 turns.

The tertiary winding requires 2 × 68 turns.

Note 1 On core-type transformers, the windings are divided into two sections, one on each limb. In this case the nearest *even* number of turns would be chosen, to make the sections equal.

Note 2 When transformer cores are supplied for the user to wind his own coils, it is usual, for convenience, to specify the magnetic limits at a particular frequency, in terms of the volts per turn. For this core it is 6/67.57 = 88.8 mV/turn or 11.26 turns per volt; at 50 Hz.

Example 2.2

A 20-kVA, 3810/230-V, 50-Hz single-phase transformer operates at a maximum flux density of 1.25 teslas, for which the iron requires a value of $H = 0.356$ At/mm. The core cross-section is 0.016 m^2 and the mean length of the magnetic path through the core is 1.4 m. The primary and secondary turns are 860 and 52 respectively. It is decided to use the transformer as an inductor and to keep the reactance substantially constant with current, the core is sawn through transversely and packed with brass to give an 'air' gap. If the secondary winding is used, find the length of the air gap so that when carrying rated current, the maximum flux density is not exceeded. What is then the inductance and the reactance at 50 Hz?

Rated current $= \dfrac{20\,000}{230} = 86.96$ A r.m.s.

The peak m.m.f. exerted by this current is: $\sqrt{2} \times 86.96 \times 52 = 6394.7$ At

At peak flux density, the m.m.f. absorbed by the iron is $H \times l = 0.356 \times 1.4 \times 10^3 = 498.4$ At.

$\therefore$ m.m.f. available for the air gap $= 6394.7 - 498.4 = 5896.3$ At.

Neglecting air-gap fringing so that the flux density is assumed to be the same as in the iron:

$$\text{Air-gap m.m.f.} = H \times l = \frac{B}{\mu_0} \times l = \frac{1.25}{4\pi/10^7} \times l = 5896.3$$

from which $\underline{l = 5.93\,\text{mm}}$.

The effective permeance of the core + gap =

$$\Lambda = \frac{\text{flux}}{\text{m.m.f.}} = \frac{1.25 \times 0.016}{6394.7}$$

$$= 3.128 \times 10^{-6}\ \text{Wb/At}$$

so inductance $= N^2\Lambda = 52^2 \times 3.128 \times 10^{-6} = \underline{8.458\,\text{mH}}$

and reactance $= 2\pi \times 50 \times 8.458 \times 10^{-3} = \underline{2.657\,\omega}$

Alternatively, the e.m.f. $= 4.44 \times (1.25 \times 0.016) \times 50 \times 52 = 230.9\,\text{V}$ and the impedance = $230.9/86.96 = \underline{2.656\,\omega}$

Note 1 Resistance has been neglected in the above calculation; it would be relatively small.

Note 2 The inductance is not quite constant with current, but at the maximum value, the iron absorbs less than 8% of the total m.m.f. At lower currents, the linearity will improve since the iron will absorb proportionately less At.

Example 2.3

A 230/6-V, single-phase transformer is tested with its secondary winding short circuited. A low voltage (20 V) is applied to the primary terminals and it then takes an input current of 1 A; the power supplied being 10 watts. Calculate the values of leakage reactance and resistance, referred to the primary side and then to the secondary side. If the magnetising impedance is neglected, calculate the secondary terminal voltage when a load impedance of value 0.12 + j0.09 ω is connected.

On s.c. test, the input power factor is: $\dfrac{P}{VI} = \dfrac{10}{20} \times 1 = 0.5 = \cos\varphi_{sc}$

and the leakage impedance (neglecting the relatively small current $\mathbf{I}_0 = I_p - jI_m$ through magnetising branch $R_m \parallel X_m$), is:

$$\frac{V}{I} = \frac{20}{1} = 20\,\omega = Z_{sc}$$

$\therefore$ impedance referred to the primary side $= Z_{sc}(\cos\varphi_{sc} - j\sin\varphi_{sc})$. ($\sin\varphi_{sc}$ is –ve).

$$= R_1 + R_2' + j(x_1 + x_2') = 20(0.5 + j0.866) = \underline{10 + j17.32\,\omega}$$

Impedance referred to the secondary side $= Z_{sc}\left(\frac{N_2}{N_1}\right)^2 = \left(\frac{6}{230}\right)^2 . (10 + j17.32)$

$= R_1' + R_2 + j(x_1' + x_2) = \underline{0.0068 + j0.0118\,\omega}.$

It is convenient to solve for the load test with the impedance referred to the secondary side. The diagram shows the equivalent circuit connected to the load, and the phasor diagram – not to scale.

The load current would be:

$$\frac{6}{0.1268 + j0.1018} = \frac{6(0.1268 - j0.1018)}{(0.1268)^2 + (0.1018)^2}$$

$$= 28.77 - j23.1 = 36.9 \underline{/-38°.8}\,\text{A}$$

Voltage at secondary terminals, i.e. across the load $= 36.9\sqrt{0.12^2 + 0.09^2} = \underline{5.535\,\text{V}}$

An alternative approach is to calculate the drop of voltage due to the transformer internal impedance; the so-called regulation. From the phasor diagram as drawn, this is approximately equal to $RI\cos\varphi - XI\sin\varphi - |E - V|$. (sin φ is –ve.)

The load power-factor is $\frac{0.12}{\sqrt{0.12^2 + 0.09^2}} = 0.8 = \cos\varphi. \therefore \sin\varphi = -0.6$ (lagging)

Regulation $= 36.9(0.0068 \times 0.8 + 0.0118 \times 0.6) = 0.462\,\text{V}$

So load voltage $= 6 - 0.462 = 5.538\,\text{V}$

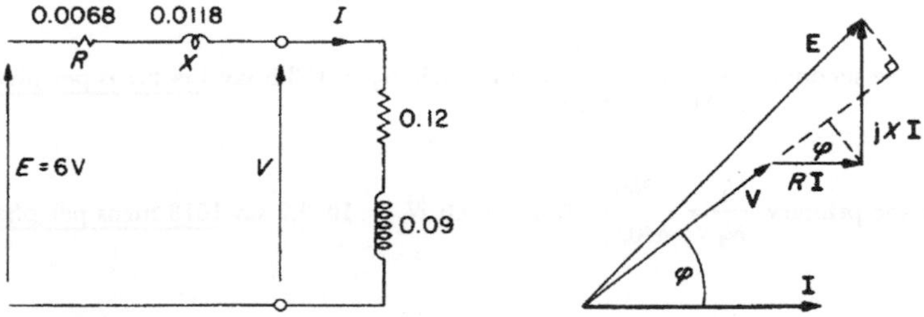

Figure E.2.3

Note 1 In this case there is enough information to calculate V from $\mathbf{E} - \mathbf{ZI}$ where, with $\mathbf{E}$ as the reference phasor, the phase angle of $\mathbf{I}$ with respect to $\mathbf{E}$ could be found from the total series impedance. If only the load power-factor φ is known however, this information is not available and the regulation expression must be used. A more accurate regulation expression can be derived.[1]

Note 2 If the load power-factor had been leading, – sin φ would have been negative and the load voltage could have been *higher* than the open-circuit voltage (E).

Example 2.4

A 3-phase, 50-Hz transformer is to have primary, secondary and tertiary windings for each phase. The specification is as follows: Primary to be 6600 V and delta connected. Secondary to be 1000 V and delta connected. Tertiary to be 440 V and star connected. Determine suitable numbers of turns to ensure that the peak flux does not exceed 0.03 Wb.

If the secondary is to supply a balanced load of 100 kVA at 0.8 p.f. lagging and the tertiary is to supply a balanced load of 50 kW at u.p.f., determine the primary line-current and power factor for this condition. Neglect all transformer imperfections.

This is the first 3-phase problem in the book and it is an excellent opportunity to become absolutely familiar with the basic circuit relationships. It is always helpful to draw a circuit diagram as on the figure, where these relationships become obvious. Note that it is standard practice to specify line voltages, line currents and total power for 3-phase devices so that, as in this case, conversion to phase values may be necessary. For example, the turns ratio is the phase-voltage ratio and is usually specified for the no-load condition.

Using the e.m.f. equation and as before, designing the low-voltage winding first:

$$440/\sqrt{3} = 4.44 \times 0.03 \times 50 \times N_3$$

from which $N_3 = 38.14$ say 39 turns per phase – to keep the flux below the specified level.

For the secondary: $\dfrac{N_2}{N_3} = \dfrac{1000}{440/\sqrt{3}}$ from which $N_2 = 153.5$ say 154 turns per phase

For the primary: $\dfrac{N_1}{N_3} = \dfrac{6600}{440/\sqrt{3}}$ from which $N_1 = 1013.2$ say 1013 turns per phase

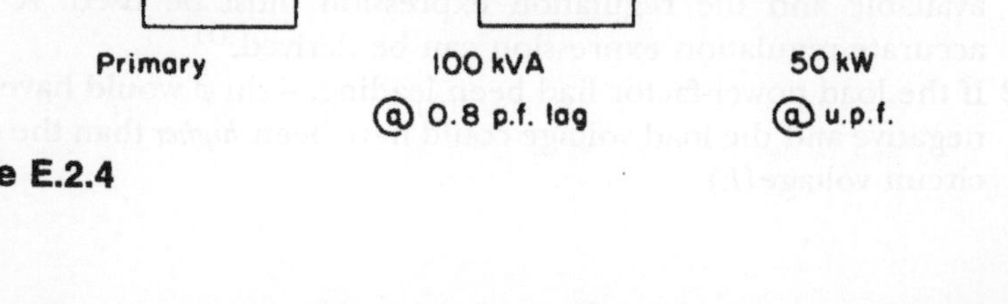

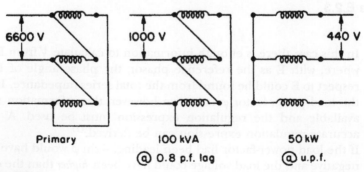

Figure E.2.4

Load condition

$$\text{Tertiary current per phase} = \frac{\text{power per phase}}{\text{voltage per phase}} = \frac{50\,000/3}{440/\sqrt{3}} = 65.6\,\text{A}$$

$$\text{Referred to primary; } I_3' = 65.6 \times \frac{39}{1013} = 2.525\ \text{A}$$

But since referred currents are required in order to calculate the total primary current, it is simpler to go directly to this by dividing the tertiary kVA by primary voltage.

i.e. $I_3' = \dfrac{50\,000/3}{6600}$ and since power factor is unity, $\mathbf{I}_3' = 2.525 + j0.$

Similarly $\mathbf{I}_2' = \dfrac{100\,000/3}{6600}(0.8 - j0.6)$ since p.f. is 0.8 lagging. $\mathbf{I}_2' = \underline{4.04 - j3.03}$

So $\mathbf{I}_1 = \mathbf{I}_2' + \mathbf{I}_3' + \mathbf{I}_0\ (= 0)$ $= \underline{6.565 - j3.03}$

and the primary is delta connected so line current $= \sqrt{3} \times \sqrt{6.565^2 + 3.03^2}$

$= \sqrt{3} \times 7.23 = \underline{12.52\,\text{A}}$

Power factor $= \cos\varphi = 6.565/7.23 = \underline{0.908 \text{ lagging.}}$

Example 2.5

On open circuit, a 3-phase, star/star/delta, 6600/660/220-V transformer takes 50 kVA at 0.15 p.f. What is the primary input kVA and power factor when, for balanced loads, the secondary delivers 870 A at 0.8 p.f. lagging and the tertiary delivers 260 line A at unity power factor? Neglect the leakage impedances.

The connections are different here from the previous example but a sketch of the appropriately modified circuit diagram would be instructive.

The data given permit I_0 to be calculated from kVA/($\sqrt{3}$ × kV)-star connection:

$$\mathbf{I}_0 = \frac{50}{\sqrt{3} \times 6.6}(\cos\varphi + j\sin\varphi) = 4.374(0.15 - j0.9887) = 0.656 - j4.324$$

$$\mathbf{I}_2' = 870(0.8 - j0.6) \times \frac{660/\sqrt{3}}{6600/\sqrt{3}} = 69.6 - j52.2$$

$$\mathbf{I}_3' = \frac{260}{\sqrt{3}}(1 + j0) \times \frac{220}{6600/\sqrt{3}} = \underline{8.667 + j0}$$

$$\mathbf{I}_1 = \mathbf{I}_0 + \mathbf{I}_2' + \mathbf{I}_3' = \underline{78.92 - j56.52}$$

$$I_1 = \sqrt{78.92^2 + 56.52^2} = 97.07\,\text{A at } \cos\varphi = 78.92/97.07 = \underline{0.813 \text{ lag}}$$

Input kVA $= \sqrt{3} \times 6.6 \times 97.07 = \underline{1109.6\,\text{kVA.}}$

Example 2.6

A 3-phase, 3-winding, delta/delta/star, 33 000/1100/400-V, 200-kVA transformer carries a secondary load of 150 kVA at 0.8 p.f. lagging and a tertiary load of 50 kVA at 0.9 p.f. lagging. The magnetising current is 4% of rated current; the iron loss being 1 kW total. Calculate the value of the primary current and power factor and input kVA when the other two windings are operating on the above loads.

Again the leakage impedances will be neglected. See Reference 2 for an exact equivalent circuit for the 3-winding transformer.

$$\text{Power component of } I_0 = \frac{1\,\text{kW}/3}{33\,\text{kV}} = 0.01\,\text{A}$$

$$\text{Magnetising component of } I_0 = \frac{4}{100} \times \frac{200/3}{33} = 0.081\,\text{A}$$

$$\therefore \mathbf{I}_0 = 0.01 - j0.081$$

$$\mathbf{I}_2' = \frac{150/3}{33}(0.8 - j0.6) = 1.212 - j0.909$$

$$\mathbf{I}_3' = \frac{50/3}{33}(0.9 - j0.436) = \underline{0.455 - j0.22}$$

$$\mathbf{I}_1 = \underline{1.677 - j1.21}$$

$$\text{Line current} = \sqrt{3} \times \sqrt{1.677^2 + 1.21^2} = \sqrt{3} \times 2.07 = \underline{3.58\,\text{A}}$$

$$\text{at power factor} \quad 1.677/2.07 = \underline{0.81 \text{ lagging}}$$

$$\text{and input} \quad \text{kVA} = \sqrt{3} \times 33 \times 3.58 = \underline{204.6\,\text{kVA}}$$

Example 2.7

The following are the light-load test readings on a 3-phase, 100-kVA, 400/6600-V, star/delta transformer:

Open circuit; supply to low-voltage side 400 V, 1250 W
Short circuit; supply to high-voltage side 314 V, 1600 W, full-load current.

Calculate the efficiencies at full load, 0.8 power factor and at half full load, u.p.f. Calculate also the maximum efficiency. What is the percentage leakage impedance based on 100% = rated V/rated I?

The losses on a transformer are: Fe loss which varies very little at constant voltage and frequency and Cu loss which is proportional to $(\text{current})^2$, or $(\text{load kVA})^2$ at constant voltage.

At any load, $\varLambda$ times rated value (load here referring to load current or load kVA):

$$\text{Efficiency} = \frac{\text{Output}}{\text{Input}} = \frac{\varLambda \times \text{kVA} \times \cos\varphi}{\varLambda \times \text{kVA} \times \cos\varphi + \text{Losses}}$$

where kVA is the rated value

$$\eta = \frac{\varLambda\,\text{kVA}\cos\varphi}{\varLambda\,\text{kVA}\cos\varphi + \text{Fe loss} + \varLambda^2\,(\text{Cu loss at full load})}$$

The maximum value of this expression is easily shown by differentiation, to occur when $\varLambda = \sqrt{(\text{Fe loss})/(\text{Cu loss at full load})}$ at any particular power factor.

The o.c. test gives the normal Fe loss since rated voltage is applied and the s.c. test gives the copper loss at full load since rated current is flowing.

For full load, 0.8 p.f.

$$\eta = \frac{1 \times 100 \times 0.8}{(1 \times 100 \times 0.8) + 1.25 + 1^2 \times 1.6} = \frac{80}{82.85} = 0.9656 = \underline{96.56\%.}$$

For half full load, u.p.f.

$$\eta = \frac{0.5 \times 100 \times 1}{(0.5 \times 100 \times 1) + 1.25 + (0.5)^2 \times 1.6} = \frac{50}{51.65} = 0.968 = \underline{96.8\%.}$$

Maximum efficiency when

$$A = \sqrt{1.25/1.6} = 0.884 \text{ and power factor is unity.}$$

$$\text{Max. } \eta = \frac{0.884 \times 100 \times 1}{88.4 + 1.25 + (0.884)^2 \times 1.6} = \frac{88.4}{90.9} = 0.9725 = \underline{97.25\%.}$$

Leakage impedance

$$\text{Rated secondary current per phase} = \frac{100\,000/3}{6600} = 5.05\,\text{A.}$$

$$\text{Short-circuit power factor} = \frac{1600/3}{314 \times 5.05} = 0.336.$$

$$Z_{sc} \text{ (referred to secondary)} = \frac{314}{5.05}(0.336 + j0.9417) = 20.89 + j58.55\,\omega.$$

$$\text{Base impedance} = 100\% = \frac{6600}{5.05} = 1306.9\,\omega.$$

$$\text{Hence percentage impedance} = \frac{1}{1306.9}(20.89 + j58.55) \times 100$$

$$= (0.016 + j0.0448) \times 100 = \underline{1.6 + j4.48\%.}$$

The value before multiplying by 100 is called the *per-unit* impedance; see Sections 3.3, 5.3 and Example (4.2d).

Example 2.8

The following light-load, line-input readings were taken on a 3-phase, 150-kVA, 6600/440-V, delta/star-connected transformer:

Open-circuit test	1900 W,	440 V,	16.5 A
Short-circuit tests	2700 W,	315 V,	12.5 A.

(a) Calculate the equivalent-circuit parameters per phase, referred to the h.v. side.

(b) Determine the secondary terminal voltages when operating at rated and half-rated current if the load power-factor is 0.8 lagging. Calculate also the efficiencies for these loads.

(c) Determine the secondary terminal voltage when operating at rated current with a load power-factor of 0.8 leading and calculate the efficiency for this condition.

Although this problem is for a 3-phase transformer, once the equivalent-circuit parameters per phase have been determined correctly, it may proceed as for single-phase, making due allowance for the fact that the calculations will give phase currents and powers. A balanced load is assumed. The figure shows the circuit and the approximate equivalent circuit which should help to avoid confusion when analysing the tests. On short circuit, I_0 may be neglected and on open circuit, the leakage impedance has little effect, but see Example 4.1.

Open-circuit test

Since iron loss is a function of voltage, this test must be taken at rated voltage if the usual parameters are required. The voltage is 440 V so it must have been taken on the low-voltage side. Referring I_0 to the primary will permit the magnetising parameters, referred to the h.v. side, to be calculated directly.

$$I_0 = 16.5 \times \frac{440/\sqrt{3}}{6600} = 0.635\,\text{A}$$

$$\cos\varphi_{oc} = \frac{1900}{\sqrt{3} \times 440 \times 16.5} = 0.151; \ \sin\varphi_{oc} = -0.9985.$$

Hence:

$$R_m = \frac{V}{I_0 \cos\varphi_{oc}} = \frac{6600}{0.635 \times 0.151} = 68.8\,\text{k}\omega. \qquad X_m = \frac{6600}{0.635 \times 0.9985} = \underline{10.4\,\text{k}\omega.}$$

Short-circuit test

This does not have to be taken at exactly rated current, since with unsaturated leakage reactance, a linear relationship between voltage and current can be assumed. Only a low voltage is required, so the readings show that the test was taken on the h.v. side.

$$\cos\varphi_{sc} = \frac{2700}{\sqrt{3} \times 315 \times 12.5} = 0.396; \ \sin\varphi_{sc} = -0.918.$$

$$\mathbf{Z}_{sc} = \frac{315}{12.5/\sqrt{3}}(0.396 + j0.918) = \underline{17.3 + j40\,\omega.}$$

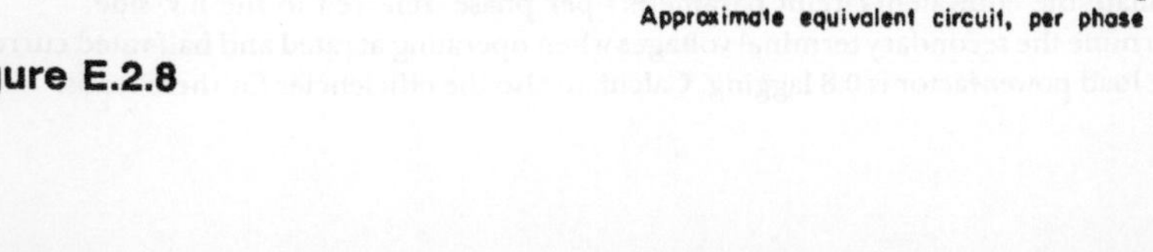

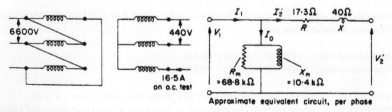

Figure E.2.8

Load conditions

$$\text{Rated current, by definition} = \frac{\text{Rated VA per phase}}{\text{Rated voltage per phase}} = \frac{150/3}{6.6} = 7.575\,\text{A}.$$

The same regulation expression can be used as in Example 2.3 and $\cos\varphi = 0.8$. $\sin\varphi = -0.6$ for (b) and +0.6 for (c)

		(a)	(b)	(c)	
H.V. Load current per phase	=	7.575	3.787	7.575	A
$I\cos\varphi$	=	6.06	3.03	6.06	A
$I\sin\varphi$	=	−4.545	−2.272	+4.545	A
Regulation = $17.3I\cos\varphi - 40I\sin\varphi$	=	286.7	143.3	−77	V
Terminal voltage = 6600 − regulation	=	6313.3	6456.7	6677	V
Referred to secondary [$\times 440/(\sqrt{3}\times 6600)$]	=	243	248.5	257	V
Secondary terminal voltage (line)	=	420.9	430.4	445.1	V
Load current referred to secondary	=	196.8	98.4	196.8	A
Secondary (output) power = $\sqrt{3}VI\cos\varphi$	=	114.8	58.7	121.3	kW
Cu loss = $2.7\times(\sqrt{3}\times I_{hv}/12.5)^2$	=	2.97	0.74	2.97	kW
Total loss (+ Fe loss = 1.9)	=	4.87	2.64	4.87	kW
Efficiency	=	95.93	95.7	96.14	%

Note that when calculating the efficiency, the changing secondary voltage has been allowed for, and this is higher than the o.c. voltage when the power factor is leading.

Example 2.9

A 3-phase transformer has a star-connected primary and a delta-connected secondary. The primary/secondary turns ratio is 2/1. It supplies a balanced, star-connected load, each phase consisting of a resistance of 4 ω in series with an inductive reactance of 3 ω.

(a) If the transformer was perfect, what would be the value of the load impedance per phase viewed from the primary terminals?

From a practical test, the following primary input-readings were taken:

	Total power	Line voltage	Line current
Secondary short circuited	12 W	7.75 V	2 A
Secondary load connected	745 W	220 V	2.35 A

(b) Deduce, from the readings, the equivalent circuit of the load viewed from the primary winding and explain why it differs from (a).

(a) The circuit diagram is shown on the figure and the first step is to transform the star-connected load to an equivalent delta; i.e. one that carries the same current as the secondary winding and has the same voltage across it, as indicated on the figure. Since the current per load-phase will thus be reduced by $\sqrt{3}$ and its voltage increased by $\sqrt{3}$, the effect will be to transform the impedance by a factor of 3 times, from that of the star-connection.

Hence $\mathbf{Z} = 12 + j9\,\omega$ per phase.

It is now possible to treat each balanced phase separately, as a single-phase problem. The effect of the 2/1 turns ratio and a perfect transformer will be to decrease the current by 1/2 and increase the voltage by 2/1, giving an impedance transformation of 2^2; i.e. $(N_1/N_2)^2$. Viewed from the primary therefore, the load impedance will appear as:

$$\underline{\mathbf{Z}'_{\text{load}} = 48 + j36\,\omega \text{ per phase.}}$$

(b) From the s.c. tests:

$$\cos\varphi_{sc} = \frac{12}{\sqrt{3} \times 7.75 \times 2} = 0.4469; \ \sin\varphi_{sc} = -0.8845.$$

Hence

$$\mathbf{Z}_{sc} = \frac{7.75/\sqrt{3}}{2}(0.4469 + j0.8845) = 1 + j2\,\omega \text{ per phase.}$$

From the load test:

$$\cos\varphi_{\text{load}} = \frac{745}{\sqrt{3} \times 220 \times 2.35} = 0.8319; \ \sin\varphi_{\text{load}} = -0.555.$$

$$\mathbf{Z}_{\text{input}} = \frac{220/\sqrt{3}}{2.35}(0.8319 + j0.555) = 45 + j30\,\omega \text{ per phase.}$$

Hence,

$$\text{referred load impedance} = \mathbf{Z}_{\text{input}} - \mathbf{Z}_{sc} = \underline{44 + j28 \text{ per phase.}}$$

The equivalent circuit per phase from these results is shown on Figure E.2.9, which includes the unknown magnetising impedance $\mathbf{Z}_m$ across the input terminals. A small part of the input current will flow through here so that the figure of 2.35 A used in calculating $\mathbf{Z}_{\text{input}}$ will be higher than the true value of I'_2. Hence, the above value of the referred load impedance will be lower than calculated from part (a).

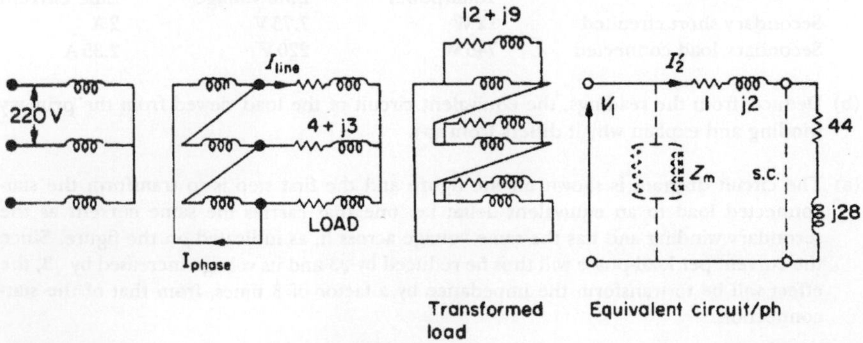

Figure E.2.9

Example 2.10

A 3-phase, 11 000/660-V, star/delta transformer is connected to the far end of a distribution line for which the near-end voltage is maintained at 11 kV. The effective leakage reactance and resistance per-phase of the transformer are respectively 0.25 ω and 0.05 ω referred to the low-voltage side. The reactance and resistance of each line are respectively 2 ω and 1 ω.

It is required to maintain the terminal voltage at 660 V when a line current of 260 A at 0.8 lagging power factor is drawn from the secondary winding. What percentage tapping must be provided on the h.v. side of the transformer to permit the necessary adjustment? The transformer magnetising current may be neglected and an approximate expression for the regulation may be used. Neglect also the changes to the impedance due to the alteration of the turns ratio.

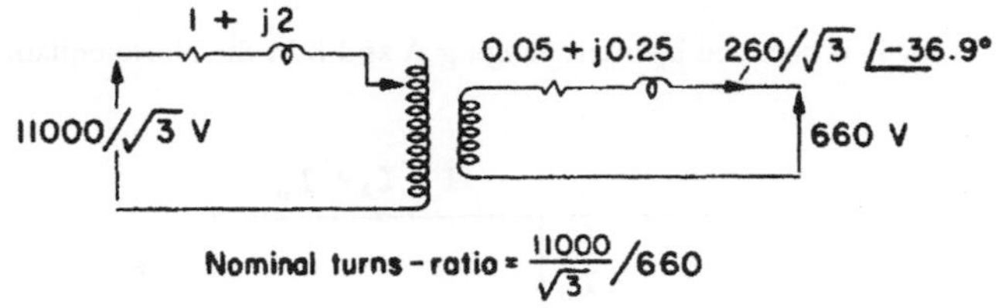

Figure E.2.10

The equivalent circuit per phase is shown on the figure. The line impedance can also be referred to the secondary side and included in the regulation expression.

$$\text{Referred line impedance} = \left(\frac{660}{11\,000/\sqrt{3}}\right)^2 \times (1 + j2) = 0.0108 + j0.0216\,\omega \text{ per phase.}$$

The total impedance referred to the secondary = 0.0608 + j0.2716 ω per phase.

$$\text{Voltage regulation per phase} = I(R\cos\varphi - X\sin\varphi) = \frac{260}{\sqrt{3}}(0.0608 \times 0.8 + 0.2716 \times 0.6)$$

$$= 7.3 + 24.46 = 31.76\,\text{V}.$$

So o.c. voltage of transformer must be 660 + 31.76

$$\text{and turns ratio must be: } \frac{11\,000/\sqrt{3}}{660 + 31.76} \text{ instead of } \frac{11\,000/\sqrt{3}}{660}$$

i.e. 9.18 instead of 9.623

$$\text{h.v. tapping must be at: } \frac{9.18}{9.623} = \underline{\underline{95.4\%}}.$$

Example 2.11

Two single-phase transformers operate in parallel to supply a load of 24 + j10 ω. Transformer A has a secondary e.m.f. of 440 V on open circuit and an internal impedance in secondary terms of 1 + j3 ω. The corresponding figures for Transformer B are 450 V and 1 + j4 ω. Calculate the terminal voltage, the current and terminal power-factor of each transformer.

The equivalent circuit is shown in the figure and yields the following equations:

$$E_A - Z_A I_A = V = (I_A + I_B)Z$$

$$E_B - Z_B I_B = V = (I_A + I_B)Z$$

Solving simultaneously gives:

$$I_A = \frac{E_A Z_B + (E_A - E_B)Z}{Z(Z_A + Z_B) + Z_A Z_B}$$

The expression for I_B is obtained by interchanging A and B in the above equation.

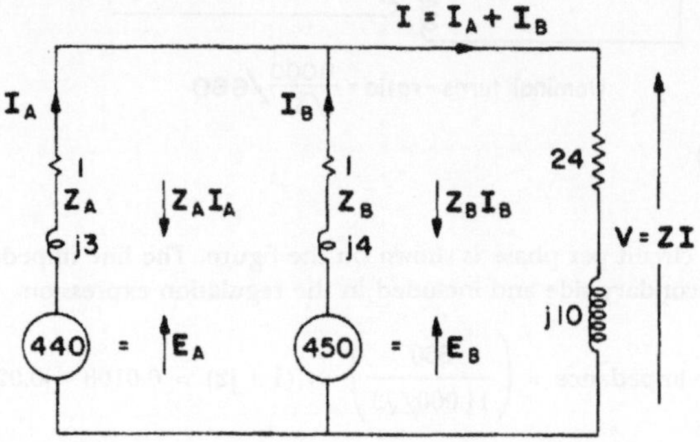

Figure E.2.11

A reference phasor must be chosen and if the two transformers have a common primary voltage, the equivalent-circuit e.m.f.s (E_{oc}) will be in phase, so will be a convenient choice.

Hence $E_A = 440 + j0$ and $E_B = 450 + j0$.

Substituting:

$$I_A = \frac{440(1 + j4) + (-10)(24 + j10)}{(24 + j10)(2 + j7) + (1 + j3)(1 + j4)} = \frac{200 + j1660}{-33 + j195} = \frac{1672\,\underline{/83°.13}}{197.8\underline{/99°.6}}$$

$$= 8.45\underline{/-16°.5}$$

$$I_B = \frac{450(1 + j3) + 10(24 + j10)}{197.8\underline{/99°.6}} = \frac{690 + j1450}{197.8\underline{/99°.6}} = \frac{1605.8\underline{/64°.6}}{197.8\underline{/99°.6}}$$

$$= 8.12\underline{/-35°}$$

$$I_A + I_B = 8.1 - j2.4 + 6.65 - j4.66 = 14.75 - j7.06 = 16.35\underline{/-25°.6}.$$

Load impedance $\mathbf{Z} = 24 + j10 = \underline{26/22°.6}$.

Terminal voltage $\mathbf{V} = \mathbf{Z}(\mathbf{I_A} + \mathbf{I_B}) = \underline{425.1/-3°}$.

Relative to V, $\mathbf{I_A} = \underline{8.45/-16°.5 + 3°} = \underline{8.45\,A \text{ at } \cos 13°.6 = 0.972\,\text{p.f. lag}}$.

Relative to V, $\mathbf{I_B} = \underline{8.12/-35° + 3°} = \underline{8.12\,A \text{ at } \cos 32° = 0.848\,\text{p.f. lag}}$.

Note 1 Although the two currents are similar, the power and reactive components of these currents are quite different. The transformers must be identical (in *per-unit* terms) with equal e.m.fs, if they are to share the load in proportion to their ratings. Calculations like the above determine whether any discrepancies are tolerable.

Note 2 To calculate the input primary currents, the secondary currents must be referred through the turns ratio and the corresponding components I_0 added.

Note 3 Two 3-phase transformers in parallel would be solved the same way, but using their *per-phase* equivalent circuits.

Note 4 The equation for $\mathbf{I_A}$ can be rearranged as:

$$\mathbf{I_A} = \frac{\mathbf{E_A Z_B}}{\mathbf{Z(Z_A + Z_B) + Z_A Z_B}} + \frac{\mathbf{E_A - E_B}}{\mathbf{Z_A + Z_B + Z_A Z_B / Z}}$$

The right-hand term is sometimes referred to as a circulating current due to the difference voltage. Strictly, it is only a mathematically expressed component of the total current, except when $Z = \infty$, i.e. when the load is open circuited. This true circulating current is then seen to be $(\mathbf{E_A - E_B})/(\mathbf{Z_A + Z_B})$, which could have been deduced directly from the circuit diagram.

2.2 Symmetrical components

A set of balanced 3-phase components (+), (–) and (0) can be combined to give an unbalanced set of 3-phase time-phasors as shown on Figure 2.1. An operator h, to give a +120° angular shift is introduced, noting that to give +240° shift, h would have to be applied twice. Thus, $h = 1/120°$, $h^2 = 1/240°$ and so $h^3 = 1/360° = 1$, showing that $h = -0.5 + j0.866$ expressed as a complex number and together with h^2 and h^3 (= 1), form the cube roots of unity. When added together, $h + h^2 + 1$ form a closed delta thus summing to zero. From Figure 2.1, using this operator and one positive,

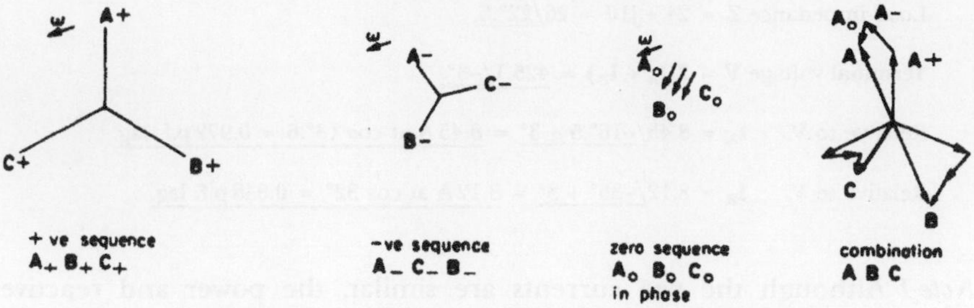

Figure 2.1 *Symmetrical components of an unbalanced system*

one negative and one zero sequence component, usually for the A phase, the sequence components and their combination for the other two phases can be expressed mathematically. The matrix form is convenient:

$$\begin{bmatrix} \mathbf{A} \\ \mathbf{B} \\ \mathbf{C} \end{bmatrix} = \begin{bmatrix} 1 & 1 & 1 \\ h^2 & h & 1 \\ h & h^2 & 1 \end{bmatrix} \begin{bmatrix} \mathbf{A}_{(+)} \\ \mathbf{A}_{(-)} \\ \mathbf{A}_{(0)} \end{bmatrix} \tag{2.2}$$

To obtain the sequence components from the actual unbalanced values we require the inverse matrix, which by standard techniques can be shown to lead to the following equation:

$$\begin{bmatrix} \mathbf{A}_{(+)} \\ \mathbf{A}_{(-)} \\ \mathbf{A}_{(0)} \end{bmatrix} = \frac{1}{3} \begin{bmatrix} 1 & h & h^2 \\ 1 & h^2 & h \\ 1 & 1 & 1 \end{bmatrix} \begin{bmatrix} \mathbf{A} \\ \mathbf{B} \\ \mathbf{C} \end{bmatrix} \tag{2.3}$$

So, under certain conditions, an unbalanced system **A, B, C** can be solved from three independent balanced systems if we know the impedance offered to positive, negative and zero-sequence currents. The phase impedances must be equal for this simple technique to be possible and the deeper implications of the method require further study, for example in References 2 and 3. Only introductory applications are being considered in this text.

Example 2.12

A 3-phase, 3-winding, $V_1/V_2/V_3$, star/star/delta transformer has a single-phase load of $\mathbf{I}_a$ amps on the secondary a-phase. The tertiary line terminals are open circuited but current may circulate within the delta. Neglecting the leakage impedances and assuming the magnetising current is so small as to be neglected, determine the currents in each phase of each winding.

This problem can be solved without recourse to symmetrical-component theory[1], but it is a conveniently simple example to introduce the application of this theory, which is also to be used later in Example 4.19. In the present instance, only the resolution of currents is required and the total m.m.f. for each sequence and for each phase may be summed to zero since $\mathbf{I}_m = 0$.

From the loading condition, the symmetrical components for the secondary follow from eqn (2.3):

$$\begin{bmatrix} \mathbf{I}_{2(+)} \\ \mathbf{I}_{2(-)} \\ \mathbf{I}_{2(0)} \end{bmatrix} = \frac{1}{3}\begin{bmatrix} 1 & h & h^2 \\ 1 & h^2 & h \\ 1 & 1 & 1 \end{bmatrix}\begin{bmatrix} \mathbf{I}_a \\ 0 \\ 0 \end{bmatrix} = \begin{bmatrix} \mathbf{I}_a/3 \\ \mathbf{I}_a/3 \\ \mathbf{I}_a/3 \end{bmatrix}$$

The three sequence components are equal for this particular loading condition. For the b and c phases there is of course an (h) or (h^2) displacement in accordance with Figure 2.1 and eqn (2.2). In the following solution, the primary and tertiary currents will be referred to the secondary, e.g. as $\mathbf{I}'_A$ and $\mathbf{I}'_{3A}$. Actually, in the tertiary winding, no positive or negative sequence currents can flow since these two balanced 3-phase systems must individually sum to zero. The only current which can flow with the line terminals open circuited is zero sequence, $\mathbf{I}'_3$ say, all in phase and equal, circulating in the closed delta. The m.m.f. for each phase and each sequence is now summed to zero. For the A phase:

+ve sequence	$0 = [\mathbf{I}'_{1A} + \mathbf{I}_a/3 + 0]\ (+)$
–ve sequence	$0 = [\mathbf{I}'_{1A} + \mathbf{I}_a/3 + 0]\ (-)$
zero sequence	$0 = [\mathbf{I}'_{1A} + \mathbf{I}_a/3 + \mathbf{I}_3']\ (0)$

Adding: $0 = \mathbf{I}'_A + \mathbf{I}_a + \mathbf{I}'_{3A}$

from which the A-phase current $\mathbf{I}'_A = -\mathbf{I}_a - \mathbf{I}'_{3A}$

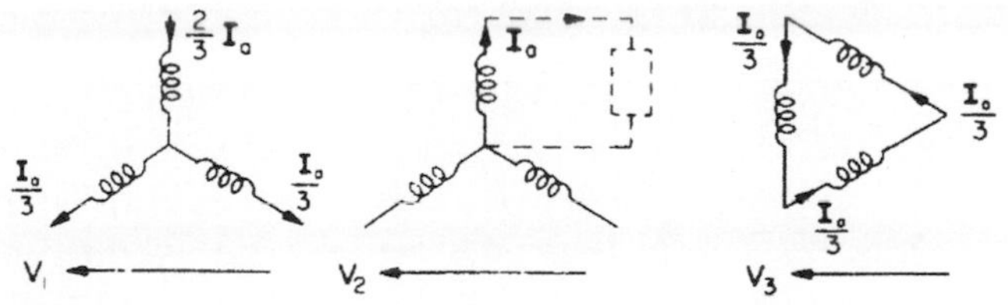

Figure E.2.12

For the B phase:

+ve sequence	$0 = [I'_{1B} + h^2 I_a/3 + 0]$ (+)
−ve sequence	$0 = [I'_{1B} + h\, I_a/3 + 0]$ (−)
zero sequence	$0 = [I'_{1B} + I_a/3 + I'_3]$ (0)

$$\text{Adding: } 0 = I'_B + 0 + I_{3A}{}'$$

from which the B-phase current $I'_B = -I'_{3A} = -I'_3$

Similarly, the C-phase current $I'_C = -I'_{3A} = -I'_3$

Combining the above answers:

$$I'_A = -I_a - I'_{3A}$$
$$I'_B = \quad - I'_{3A}$$
$$I'_C = \quad - I'_{3A}$$

$$\text{Adding} \quad 0 = -I_a - 3I'_{3A} \quad \text{so } I'_{3A} = -I_a/3$$

Hence $I'_A = -\frac{2}{3} I_a$ and $I'_B = I'_C = \frac{1}{3} I_a$

The solution is shown on the figure, the actual currents being obtainable using the turns ratios derived from V_1, V_2 and V_3.

The sequence currents can be checked from eqn (2.3) as:

$$I'_{A(+)} = -\tfrac{1}{3} I_a = I'_{A(-)} \text{ and } I'_{A(0)} = 0.$$

$$I'_{3(+)} = 0 = I'_{3(-)} \text{ and } I'_{3(0)} = -\tfrac{1}{3} I_a.$$

3 D.C. machines

For a first approach to the subject of electrical drives, the d.c. machine provides a simple introduction to the problems encountered since, for steady-state operation, it can be represented with reasonable accuracy, as a variable e.m.f. E (α speed × flux) behind the armature-circuit resistance R_a. The field m.m.f. may be provided in various ways and with more than one field winding. Non-linearities, especially saturation effects, may be involved in the calculations. The first three of the following examples bear somewhat lightly on machine-design aspects but the remainder are concerned with motoring; regenerating and braking. It is not the intention in this book to cover machine-winding design since this is a specialised study, for which the essential background is explained in Reference 1 for example. Here the emphasis is on electromechanical performance. Chapter 6 deals with transient operation and closed-loop control and Chapters 7 and 8 with d.c. machines in power-electronic circuits.

3.1 Revision of equations

The average (d.c.) e.m.f. generated in a winding with z_s conductors connected in series, of active length l, rotating at velocity $v = \pi dn$ where d is the armature diameter and n the speed of rotation in rev/sec, is obtained from:

$$E = B_{av} l v z_s \text{ and } B_{av} = \frac{\text{Flux per pole}}{\text{Cylindrical area/No. of poles}} = \frac{\phi}{\pi dl/2p}$$

$$E = \frac{2p\phi}{\pi dl} \times l \times \pi dn \times z_s = 2p\phi n z_s \qquad (3.1)$$

A d.c. machine armature-winding always has parallel paths and z_s is the total number of armature conductors Z, divided by the number of parallel paths. A simple wave winding has the smallest number, 2, and $z_s = Z/2$. For a simple lap winding, $z_s = Z/2p$.

Reference back to Sections 1.3 and 1.4 will be helpful in understanding the following development of the power-balance equation. For simplicity at this stage, the brush loss – typically $2 \times I_a$ watts – will be assumed to be included in the R_a effect.

From the basic circuit equation for a motor:

$$V = E + RI_a \tag{3.2}$$

Multiplying by I_a gives the power-balance equation:

$$VI_a = EI_a + RI_a^2 \tag{3.3}$$

Terminal power = Converted (air-gap) power + Electrical loss

$$P_{elec} = P_{gap} + \text{Electrical loss}$$

The power converted, EI_a, can be expressed in mechanical terms as $\omega_m T_e$ (rad/s × Nm) and hence by equating these two expressions:

$$\frac{T_e}{I_a} = \frac{E}{\omega_m} = \frac{2p\phi n z_s}{2\pi n} = \frac{pz_s}{\pi}\phi = k_\phi \tag{3.4}$$

Note that k_ϕ, the electromechanical conversion coefficient, is directly proportional to the flux per pole and has alternative units of e.m.f. per rad/s, or electromagnetic torque per ampere. It is sometimes called the speed constant or the torque constant and when SI units are used, these have the same numerical value, k_ϕ.

Hence:

$$E = k_\phi \omega_m \tag{3.5}$$

$$T_e = k_\phi I_a \tag{3.6}$$

and: $k_\phi = f(F_f, F_a)$

where F_a, the armature m.m.f. per pole, has some demagnetising effect usually, and this may be even greater than 10% of F_f on uncompensated machines. It will be neglected in the following examples except where special magnetisation curves are taken for series motors in Section 3.4. For a single field winding then:

$$k_\phi = f(I_f) \tag{3.7}$$

Several other equations may be derived from the above for the purpose of expressing a particular variable as the unknown. The most important is the speed/torque equation from:

$$\omega_m = \frac{E}{k_\phi} = \frac{V - RI_a}{k_\phi} = \frac{V}{k_\phi} - \frac{RT_e}{k_\phi^2} \tag{3.8}$$

I_a could be obtained from $(V-E)/R$ and k_ϕ from E/ω_m, but E (and I_a) are dependent variables, determined from power considerations and the armature current may have to be obtained from:

$$I_a = \frac{V - \sqrt{V^2 - 4R\omega_m T_m}}{2R} \tag{3.9}$$

and the flux from:

$$k_\phi = \frac{V + \sqrt{V^2 - 4R\omega_m T_m}}{2\omega_m} \tag{3.10}$$

which are derived from eqns (3.3, 3.4 and 3.8). The choice of sign before the radical gives the lower current and higher flux, which is correct for the normal, low-resistance case. The opposite choice is applicable to high-resistance circuits; see Reference 1 and Example 3.12. Note that on steady state, $T_e = T_m$.

3.2 Solution of equations

The following flow diagram has been prepared to act as a guide to thought while ascertaining a solution procedure. Not every possibility has been covered but the systematic approach indicated should be helpful in all cases. Non-linearities are taken into account either at the beginning or at the end of the solution. For example, if I_f is given, k_ϕ follows and hence leads to a solution of the linear equations for torque and speed, say. Alternatively, if k_ϕ is calculated from the linear equations, I_f follows from the magnetisation curve. On the mechanical side, if the non-linear relationship $T_m = f(\omega_m)$ is available, then any particular speed will yield the corresponding torque T_m which is the same as T_e on steady state. Alternatively, if $\omega_m = f(T_e)$ is calculated from the electromagnetic equations, then the intercept with the $T_m = f(\omega_m)$ curve determines the steady-state speed. Relationships for transient speed-calculations are also available from these two curves; see Chapter 6.

D.C. machine solution programme

Input data from:

V, I_a, Power, Efficiency η, ω_m, k_ϕ, I_f, Resistances, T_{loss}, $T_m = f(\omega_m)$

↓

Magnetisation curve given, or from test values of $(V, I_a, R, \omega_m, I_f)_{test}$

we can obtain $k_\phi = \left(\dfrac{V - RI_a}{\omega_m}\right)_{test}$ **as $f(I_f)$.**

Alternatively, rated $k_{\phi R}$ may be established from rated values of the above. The subsequent calculation(s) of k_ϕ could be expressed as *per-unit* flux values; namely $k_\phi / k_{\phi R}$.

↓ ↓ ↓ ↓

Given: V, R $I_f \rightarrow k_\phi$ $T_m \rightarrow T_e \rightarrow k_\phi I_a$ Find: $\omega_m = \dfrac{V - RI_a}{k_\phi}$	Given: V, R, η, ω_m, Output power Find: I_a, k_ϕ, I_f, T_e, T_{loss}	Given: V, R and $T_m = f(\omega_m)$ Find: I_a, k_ϕ, I_f	Given: V, Parameters and circuit configuration Find: $\omega_m = f(T_e)$ throughout all modes
↓	↓	↓	↓
Speeds	Currents	Field-circuit resistance	Speed/Torque curves

Example 3.1

A 4-pole d.c. armature wave winding has 294 conductors:

(a) What flux per pole is necessary to generate 230 V when rotating at 1500 rev/min?

(b) What is the electromagnetic torque at this flux when rated armature current of 120 A is flowing?

(c) How many interpole ampere turns are required with this current if the interpole gap density is to be 0.15 tesla and the effective radial air gap is l_g = 8 mm? Neglect the m.m.f. absorbed by the iron.

(d) Through what mechanical angle must the brushes be moved away from the quadrature axis if it is required to produce a direct-axis magnetisation of 200 At/pole?

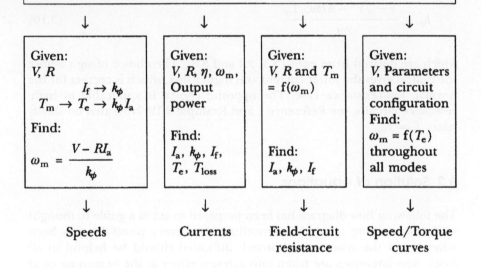

(a) From eqn (3.1) $E = 2 \times p \times \phi \times n \times z_s$

substituting: $230 = 2 \times 2 \times \phi \times \frac{1500}{60} \times \frac{294}{2}$

from which: $\underline{\phi = 0.0156\,\text{Wb.}}$

(b) From eqns (3.4) and (3.6) $T_e = k_\phi \cdot I_a = \frac{p.z_s}{\pi} \cdot \phi \cdot I_a$

$$= \frac{2 \times 147}{\pi} \times 0.0156 \times 120 = \underline{175.7\,\text{Nm.}}$$

(c) On the interpolar (quadrature) axis, maximum armature At per pole F_a occurs, and the interpole m.m.f. must cancel this and also provide sufficient excess to produce the required commutating flux opposing that of the armature.

$$F_a = \frac{\text{Total ampere turns}}{\text{No. of poles}} = \frac{\text{amps/conductor} \times \text{conductors}/2}{2p}$$

$$= \frac{120/2 \times 294/2}{4} = 2205\,\text{At/pole}$$

Required

$$H = \frac{\text{At}}{l_g} = \frac{B}{\mu_0}. \text{ Hence At} = \frac{B \cdot l_g}{\mu_0} = \frac{0.15 \times 8 \times 10^{-3}}{4\pi/10^7} = 955\,\text{At/pole}$$

Hence total interpole m.m.f. required = 2205 + 955 = $\underline{3160\,\text{At/pole}}$

This would typically be obtained with 27 turns per pole carrying $I_a = 120\,\text{A}$, with small adjustments to the air-gap length if necessary, following commutation tests.

(c) From the diagram, a brush axis shift of α produces a demagnetising (or magnetising) m.m.f. of $F_a \times 2\alpha/180$ At/pole. Hence:

$$200 = 2205 \times 2\alpha/180$$

so: $\alpha = 8.16$ $\underline{\text{electrical degrees}}$

$= 8.16/p = \underline{4.08 \text{ mechanical degrees}}$

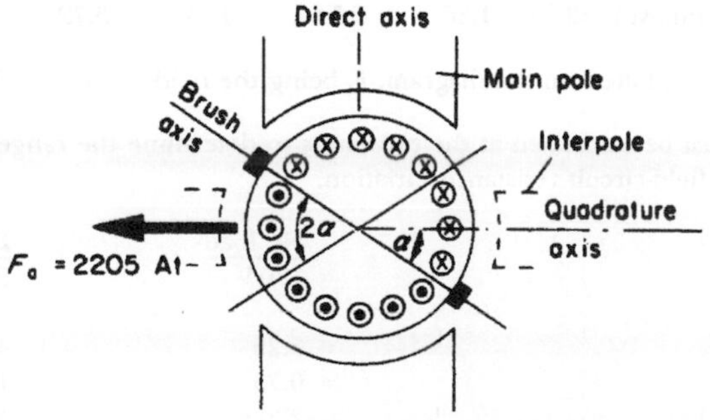

Figure E.3.1 *(2-pole machine shown for simplicity)*

Example 3.2

A d.c. shunt-wound, (self-excited) generator rated at 220 V and 40 A armature current has an armature resistance of 0.25 Ω. The shunt field resistance is 110 Ω and there are 2500 turns per pole. Calculate:

(a) the range of external field-circuit resistance necessary to vary the voltage from 220 V on full load to 170 V on no load when the speed is 500 rev/min;
(b) the series-winding m.m.f. required to give a level-compound characteristic at 220 V when running at 500 rev/min;
(c) the maximum voltage on o.c. if the speed is reduced to 250 rev/min and all external field resistance is cut out.

Armature reaction and brush drops may be neglected.

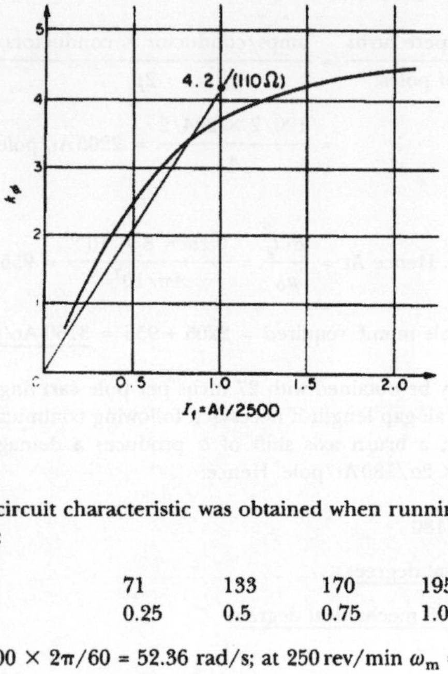

Figure E.3.2

The following open-circuit characteristic was obtained when running at 500 rev/min with the shunt field excited:

E.M.F.	71	133	170	195	220	232 V
Field current	0.25	0.5	0.75	1.0	1.5	2.0 A

ω_m at 500 rev/min = 500 × 2π/60 = 52.36 rad/s; at 250 rev/min ω_m = 26.18 rad/s.

Hence k_ϕ = e.m.f. (above)/52.36 = 1.36 2.54 3.25 3.72 4.2 4.43

The k_ϕ/I_f curve is plotted on the diagram, F_f being the field current × 2500.

(a) The e.m.f. must be calculated at the two limits to determine the range of field current and hence of field-circuit resistance variation.

	No load	Full load
$E = V + RI_a$	= 170 V	220 + 40 × 0.25 = 230 V
$k_\phi = E/52.36$	= 3.25	4.39
I_f from curve	= 0.75	1.86
$R_F = (E - RI_a)/I_f$	= 226.4	118.3 Ω
External field-circuit resistance = R_F − 110	= 116.4 Ω	8.3 Ω

(b) The terminal voltage must be 220 V on no load and on full load.

		No load	Full load
V	=	220	220
$E = V + R_a I_a$	=	220	230
(On no load, $I_a = I_f$ is small).			
Required k_ϕ at $\omega_m = 52.36$	=	4.2	4.39
Field m.m.f. required, from curve	=	1.5×2500	1.86×2500
Difference = series m.m.f. required	=	$(1.86 - 1.5) \times 2500$ =	840 At/pole

(c) The open-circuit curve could be redrawn in terms of e.m.f. against field current with the e.m.f. reduced in the ratio 250/500. Alternatively, the k_ϕ curve can be used, since it is the e.m.f. at 250 rev/min scaled down by the divisor 26.18 rad/s. The slope of the resistance line (V/I_f) for 110 ω must also be reduced, to 110/26.18 (= 4.2) as shown. The field line intersects the characteristic at a k_ϕ (= e.m.f./26.18) of 3.38. Hence the voltage on open circuit – which is the terminal voltage neglecting the very small $R_a I_f$ drop – is 3.38×26.18 = 88.5 V.

Example 3.3

The machine of the last question is to be run as a motor from 220 V. A speed range of 2/1 by field control is required. Again neglecting the effect of armature reaction and brush drop and assuming $I_a = 0$ on no load, calculate:

(a) the range of external field-circuit resistance required, as a shunt motor, to permit speed variation from 500 rev/min on no load, to 1000 rev/min with the armature carrying its rated current of 40 A;
(b) the value of the series-field ampere turns required to cause the speed to fall by 10% from 500 rev/min on no load, when full-load current is taken;
(c) the speed regulation (no load to 40 A load) with this series winding in circuit and the shunt field set to give 1000 rev/min on no load.
(d) By how much would this series winding increase the torque at 40 A compared with condition (a) at the minimum field setting?

Parts (a) and (b) require calculation of the k_ϕ range to find the excitation needs.

(a)		I_a = 0 A, no load	I_a = 40 A
Speed		500	1000 rev/min
ω_m		52.36	104.7 rad/s
$E = 220 - 0.25 \times I_a$		220	210 V
$k_\phi = E/\omega_m$		4.2	2.0 Nm/A
I_f from mag. curve		1.5	0.38 A
$R_F = 220/I_f$		146.7	579 ω
External resistance = $R_F - 110$	=	36.7	469 ω

(b)			
Speed		500	450 rev/min
ω_m		52.36	47.12 rad/s
$E = 220 - 0.25 \times I_a$		220	210 V
$k_\phi = E/\omega_m$		4.2	4.46 Nm/A
Field ampere turns from mag. curve	=	1.5×2500	2.05×2500
Difference is required series m.m.f.	=	$(2.05 - 1.5) \times 2500$ =	1375 At/pole

With each series turn carrying 40 A this would require 1375/40 = 34 + turns. With 40 turns say and a diverter, the test performance could be adjusted to give the 10% speed regulation specified.

(c) The speed will be obtained from $\omega_m = E/k_\phi$.
On no load, k_ϕ = 220 V/104.7 rad/s = 2.1.
From mag. curve this requires 0.4 A × 2500 = 1000 At/pole shunt excitation.
On load, the total excitation is therefore 1000 + 1375 = 2375 At/pole.
k_ϕ will therefore correspond to 2375/2500 = 0.95 A giving: 3.65 Nm/A.

Hence, ω_m on load will be $\frac{E}{k_\phi} = \frac{210}{3.65}$ = 57.5 rad/s = 549 rev/min

Speed regulation from 1000 rev/min is therefore $\frac{1000 - 549}{1000} \times 100$ = 45.1%.

Note the great increase from the 500 rev/min condition because of the weak shunt field.

(d) $T_e = k_\phi, I_a$. At minimum field, 1000 rev/min $T_e = 2 \times 40 = 80$ Nm

With additional series excitation $T_e = 3.65 \times 40 = 146$ Nm

So although the speed has fallen considerably, due to the series winding, the electromagnetic torque has increased by 66/80 = 82% for the same reason and the air-gap power is the same.

Example 3.4

In the shunt motor of the last question, the no-load armature current was neglected. In fact, the total no-load input current is 5 A when both field and armature are directly connected across the 220 V supply, the output (coupling) torque being zero, so that the only torque is that due to the friction, windage and iron losses. Calculate the speed, output power and efficiency when the load has increased to demand rated armature current, 40 A.

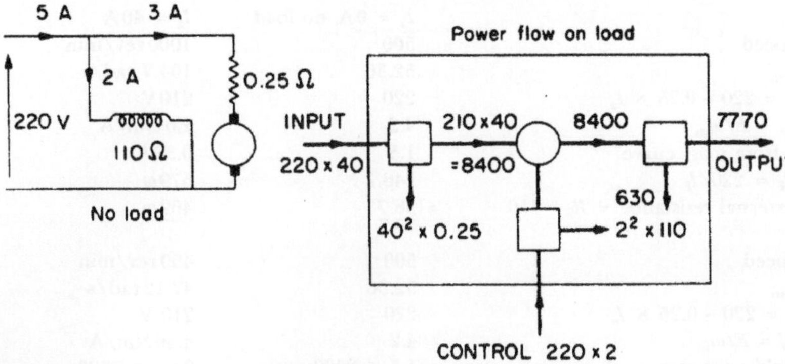

Figure E.3.4

The no-load condition is shown on the first diagram where it is seen that the armature current is 3 A, the field taking 220/110 = 2 A.
The value of k_ϕ from the o.c. curve at 2 A is 4.43 Nm/A.
The no-load e.m.f. is 220 − 0.25 × 3 = 219.25 V.
Hence speed $\omega_m = E/k_\phi$ = 219.25/4.43 = 49.49 rad/s = 473 rev/min.
The air-gap power $P_{gap} = \omega_m \cdot T_e = E \cdot I_a$ = 219.25 × 3 = 658 watts is consumed in friction, windage and iron losses and corresponds to T_{loss} = 658/49.49 = 13.3 Nm.

For the load condition, with 40 A in the armature, the second diagram is a useful representation of the power flow. The explanation of the numerical values involves a few minor calculations which can be understood by reference back to Figure 1.10.

Speed = E/k_ϕ = 210/4.43 = 47.4 rad/s = 453 rev/min.
If we neglect any small change in loss torque with this speed fall then:
Mechanical loss = $\omega_m \cdot T_{loss}$ = 47.4 × 13.3 = 630 watts.
Output power = P_{gap} − mechanical loss = 210 × 40 − 630 = 7.77 kW = 10.4 hp.
Input power = 220(40 + 2) = 9.24 kW so efficiency = 7.77/9.24 = 84.1%.

Example 3.5

In the back-to-back test circuit shown, Machine 1 is a motor driving Machine 2 which is a generator. The generated power is fed back into the common 250-V line so that only the machine losses have to be supplied. Currents in various parts of the circuit, together with the

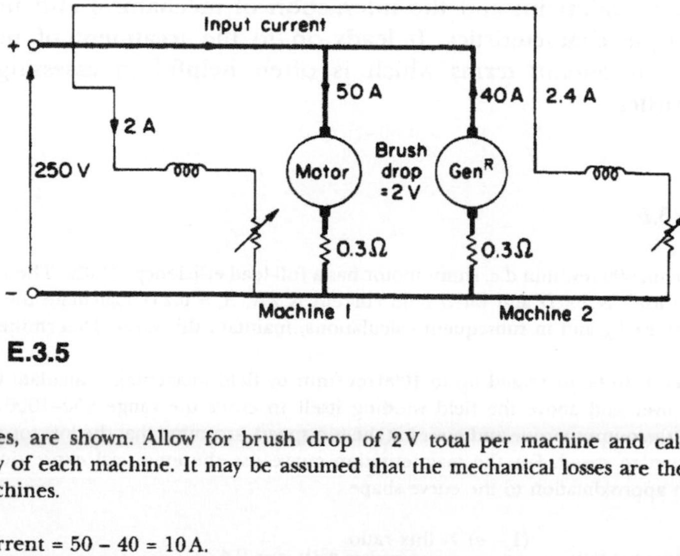

Figure E.3.5

resistances, are shown. Allow for brush drop of 2 V total per machine and calculate the efficiency of each machine. It may be assumed that the mechanical losses are the same for both machines.

Input current = 50 − 40 = 10 A.
Input power to armature circuits = 250 × 10 = 2500 watts.
Total armature losses, excluding friction, windage and iron losses
$= 50^2 \times 0.3 + 2 \times 50 + 40^2 \times 0.3 + 2 \times 40 = 1410$ watts.
∴ Total mechanical loss = 2500 − 1410 = 1090 watts = 545 watts per machine

$$\text{Motor efficiency} = \frac{\text{Input} - \text{losses}}{\text{Input}} = 1 - \frac{(50^2 \times 0.3) + (2 \times 50) + 545 + (250 \times 2)}{250(50 + 2)}$$

$$= 1 - \frac{1895}{13\,000}$$

$$= \underline{85.42\%}$$

$$\text{Generator efficiency} = \frac{\text{Output}}{\text{Output} + \text{losses}}$$

$$= \frac{250 \times 40}{(250 \times 40) + (40^2 \times 0.3) + (2 \times 40) + 545 + (250 \times 2.4)}$$

$$= \frac{10\,000}{11\,705} = \underline{85.43\%}$$

The most convenient expressions to suit the data have been chosen. It is a coincidence that efficiencies are the same. The motor has the higher copper loss and the generator the higher flux and hence field loss and, in practice, a slightly higher iron loss also. But note also that efficiency is a function of output and for the generator this is 10 kW whereas for the motor it is 11.105 kW.

The next few examples illustrate the consequences of changing the machine parameters, sometimes with the object of achieving a certain speed against a specified mechanical load characteristic. This brings in the overall drive viewpoint and the interaction of mechanical and machine speed/torque characteristics. It leads on to the treatment of machine equations in *per-unit* terms which is often helpful in assessing drive characteristics.

Example 3.6

A 500-V, 60-hp, 600-rev/min d.c. shunt motor has a full-load efficiency of 90%. The resistance of the field itself is 200 Ω and rated field current is 2 A. $R_a = 0.2\,\Omega$. Calculate the full-load (rated) current I_{aR} and in subsequent calculations, maintain this value. Determine the loss torque.

The speed is to be increased up to 1000 rev/min by field weakening. Calculate the extra resistance, over and above the field winding itself to cover the range 600–1000 rev/min. Determine the output torque and power at the top speed, assuming that the loss torque varies in proportion to speed. For the magnetisation curve use the empirical expression below, which is an approximation to the curve shape.

$$\text{Field-current ratio} = \frac{(1 - a) \times \text{flux ratio}}{1 - a \times \text{flux ratio}} \quad \text{with } a = 0.4$$

where the flux ratio is that between a particular operating flux (E/ω_m) and rated flux ($k_{\phi R}$). The field-current ratio is that of the corresponding field currents.

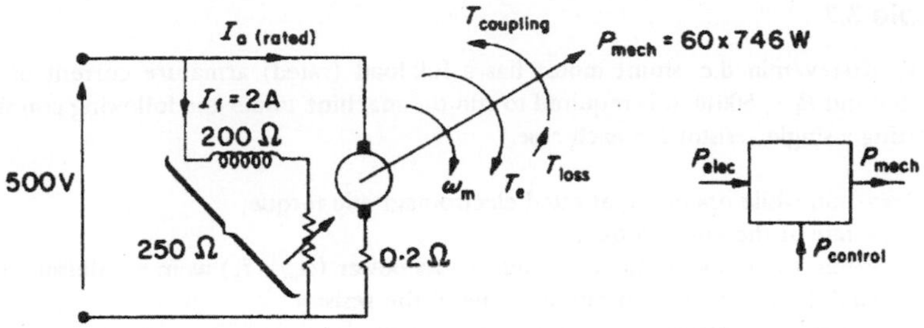

Figure E.3.6

The data are assembled on the figure as a convenient *aide-memoire*, together with a skeleton power-flow diagram from which:

Full-load efficiency $\eta_R = \dfrac{P_{mech}}{P_{elec} + P_{control}} = \dfrac{60 \times 746}{500 \times I_{aR} + 500 \times 2} = \dfrac{90}{100}$

from which: $I_{aR} = \underline{97.5\,\text{A}}$

Hence: $k_{\phi R} = \dfrac{E_R}{\omega_{mR}} = \dfrac{500 - 0.2 \times 97.5}{20\pi} = 7.65\,\text{Nm/A}$

and: $T_{eR} = k_{\phi R} \cdot I_{aR} = 7.65 \times 97.5 = 745.9\,\text{Nm}$

$T_{coupling} = \dfrac{60 \times 746}{20\pi} = \underline{712.4\,\text{Nm}}$

$T_{loss} = T_{eR} - T_{coupling} = \underline{33.5\,\text{Nm}}$

At 1000 rev/min, $T_{loss} = 33.5 \times 1000/600 = 56\,\text{Nm}$

At 1000 rev/min, $k_\phi = \dfrac{500 - 0.2 \times 97.5}{1000 \times 2\pi/60} = 4.59\,\text{Nm/A}$

$\therefore$ flux ratio $= 4.59/7.65 = 0.6.$

Hence, field-current ratio $= \dfrac{I_{f(1000)}}{2\,\text{A}} = \dfrac{0.6 \times 0.6}{1 - 0.4 \times 0.6}$

from which: $I_{f(1000)} = 0.947\,\text{A}$

and: $R_F = 500/0.947$

$= 528\,\Omega$ so external resistance $= \underline{328\,\Omega}$

Coupling torque $= T_e - T_{loss} = 4.59 \times 97.5 - 56 = \underline{392\,\text{Nm}}$

Mechanical output power $= \omega_m \cdot T_{coupling} = 104.7 \times 392 = \underline{41\,\text{kW}} = 55\,\text{hp}$

Example 3.7

A 500-V, 500-rev/min d.c. shunt motor has a full-load (rated) armature current of 42 A. $R_a = 0.6\,\omega$ and $R_f = 500\,\omega$. It is required to run the machine under the following conditions by inserting a single resistor for each case.

(a) 300 rev/min while operating at rated electromagnetic torque;
(b) 600 rev/min at the same torque;
(c) 800 rev/min while operating at the same gross power ($\omega_m \cdot T_e$) as in condition (b).
For each condition, find the appropriate value of the resistor.
The following magnetisation curve was taken on open circuit at 500 rev/min:

Field current	0.4	0.6	0.8	1.0	1.2 A
Generated e.m.f.	285	375	445	500	540 V

The test speed $\omega_{m(test)} = 500 \times 2\pi/60 = 52.36\,\text{rad/s}.$

$k_\phi = E/\omega_{m(test)} = E/52.36 =$ 5.44 7.16 8.5 9.56 10.3 Nm/A

$$\text{Rated } k_\phi\ (k_{\phi R}) = \frac{V_R - R_a \cdot I_{aR}}{\omega_{mR}} = \frac{500 - 0.6 \times 42}{52.36} = 9.07\,\text{Nm/A}$$

Rated $T_e(T_{eR}) = k_{\phi R} \cdot I_{aR} = 9.07 \times 42 = 381\,\text{Nm}.$
Rated field current from curve at $k_\phi = 9.07$, is 0.9 A $\therefore R_F = 500/0.9 = 555\,\Omega$

Now consider the speed/torque equation (3.8): $\omega_m = \dfrac{V}{k_\phi} - \dfrac{R \cdot T_e}{k_\phi^2}$

It can be seen that at a fixed terminal voltage V, any increase of speed beyond the rated value can only be obtained by reduction of flux; i.e. by inserting extra resistance in the field circuit.

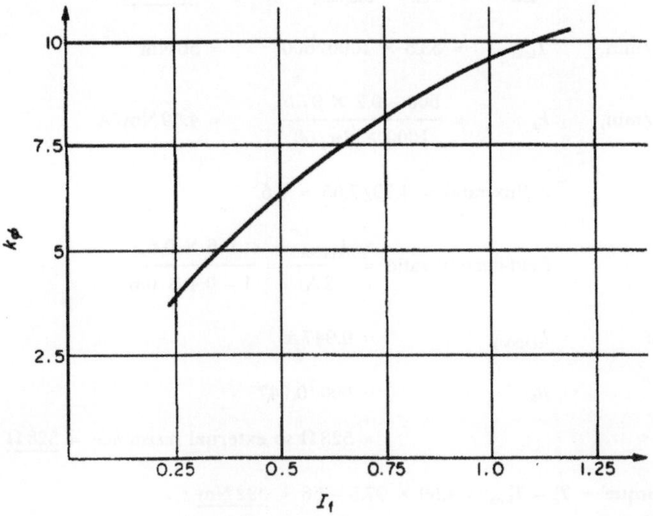

Figure E.3.7

A reduction of speed, without exceeding the flux limit imposed by saturation, can only be obtained by operating on the second term – in practice this means increasing R since reduction of k_ϕ would give rise to excessive armature currents unless R is relatively high – see Reference 1 and speed/flux curves of Example 3.12.

Hence, for (a), rearranging the equation: $R = \dfrac{(V - k_\phi \cdot \omega_m)}{T_e/k_\phi} \left(= \dfrac{V-E}{I_a} \right)$

Since $\omega_m = 300 \times 2\pi/60 = 10\pi$: $\quad R = \dfrac{500 - 9.07 \times 10\pi}{381/9.07}$

$= 5.12\,\Omega$, an extra $5.12 - 0.6 = \underline{4.52\,\Omega}$

For (b) we require k_ϕ, not knowing the armature current and hence eqn (3.10) will have to be used. At 600 rev/min $\omega_m = 20\pi$ so:

$$k_\phi = \frac{V + \sqrt{V^2 - 4 \times R \times \omega_m \times T_e}}{2\omega_m} = \frac{500 + \sqrt{500^2 - 4 \times 0.6 \times 20\pi \times 381}}{2 \times 20\pi}$$

$= 7.47$ Nm/A

From the magnetisation curve, this requires $I_f = 0.64$ A and $R_F = 500/0.64 = 781\,\Omega$.

Hence extra field-circuit resistance is $781 - 555 = \underline{226\,\Omega}$

For (c), since the same $\omega_m T_e$ product is specified, the only difference in the k_ϕ equation above is to the denominator which becomes $2 \times 800 \times 2\pi/60 = 167.6$.

k_ϕ is therefore 5.602 requiring an I_f of 0.41 A and $R_F = 500/0.41 = 1219\,\Omega$.

$\therefore$ extra resistance is $1219 - 555 = \underline{664\,\Omega}$.

Example 3.8

A 220-V, 1000-rev/min, 10-hp, d.c. shunt motor has an efficiency of 85% at this rated, full-load condition. The total field-circuit resistance R_F is then 100 Ω and $R_a = 0.4\,\Omega$. Calculate the rated values of current, flux and electromagnetic torque; (I_{aR}, $k_{\phi R}$ and T_{eR}). Express your answers to the following questions in *per unit* where appropriate by dividing them by these reference or base values, which are taken as 1 *per unit*. Take 220 V (V_R) as 1 *per unit* voltage.

(a) Find the applied voltage to give half rated speed if T_m is proportional to $\omega_m{}^2$; k_ϕ and R_a unchanged.
(b) Find the extra armature-circuit resistance to give half rated speed if T_m is proportional to speed; k_ϕ and V being 1 *per unit*.
(c) Find the *per-unit* flux to give 2000 rev/min if $T_m \propto 1/\omega_m$ (constant power) and also the armature current, if $V = 1$ *per unit* and R_a is the normal value.
(d) What electromagnetic torque is developed if the voltage, flux and speed are at half the rated values and there is an extra 2 Ω in the armature circuit?

The data could usefully be assembled on a diagram as for Example 3.6. Referring also to the associated power-flow diagram:

$$\text{Motor efficiency } \eta_R = \frac{P_{mech}}{P_{elec} + P_{control}}$$

$$0.85 = \frac{10 \times 746}{220 \times I_{aR} + 220^2/100}$$

from which: $I_{aR} = \underline{37.7\,\text{A}}$.

$$\text{Hence: } k_{\phi R} = \frac{220 - 0.4 \times 37.7}{1000 \times 2\pi/60} = \underline{1.957} \text{ and therefore } T_{eR} = 1.957 \times 37.7 = \underline{73.8\,\text{Nm}}$$

The remaining questions can all be answered from the speed/torque equation:

$$\omega_m = \frac{V}{k_\phi} - \frac{R \cdot T_e}{k_\phi{}^2} \text{ where } T_e = T_m \text{ on steady state.}$$

(a) At half speed, $\omega_m = 52.36\,\text{rad/s}$ and $T_m = (\tfrac{1}{2})^2 \times 73.8$

$$\text{From the equation: } V = k_\phi \cdot \omega_m + \frac{R \cdot T_m}{k_\phi} = 1.957 \times 52.36 + \frac{0.4}{1.957} \times \frac{73.8}{4}$$

$$= 106.2\,\text{V} = \underline{0.48 \textit{ per unit}}$$

(b) A different mechanical characteristic applies so $T_m = \tfrac{1}{2} \times 73.8 = 36.9\,\text{Nm}$

$$\text{From the equation: } R = \frac{V - k_\phi \cdot \omega_m}{T_e/k_\phi} \left(= \frac{V - E}{I_a} \right) = \frac{220 - 1.957 \times 52.36}{36.9/1.957}$$

$$= 6.23\,\Omega, \text{ i.e. an extra } 6.23 - 0.4 = \underline{5.83\,\Omega}$$

$$\text{(c) } k_\phi = \frac{V + \sqrt{V^2 - 4 \cdot R \cdot \omega_m \cdot T_m}}{2\omega_m} = \frac{220 + \sqrt{220^2 - 4 \times 0.4 \times 104.7 \times 73.8}}{2 \times 2000 \times 2\pi/60}$$

$$= 0.9784\,\text{Nm/A} = 0.9784/1.957 = \underline{0.5 \textit{ per unit}}$$

Note that rated power and speed have been used under the radical because for this mechanical load, the power is stated to be constant. Check $I_a = I_{aR}$, eqn (3.9).

$$\text{(d) } T_e = k_\phi \cdot \frac{(V - k_\phi \cdot \omega_m)}{R} = \frac{1.957}{2} \frac{(110 - 0.978 \times 52.36)}{2 + 0.4} = 24\,\text{Nm}$$

$$= 24/73.8 = \underline{0.325 \textit{ per unit}}$$

and $I_a = 24/0.978 = 24.5\,\text{A}$

3.3 *Per-unit* notation

The last question introduced the idea of expressing quantities in *per unit*, i.e. as fractions of some base or reference quantity. It is possible to solve the question throughout in *per-unit* notation and after the following explanation, it would be a good exercise to try this. The method is sometimes convenient, especially when integrating the mechanical system parameters into the drive and in simplifying scaling factors for computer solutions. For a d.c. machine, with the appropriate choice of base values, the equations, apart from being dimensionless, are the same as those used for actual values as explained in Reference 1. The most convenient base values are:

Rated voltage	V_R	= 1 *per-unit* voltage
Rated current	I_{aR}	= 1 *per-unit* current
Rated flux	$k_{\phi R}$	= 1 *per-unit* flux

From these:

Rated torque = $k_{\phi R} \cdot I_{aR}$ is also 1 *per-unit* torque
Rated power = $V_R \cdot I_{aR}$ is also 1 *per-unit* power

and 1 *per-unit* resistance = V_R/I_{aR}, since only three of the seven practicable quantities, derived from the products and quotients of the first three, can be defined independently. It follows also that 1 *per-unit* speed is predetermined as $V_R/k_{\phi R}$, since these two parameters have been chosen. Therefore rated speed is not 1 *per unit* but:

$$\frac{E_R}{k_{\phi R}} = \frac{V_R - R_a \cdot I_{aR}}{k_{\phi R}} = \frac{1 - R_a \times 1}{1} = 1 - R_a \textit{ per unit.}$$

Per-unit resistance is:

$$\frac{\text{ohmic value}}{\text{base value}} = \frac{R_a}{V_R/I_{aR}} = \frac{R_a \cdot I_{aR}}{V_R},$$

which is the fraction of base voltage, which is absorbed across the armature-circuit resistance at base current.

In the following example, the field current is also expressed in *per unit* using the same empirical expression as in Example 3.6.

Example 3.9

A d.c. shunt motor runs at 1000 rev/min when supplied from rated voltage, at rated flux and drives a total mechanical load, including the loss torque, which has coulomb friction, viscous friction and square-law components given by the following expression:

$$T_m = 30 + 30\left(\frac{\text{rev/min}}{1000}\right) + 30\left(\frac{\text{rev/min}}{1000}\right)^2 \text{ Nm}$$

The armature resistance is 0.06 *per unit* and the magnetisation curve can be approximated by the empirical expression:

$$I_f \text{ per unit} = \frac{0.6 \times \phi \text{ per unit}}{1 - 0.4 \times \phi \text{ per unit}}$$

Calculate:

(a) The values of 1 *per-unit* torque in Nm and 1 *per-unit* speed in rev/min;
(b) the required I_f and the value of I_a in *per unit*, if the speed is to be 600 rev/min with the terminal voltage set at 0.5 *per unit*;
(c) the required I_f and the value of I_a in *per unit* when the terminal voltage is set at the rated value and the speed is adjusted to (i) 1200 rev/min; (ii) 0.8 *per unit*;
(d) the required terminal voltage in *per unit* if the resistance is increased to 0.2 *per unit*, the field current is reduced to 0.6 *per unit* and the speed is to be set at the rated value.

In this comprehensive question, since the non-linear $T_m = f(\omega_m)$ relationship is given and $T_e = T_m$ in the steady state, then T_e follows if the speed is specified and conversely, any particular torque will correspond to a particular speed. Thereafter, the solution is just applying the various equations developed at the beginning of this chapter. The quadratic expression for I_a eqn (3.9), must be used because the power is given, not the value of e.m.f. or flux.

(a) 1 *per-unit* torque, from the question, must occur at a speed of 1000 rev/min

$$\text{viz.}\quad 30\left(1 + \frac{1000}{1000} + \frac{1000^2}{1000^2}\right) = \underline{90\,\text{Nm}}$$

Rated speed = 1 − R_a *per unit* is 1000 rev/min.

$$\therefore 1 \text{ per-unit speed} = \frac{1000}{1 - 0.06} = \underline{1064\,\text{rev/min}}$$

Preliminary calculation of T_e (= T_m) at stated speeds:

Part	Rev/min	*p.u.* speed $= \frac{\text{rev/min}}{1064}$	Torque, Nm	*p.u.* torque $= \frac{T_m}{90}$
	1064	1		
(d)	1000	0.94	90	1
(b)	600	0.564	$30(1 + 0.6 + 0.6^2) = 58.8$	0.653
(c) (i)	1200	1.128	$30(1 + 1.2 + 1.2^2) = 109.2$	1.213
(c) (ii)	851	0.8	$30(1 + 0.851 + 0.851^2) = 77.3$	0.858

Calculations for parts:	(b)	(c) (i)	(c) (ii)
ω_m	0.564	1.128	0.8
T_e	0.653	1.213	0.858
Power $= \omega_m T_e$	0.368	1.368	0.686
$X = V^2 - 4R \cdot \omega_m T_e$	$0.5^2 - 0.24 \times 0.368 = 0.1617$	$1^2 - 0.24 \times 1.368 = 0.6717$	$1^2 - 0.24 \times 0.686 = 0.835$
$I_a = \dfrac{V - \sqrt{X}}{2R}$	0.816	1.504	0.717
$k_\phi = T_e/I_a$	0.8	0.807	1.196
$I_f = \dfrac{0.6k_\phi}{1 - 0.4k_\phi}$	0.706	0.715	1.376

(d) The field current is set at 0.6 *per unit* $= \dfrac{0.6k_\phi}{1 - 0.4k_\phi}$ hence $k_\phi = 0.714$

The speed is to be the rated value (0.94 *per unit*)
The torque will therefore be 1 *per unit*.
Hence $I_a = T_e/k_\phi = 1/0.714 = 1.4$ *per unit*
Required $V = k_\phi \cdot \omega_m + R \cdot I_a = 0.714 \times 0.94 + 0.2 \times 1.4 = \underline{0.951}$ *per unit*

Note that the answers to part (c) show that field control of speed is not satisfactory with this mechanical load because the armature current becomes excessive at high speeds and the field current is excessive at speeds lower than rated. Part (d) shows a similar situation.

Example 3.10

A d.c. shunt motor is being considered as a drive for different mechanical loads having the following characteristics: (a) Constant power ($\omega_m T_m$); (b) constant torque and (c) torque proportional to speed. It is desired to know the effects on armature current and speed of making various changes on the electrical side. Taking as a basis that rated voltage, rated armature current and field current give rated speed and torque, express armature current and speed in *per unit* when the following changes are made:

(i) field current reduced to give half flux;
(ii) armature supply-voltage halved;
(iii) armature voltage and field flux both halved.

Consider loads (a), (b) and (c) in turn and neglect all machine losses.

The required equations for current and speed are: $I_a = T_e/k_\phi$ and $\omega_m = V/k_\phi$ and all calculations are in *per unit.*

Mechanical load characteristic	(a) $\omega_m T_m$ const. i.e. $T_m \propto 1/\omega_m$	T_m const. (b)	$T_m \propto$ speed (c)
(i) $k_\phi = 0.5$; $V = 1$.			
$\omega_m = V/k_\phi$	2	2	2
$T_e = T_m$	0.5	1	2
$I_a = T_e/k_\phi$	1	2	4
(ii) $V = 0.5$; $k_\phi = 1$.			
ω_m	0.5	0.5	0.5
T_m	2	1	0.5
I_a	2	1	0.5
(iii) $V = 0.5$; $k_\phi = 0.5$			
ω_m	1	1	1
T_m	1	1	1
I_a	2	2	2

Again, this example shows, in a simple manner, what is, and what is not a feasible strategy in the control of d.c. machines and how the nature of the mechanical load determines this; one armature-current overload is as high as four times the rated value.

Example 3.11

A d.c. motor has a *per-unit* resistance of 0.05. Determine the two values of current and of flux at which rated torque can be developed at rated speed when supplied from rated voltage.

$$I_a = \frac{V \mp \sqrt{V^2 - 4R_a \cdot \omega_m \cdot T_e}}{2R_a} = \frac{1 \mp \sqrt{1^2 - 4 \times 0.05 \times (1 - 0.05) \times 1}}{2 \times 0.05}$$

$$= \frac{1 \mp \sqrt{0.81}}{0.1} = \underline{1 \; per \; unit} \text{ or } \underline{19 \; per \; unit}$$

The numerator is the same for $k_\phi = \dfrac{1 \mp \sqrt{0.81}}{2 \times (1 - 0.05)} = \underline{1 \; per \; unit} \text{ or } \dfrac{1}{19} \; per \; unit$

Clearly, the only practical solution is the first one with $k_\phi = I_a = 1$ *per unit,* even though the same torque of 1 *per unit* is given by the second solution. This is a relatively low-resistance machine. The next example shows the effect of an increased armature-circuit resistance, when, as on some small servo motors and with 'constant' current supplies, speed increase is obtained by increasing the field current, working on the rising part of the speed/flux characteristic; see Reference 1.

Example 3.12

For a separately excited d.c. motor which at rated voltage, flux and armature current delivers rated torque at rated speed $(1 - R_a)$ *per unit*, show that the maximum speed which can be obtained by field weakening is:

(a) $V^2/4R$ *per unit* for a constant-torque load equal to rated torque and:
(b) $V \times \sqrt{(1 - R_a)/4R}$ *per unit* if rated torque is the same, but is proportional to speed and the circuit resistance is R which is not necessarily equal to R_a.
(c) Calculate for resistances of $R_a = 0.05$ and for $R = 0.5$, the values of ω_{max} in *per unit* and the values of armature current and flux at this speed, for the constant-torque load. Repeat the calculation, but this time for the case of load torque proportional to speed.
(d) For the same motor determine the required circuit resistance to permit continuous speed increase by field increase up to rated flux with rated voltage applied. Consider both mechanical load-characteristics as before.

(a) For the constant-torque load, $T_m = 1$ and the equation is $\omega_m = \dfrac{V}{k_\phi} - \dfrac{R}{k_\phi^2}$

$$\frac{d\omega_m}{dk_\phi} = -\frac{V}{k_\phi^2} + 2 \times \frac{R}{k_\phi^3}$$

and for maximum speed, this must be zero; i.e. $V = 2R/k_\phi$ or $k_\phi = 2R/V$. Substituting in the speed equation:

$$\omega_{max} = \frac{V^2}{2R} - \frac{V^2}{4R} = \underline{\frac{V^2}{4R}}.$$

(b) For the case of torque proportional to speed, by considering the identical ratios of torque and speed to their rated values:

$$\frac{T_m}{1} = \frac{\omega_m}{\omega_{mR}}; \text{ so } T_m = \frac{\omega_m}{1 - R_a}$$

and substituting in the speed equation:

$$\omega_m = \frac{V}{k_\phi} - \frac{R \times [\omega_m/(1 - R_a)]}{k_\phi^2}$$

and by rearrangement:

$$\omega_m = \frac{k_\phi \cdot V}{k_\phi^2 + \dfrac{R}{1 - R_a}}$$

differentiating:

$$\frac{d\omega_m}{dk_\phi} = \frac{V \cdot \left(k_\phi^2 + \dfrac{R}{1 - R_a}\right) - k_\phi \cdot V \cdot (2k_\phi)}{(\text{denominator})^2}$$

and this is zero when $k_\phi^2 = R/(1 - R_a)$ and this is the condition for maximum speed.

Substituting in the re-formed speed expression for this load;

$$\omega_{max} = \frac{V\sqrt{\left(\frac{R}{1-R_a}\right)}}{\frac{R}{1-R_a}+\frac{R}{1-R_a}}$$

so:

$$\omega_{max} = V \cdot \sqrt{\frac{1-R_a}{4R}}$$

Before dealing with the numerical part of this question, it is worth noting that the point of maximum speed is the changeover between the rising and falling parts of the speed/flux characteristic. If this changeover is required to occur at rated flux, so that speed increase by increasing flux can be obtained, the expressions derived for k_ϕ to give ω_{max} will also yield the required resistance to meet this condition; by substituting $k_\phi = 1$. For the constant-torque load, required $R = V/2$ and for $T_m \propto \omega_m$, $R = 1 - R_a$. This information is relevant to the final part (d) of the question but generally it will be found that at ω_{max}, $E = k_\phi\omega_{max} = V/2$; i.e. for maximum speed the apparent 'load' resistance, E/I_a, is equal to the source (series), resistance $(V/2)/I_a$, cf. maximum power-transfer theorem.

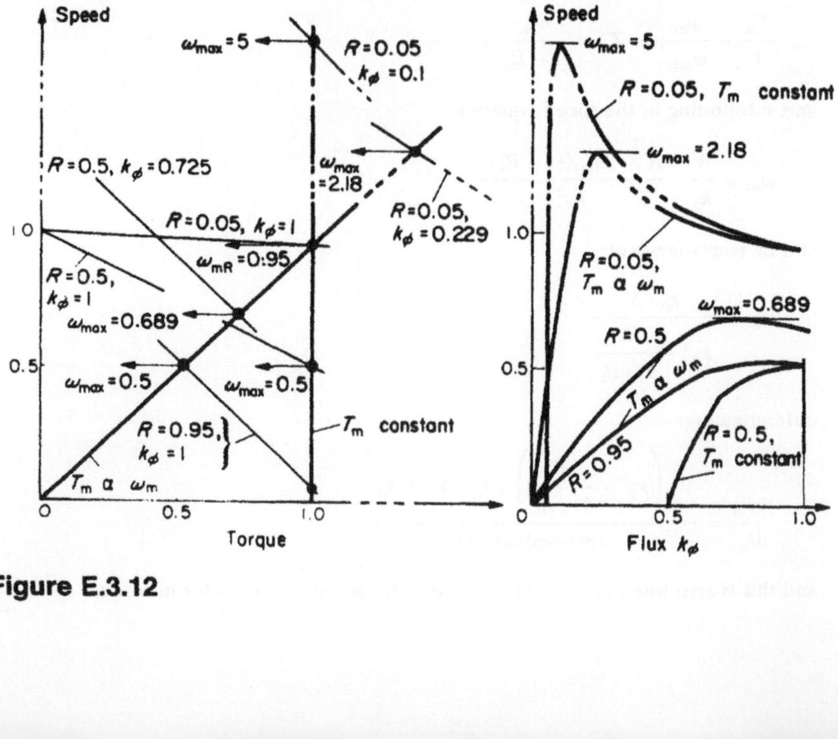

Figure E.3.12

(c) *Numerical solution*

	T_m constant			$T_m \propto \omega_m$		
		$R = 0.05$	$R = 0.5$		$R = 0.05$	$R = 0.5$
Speed at rated torque = $(1 - R)$		0.95	0.5		0.95	0.5
ω_{max}	$\dfrac{V^2}{4R} =$	5	0.5	$V\sqrt{\dfrac{1-R_a}{4R}} =$	2.18	0.689
T_m at ω_{max}		1	1	$\dfrac{\omega_{max}}{1-R_a} =$	2.29	0.725
Power = $\omega_{max} T_m$		5	0.5		5	0.5
$\sqrt{1 - 4R \times \text{Power}}$		0	0		0	0
$I_a = \dfrac{1 \pm \sqrt{}}{2R}$		10	1		10	1
$k_\phi = \dfrac{1 \mp \sqrt{}}{2\omega_{max}}$		0.1	1		0.229	0.725
$T_e = k_\phi \cdot I_a = T_m$		1	1		2.29	0.725

The above results are shown in outline on the accompanying speed/torque and speed/flux curves. It can be seen that the high-resistance circuit keeps the current and speed within rated limits though the power and the speed and/or torque cannot reach rated values. The zero for the square-root term confirms that the maximum conditions have been reached, with only one solution for I_a and k_ϕ.

In the solution of the final part of the question, maximum speed ω_{max} will be reached at maximum flux by suitable adjustment of the resistance. The solution is also shown on the accompanying diagram.

(d) For constant-torque load, required resistance for continuous increase of speed with flux is $V/2 = 0.5$ *per unit* for rated voltage. The solution has already been covered for this value of R on p. 56 and the curves are on the figures.

For the case of $T_m \propto \omega_m$, the required resistance is $1 - R_a = 1 - 0.05 = 0.95$ *per unit*, see table on p. 58 and the appropriate curves on Figure E.3.12.

$\omega_{max} = V\sqrt{\dfrac{1-R_a}{4R}}$	$= \sqrt{(1-0.05)/(4\times 0.95)}$	$= 0.5$
T_m at this speed	$= \omega_{max}/(1-R_a) = 0.5/(0.95)$	$= 0.526$
Power	$= \omega_{max}\cdot T_m = 0.526\times 0.5$	$= 0.263$
$\sqrt{1-4R\times \text{Power}}$	$= 1-4\times 0.95\times 0.526$	$= 0$
$\therefore I_a = 1/2R$	$= 1/1.9$	$= 0.526$
$k_\phi = 1/2\omega_{max}$	$= 1/(2\times 0.5)$	$= 1$
$T_e = k_\phi\cdot I_a = T_m$	$= 1\times 0.526$	$= 0.526$

3.4 Series motors

The special characteristics of the series motor make it suitable for many applications requiring high overload-torque per ampere as in traction, together with its falling speed/torque curve limiting the power demand, this being suitable also for crane and fan drives with simple resistance control of the characteristic. The motor can also be designed to run on a.c. (the universal motor), but there is a deterioration of performance due to the loss of voltage in the machine reactance which reduces the machine e.m.f. and speed, for a given flux and supply voltage. The field iron must be laminated to reduce eddy-current effects which otherwise would not only increase the iron losses but also, in delaying the flux response, would prevent the flux from being in phase with the excitation m.m.f. If these are in phase, the torque is still given by $k_\phi \cdot i_a$ and neglecting saturation this is $k_f \cdot i_a^2$ instantaneously, which is unidirectional and for a sinusoidal r.m.s. current I_a, oscillates at twice supply frequency, from zero to $k_f(\sqrt{2}I_a)^2$. The mean torque is thus $k_f I_a^2$ as for a d.c. current I_a, but this assumes that k_f does not saturate as it peaks to $\sqrt{2}$ times the value corresponding to I_a. Quite apart from saturation effects which occur even on d.c. operation at the higher currents, it can be seen that on a.c., the high peak current necessarily causes further flux reduction. However, the good starting torque and the variable-speed facility results in this single-phase series motor tending to dominate the mains-supplied domestic appliance and portable machine-tool fields of application. High speeds up to and beyond 10 000 rev/min give a good power/weight ratio and reasonable efficiencies. Applications beyond the small-power range, > 1 kW, are no longer common. Commutation difficulties required design refinements and compensating windings to oppose the armature m.m.f. and thus reduce the reactive voltage loss, when these motors were used for traction, usually at 25 or $16\frac{2}{3}$ Hz.

The first three examples to follow neglect saturation, one of them illustrating a.c. operation. Further examples calculate d.c. machine speed/torque curves allowing for the non-linearities due to armature reaction, saturation and the mechanical load.

Example 3.13

A 220-V d.c. series motor runs at 700 rev/min when operating at its full-load current of 20 A. The motor resistance is 0.5 Ω and the magnetic circuit may be assumed unsaturated. What will be the speed if:

(a) the load torque is increased by 44%?
(b) the motor current is 10 A?

Speed is given by $E/k_\phi = (V - RI_a)/k_\phi$ and for the series machine, neglecting saturation $k_\phi \propto I_a$ since $I_a = I_f$. Hence, $T_e = k_\phi \cdot I_a \propto I_a^2$.

If the torque is increased to 1.44, this will be achieved by an increase of current by a factor of $\sqrt{1.44} = 1.2$. The same increase of flux will occur.

$$\text{Thus } \frac{\omega_{m1}}{\omega_{m2}} = \frac{E_1}{E_2} \cdot \frac{k_{\phi 2}}{k_{\phi 1}} = \frac{220 - 0.5 \times 20}{220 - 0.5 \times (20 \times 1.2)} \times \frac{1.2}{1} = 1.212$$

$\therefore$ new speed $= 700/1.212 =$ <u>578 rev/min</u>

$$\text{For a current of 10 A: } \frac{\omega_{m1}}{\omega_{m2}} = \frac{220 - 0.5 \times 20}{220 - 0.5 \times 10} \times \frac{10}{20} = 0.488$$

$\therefore$ new speed $= 700/0.488 =$ <u>1433 rev/min</u>

Example 3.14

The series motor of the previous example is to be supplied from a 220-V, 50-Hz, a.c. supply and it can be assumed that the field iron is correctly laminated. If the inductance between terminals is 15 mH, at what speed will the machine run when producing the same average torque as in part (a) of Example 3.13? Assuming the torque loss is 7.5 Nm, compare the efficiencies for d.c. and a.c. operation at this load condition.

On a.c. the rotational e.m.f. (from *Blv*) is in phase with the flux density pulsation, which, with correct lamination, will be in phase with the exciting m.m.f. The phasor diagram can be drawn in terms of the r.m.s. values of the various components. The voltage drops are:

$$R \cdot I_a = 0.5 \times 24 = 12\,\text{V}$$

$$X \cdot I_a = 2\pi \times 50 \times 0.015 \times 24 = 113.1\,\text{V}$$

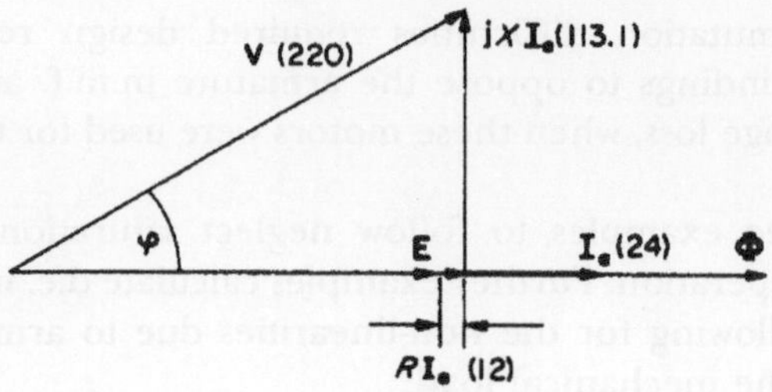

Figure E.3.14

From the phasor diagram $E = \sqrt{220^2 - 113.1^2} - 12 = 176.7\,\text{V}$

The value of k_φ can be obtained from operation on d.c.

At 20 A and 700 rev/min, $k_\varphi = \dfrac{220 - 0.5 \times 20}{2\pi \times 700/60} = 2.865$

so at 24 A, $k_\varphi = 1.2 \times 2.8647 = 3.438$ which on a.c. is an r.m.s. value and will give the r.m.s. e.m.f. hence:

$$\omega_m = \frac{176.7}{3.438} = 51.4 = \underline{491\ \text{rev/min}}$$

The power factor is $\cos\varphi = 188.7/220 = 0.858$

The output torque is $k_\varphi I_a - T_{loss} = 3.438 \times 24 - 7.5 = 75\,\text{Nm}$

Hence the efficiency $= \dfrac{75 \times 51.4}{220 \times 24 \times 0.858} = \underline{0.851}$

On d.c. the efficiency $= \dfrac{75 \times 2\pi \times 578/60}{220 \times 24} = \underline{0.86}$

On a.c. therefore, there is a slight fall in efficiency but in practice the efficiency would be rather lower because of the additional field-iron losses and more pronounced saturation effects.

Example 3.15

A d.c. series motor has a *per-unit* resistance of 0.05 based on rated voltage, rated current and rated flux as reference quantities. Assuming the machine is unsaturated, i.e. k_φ in *per unit* = I_a in *per unit*, calculate:

(a) The *per-unit* speed and current when the torque is 0.5 *p.u.*
(b) The *per-unit* speed and torque when the current is 0.5 *p.u.*
(c) The *per-unit* current and torque when the speed is 0.5 *p.u.*

(a) Since torque = $k_\phi \cdot I_a$ and the machine is unsaturated, torque in *per unit* = I_a^2

Hence $I_a^2 = 0.5$ so $I_a = \underline{0.707}$

$$\omega_m = \frac{E}{k_\phi} = \frac{1 - 0.05 \times 0.707}{0.707} = \underline{1.36}$$

(b) $I_a = 0.5$ so $T_e = 0.5^2 = \underline{0.25}$

$$\omega_m = \frac{E}{k_\phi} = \frac{1 - 0.05 \times 0.5}{0.5} = \underline{1.95}$$

(c) $I_a = \dfrac{V - E}{R} = \dfrac{V - k_\phi \cdot \omega_m}{R}$ and since $k_\phi = I_a$:

$$I_a = \frac{1 - I_a \times 0.5}{0.05} \text{ from which } I_a = \underline{1.82}$$

$$T_e = I_a^2 = 1.82^2 = \underline{3.3}$$

Series machine speed/torque curves

To allow for non-linearities in the magnetic circuit, these curves must be worked out point by point. The general method is to get expressions for E, I_a and k_ϕ as $f(I_f)$. Hence, for any particular value of I_f, speed $\omega_m = E/k_\phi$ and torque $T_e = k_\phi \cdot I_a$.

The magnetisation curve necessary to determine the k_ϕ/I_f relationship can readily be obtained – with allowance for armature reaction included – by loading the machine as a motor, with provision for varying the terminal voltage whilst the speed is held constant preferably by adjusting the mechanical load. Alternatively, if the series field can be separately excited and the machine is loaded as a generator, the same information can be obtained if the armature current is maintained at the same value as the field current, as this is increased. Hence $E_{test} = V \pm R \cdot I_a$ and k_ϕ at each value of I_f is E_{test}/ω_{test}, where $\omega_{test} = (2\pi/60) \times$ test rev/min. A test on open-circuit would not of course include armature reaction effects but would be a good approximation to the true curve. In the following examples, for convenience, the magnetisation curve data are given at the end of the question and the k_ϕ/I_f curve is derived at the beginning of the solution.

Example 3.16

A 250-V d.c. series motor has an armature-circuit resistance $R = 1.2\,\Omega$. Plot its speed/torque and speed/power curves from the following data and determine the torque and mechanical power developed at 600 rev/min. Also calculate the value of additional series resistance to limit the starting torque at full voltage to 120 Nm. The following magnetisation curve was taken when running as a motor from a variable terminal voltage and rotating at a constant speed of 500 rev/min.

Terminal voltage	114	164	205	237	259	278	V
Field current	8	12	16	20	24	28	A

Test speed $\omega_{test} = 500 \times 2\pi/60 = 52.36$ rad/s

$E_{test} = V - 1.2I_f$	104.4	149.6	185.8	213	230.2	244.4	V
$k_\phi = E_{test}/\omega_{test}$	2	2.86	3.55	4.07	4.4	4.67	Nm/A

Having determined the $k_\phi = f(I_f)$ characteristic, speed and torque will be calculated for the specified terminal voltage. The e.m.f. will be $250 - 1.2I_f$ and $I_a = I_f$.

E	240.4	235.6	230.8	226	221.2	216.4	V
$\omega_m = E/k_\phi$	120.2	82.4	65	55.5	50.3	46.3	rad/s
$N = \omega_m \times 60/2\pi$	1148	787	621	530	480	442	rev/min
$T_e = k_\phi \cdot I_f$	16	34.3	56.8	81.4	105.6	130.8	Nm
Power $= \omega_m \cdot T_e$	1.93	2.83	3.7	4.52	5.31	6.06	kW

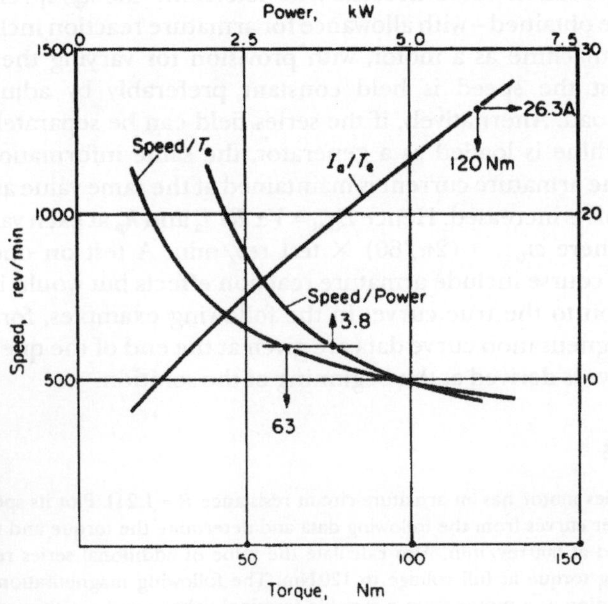

Figure E.3.16

Speed/torque and speed/power curves are plotted from the above results. At 600 rev/min:

Torque = 63 Nm and Power = 3.8 kW

For the second part of the question, it will be necessary to plot the T_e/I_a curve noting that any particular value of T_e occurs at a unique value of I_a. These data are available in the above table. From the curve at 120 Nm, $I_a = 26.3$ A. Hence, the required series resistance to limit the starting current to this value at full voltage, with e.m.f. zero = 250/26.3 = 9.51 Ω, an extra 9.51 − 1.2 = 8.31 Ω.

Example 3.17

A d.c. series motor has an armature resistance of 0.08 Ω and the field resistance is the same value.

(a) Find the speed at which a torque of 475 Nm will be developed when supplied at 250 V.
(b) The motor is driving a hoist and the load can 'overhaul' the motor so that its speed can be reversed to operate in the positive-torque, negative-speed quadrant. How much external armature-circuit resistance will be necessary to hold the speed at −400 rev/min when the torque is 475 Nm? Note that this will demand a current of the same value as in (a), but the e.m.f. is now reversed, supporting current flow; the machine is generating and the total resistance is absorbing $V + E$ volts. Calculate and draw the speed/torque curve to check that the chosen resistance is correct.

The magnetisation curve was taken by running the machine as a separately excited generator, field and armature currents being adjusted together to the same value. The following readings were obtained at a constant test speed of 400 rev/min:

Terminal voltage	114	179	218	244	254	V
Field current	30	50	70	90	110	A

$\omega_{test} = 400 \times 2\pi/60 = 41.89$ rad/s. $E_{test} = V + 0.08I_f$ ($I_a = I_f$ but field is not in series)

E_{test}	116.4	183	223.6	251.2	262.8	
$k_\phi = E_{test}/41.89$	2.79	4.37	5.34	6	6.27	Nm/A
$T_e = k_\phi \cdot I_a = k_\phi \cdot I_f$	83.7	218.5	373.8	540	689.7	Nm

(a)

$E = 250 - 0.16I_f$	245.2	242	238.8	235.6	232.4	V
$\omega_m = E/k_\phi$	87.9	55.4	44.7	39.3	37.1	rad/s
$N = \omega_m \times 60/2\pi$	839	529	427	375	354	rev/min

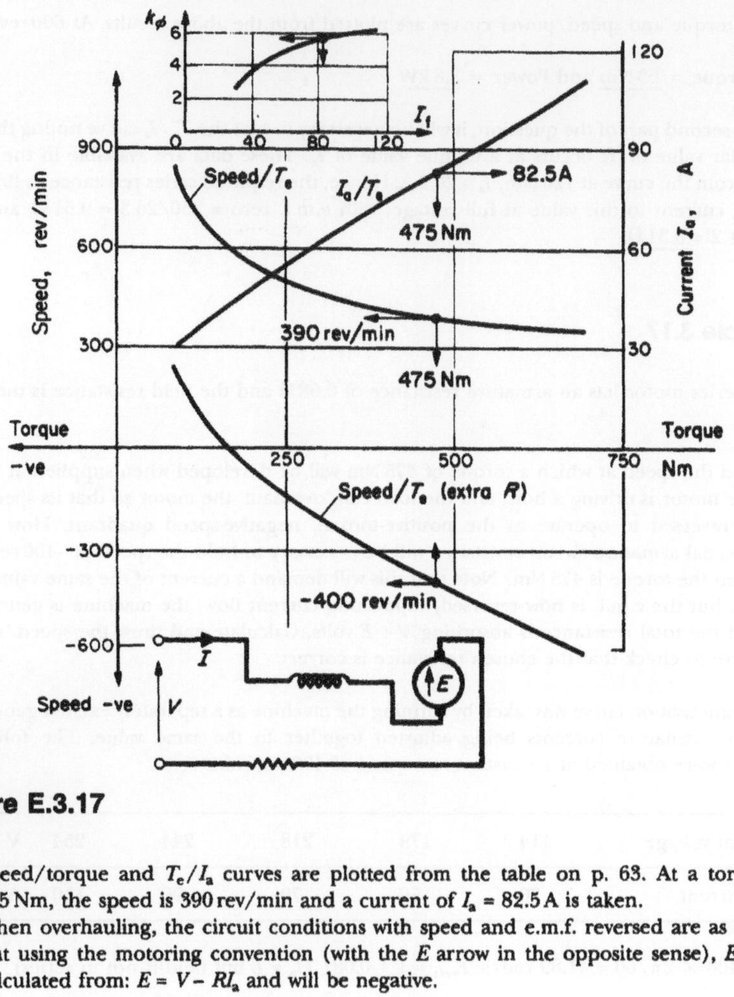

Figure E.3.17

(a) Speed/torque and T_e/I_a curves are plotted from the table on p. 63. At a torque of 475 Nm, the speed is 390 rev/min and a current of I_a = 82.5 A is taken.

(b) When overhauling, the circuit conditions with speed and e.m.f. reversed are as shown. But using the motoring convention (with the E arrow in the opposite sense), E is still calculated from: $E = V - RI_a$ and will be negative.

From the k_ϕ/I_f curve at 82.5 A, k_ϕ = 5.78.

$$E = k_\phi \cdot \omega_m = 5.78 \times (-400) \times 2\pi/60 = -242\ V$$

so $-242 = 250 - 82.5R$, from which R = 5.96; an extra 5.8 Ω.

For the field currents in the table on p. 63, the torque will be the same but the e.m.f. is now $250 - 5.96I_f$ and this permits the new speed points to be found:

E	71.2	−48	−167.2	−286.4	−405.6 V
$\omega_m = E/k_\phi$	25.5	−11	−31.3	−47.7	−64.7 rad/s
N	244	−105	−299	−456	−618 rev/min

From the plotted curve, the speed at 475 Nm is in fact −400 rev/min. Note that the machine is really operating in a braking mode, the machine is generating but the circuit as a whole is dissipative. The mode will be met again in Examples 3.20, 4.8, 6.7 and 6.8.

Example 3.18

For the same machine as in Example 3.16, calculate the speed/torque curves for the following circuit conditions, the supply voltage being 250 V throughout:

(a) with a 5 Ω series resistor and a 10 Ω diverter resistor across the machine terminals;
(b) with a 5 Ω series resistor and with the 10 Ω diverter across the armature terminals only;
(c) with a single resistor of 1.8 Ω, diverting current from the field winding;
(d) without diverters but with the series winding tapped at 75% of the full series turns. Allow for the reduced circuit resistance and assume $R_a = R_f = 0.6\,\Omega$.

These circuits have all been used in practice to change the characteristic for various control purposes, but the problem is also a good exercise in simple circuit theory. The various

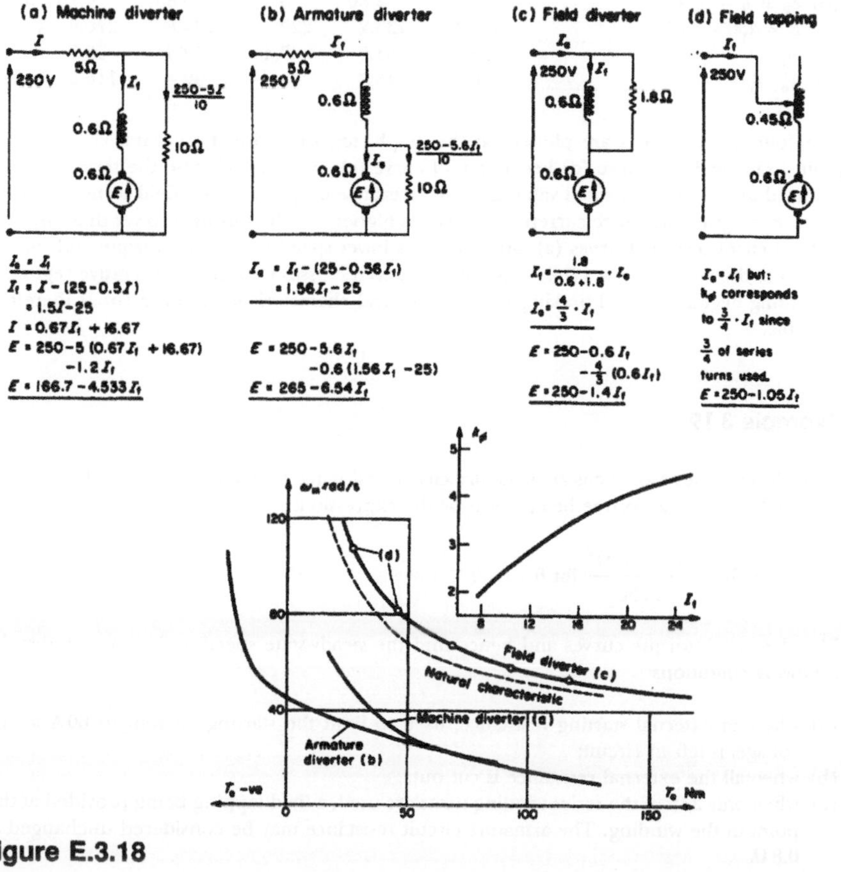

Figure E.3.18

configurations are shown on the figure on p. 65, together with the derivations of the required equations relating E and I_a to the field current I_f, which is not always the same as I_a. The magnetisation data are transferred from Example 3.16.

Field current	8	12	16	20	24	28	A
k_ϕ	2	2.86	3.55	4.07	4.4	4.67	Nm/A
(a) $E = 166.7 - 4.533I_f$	130.4	112.3	94.1	76	57.9	39.7	V
$\omega_m = E/k_\phi$	65.2	39.3	26.5	18.7	13.2	8.5	rad/s
$T_e = k_\phi \cdot I_f$	16	34.3	56.8	81.4	105.6	130.8	Nm
(b) $I_a = 1.57I_f - 25$	−12.5	−6.3	−0.04	6.2	12.4	18.7	A
$E = 265 - 6.54I_f$	212.7	186.5	160.4	134.2	108	81.9	V
ω_m	106	65.2	45.1	33	24.6	17.5	rad/s
$T_e = k_\phi \cdot I_a$	−25	−18	−0.1	25.2	54.6	87.3	Nm
(c) $I_a = 1.333I_f$	10.7	16	21.3	26.7	32	37.3	A
$E = 250 - 1.4I_f$	238.8	233.2	227.6	222	216.4	210.8	V
ω_m	119.4	81.5	64.1	54.5	49.1	45.1	rad/s
T_e	21.4	45.8	75.6	108.7	140.8	174.2	Nm
(d) k_ϕ at $I_f \times 3/4$		2.2	2.86	3.4	3.85	4.15	Nm/A
$E = 250 - 1.05I_f$		237.4	233.2	229	224.8	220.6	V
ω_m		108	81.5	67.4	58.4	53.2	rad/s
T_e		26.4	45.8	68	92.4	116.2	Nm

The four characteristics are plotted on the graph, together with the natural characteristic from Example 3.16. Curve (d) lies on top of curve (c) since in each case the series m.m.f. is reduced to 3/4 of the normal value and the resistance drop across the field terminals is the same. For a given armature current, the speed is higher and the torque is lower than with the natural characteristic. Curves (a) and (b) give a lower speed for a given torque and for the armature diverter, there is a finite no-load speed as it crosses into the regenerative region, I_a becoming negative. Rapid braking from the natural characteristic to curve (b) is therefore possible.

Example 3.19

A 500-V d.c. series motor has an armature circuit resistance of 0.8 Ω. The motor drives a fan, the total mechanical torque being given by the expression:

$$T_m = 10 + \frac{(\text{rev/min})^2}{2250} \text{ lbf ft.}$$

Plot the speed/torque curves and hence find the steady-state speed and torque under the following conditions:

(a) when an external starting resistance, used to limit the starting current to 60 A at full voltage, is left in circuit;
(b) when all the external resistance is cut out;
(c) when only 2/3 of the series winding turns are used, a field tapping being provided at this point in the winding. The armature-circuit resistance may be considered unchanged at 0.8 Ω.

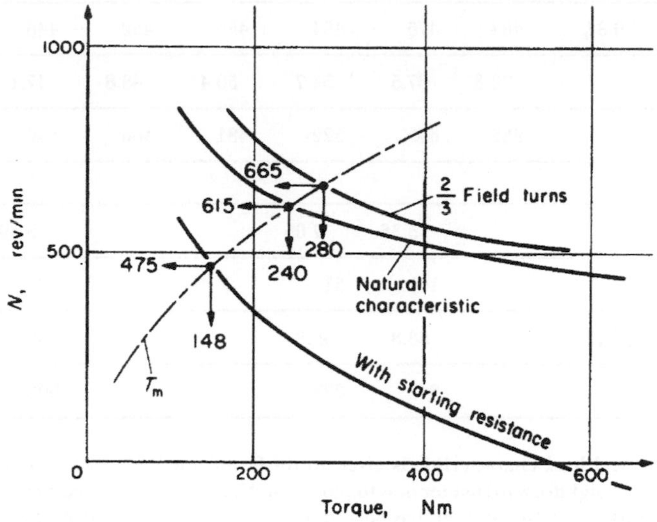

Figure E.3.19

The magnetisation curve at 550 rev/min is as follows:

Field current	20	30	45	55	60	67.5	A
Generated e.m.f.	309	406	489	521	534	545	V

Test speed $\omega_{test} = 550 \times 2\pi/60 = 57.6$ rad/s.
Required starting resistance for 60 A = 500/60 = 8.33 Ω; i.e. 8.33 – 0.8 = 7.53 Ω extra.

k_ϕ = gen. e.m.f./57.6	5.36	7.05	8.49	9.05	9.27	9.46	Nm/A

For case (a) $I_a = I_f$ and $E = 500 - 8.33I_f$
For cases (b) and (c) $I_a = I_f$ and $E = 500 - 0.8I_f$

But for (c), the value of k_ϕ must be reduced to that corresponding to $2/3\,I_f$.
Hence, for the various operating conditions:

(a) $E = 500 - 8.33I_f$	333.4	250.0	125.1	41.85	0.2	–62.3	V
$\omega_m = E/k_\phi$	62.2	35.47	13.82	4.62	0.02	–6.58	rad/s
$N = \omega_m \times 60/2\pi$	594	339	141	44	0.2	–63	rev/min
T_e for (a) and (b)	107	212	382	498	556	639	Nm

(b) $E = 500 - 0.8I_f$	484	476	464	456	452	446	V
ω_m	90.3	67.5	54.7	50.4	48.8	47.1	rad/s
N	862	645	522	481	466	450	rev/min
(c) k_ϕ at $2/3I_f$		5.36	7.05			8.49	Nm/A
$T_e = k_\phi \cdot I_a$		161	317			573	Nm
$\omega_m = E(b)/k_\phi$		88.8	65.8			52.5	rad/s
N		848	628			502	rev/min

The mechanical load characteristic is given in lbf ft and the conversion factor to Nm is 746/550. It is a straightforward matter now to plot T_m in Nm for the converted expression: $T_m = 13.6 + (\text{rev/min})^2/1659$ Nm and this has been done along with the speed/T_e curves from the above table. The intersections of the T_m and T_e curves give the steady-state points as:

(a) 475 rev/min, 148 Nm; (b) 615 rev/min, 240 Nm; (c) 665 rev/min, 280 Nm.

3.5 Braking circuits

If when motoring, the circuit conditions are so changed that current I_a (or flux k_ϕ) reverses polarity, the torque T_e will reverse. Being then in the same sense as T_m, which opposes rotation, the speed will fall, T_m being assisted by this electrical braking action. Depending on the circuit and the nature of T_m, the rotation may reverse after falling to zero, to run up again as a motor in this changed direction. Reversal of flux is sometimes employed with certain power-electronic drives but the time constant for ϕ is relatively long by comparison with the armature-current time-constant. Further, since I_a would increase excessively as k_ϕ was reduced for ultimate reversal, armature current must first be zeroed for the reversal period, usually less than a second, and during this 'dead time', motor control is lost. Consider the expression for I_a:

$$I_a = \frac{V - E}{R} = \frac{V - k_\phi \omega_m}{R}.$$

Reversal of I_a and T_e can be achieved by four different methods:

(1) Increase of $k_\phi \omega_m$, the motor becoming a generator pumping power back into the source. This regeneration would only be momentary if k_ϕ was increased, being limited by saturation and the fall of speed. If T_m

is an active load, e.g. a vehicle drive, then gravity could cause speed to increase, and be controlled by controlling the regenerated power through flux adjustment.

(2) Reversal of V. This would have to include a limiting resistor to control the maximum current. Such reverse-current braking (plugging) is very effective but consumes approximately three times the stored kinetic energy of the system in reducing the speed to zero, and would run up as a motor in reverse rotation unless prevented. See Tutorial Example T6.6.

(3) Short circuiting the machine, making $V = 0$, would also require a limiting resistor. Again the machine is generating in what is called a dynamic (or rheostatic) braking mode and this time, the resistor and the machine losses only dissipate 1 × the stored kinetic energy. Braking is slower, especially if T_m is small.

(4) Far superior to any of the above methods is to provide the relatively expensive facility of controlling V using a separate generator or power-electronic circuit in what is called the Ward-Leonard system, after its inventor. Rapid control of current, torque and speed in any of the four quadrants is made available. The next example illustrates all the methods above, showing them on the 4-quadrant diagram.

Example 3.20

A 250 V, 500 rev/min d.c. separately excited motor has an armature resistance of 0.13 Ω and takes an armature current of 60 A when delivering rated torque at rated flux. If flux is maintained constant throughout, calculate the speed at which a braking torque equal in magnitude to the full-load torque is developed when:

(a) regeneratively braking at normal terminal voltage;
(b) plugging, with extra resistance to limit the peak torque on changeover to 3 *per unit*;
(c) dynamically braking, with resistance to limit the current to 2 *per unit*;
(d) regeneratively braking at half rated terminal voltage.
(e) What terminal voltage would be required to run the motor in reverse rotation at rated torque and half rated speed?

It is first necessary to calculate rated flux and thereafter the speed is given by:

$$\omega_m = \frac{V - RI_a}{k_\phi} \text{ or } \frac{V}{k_\phi} - \frac{RT_e}{k_\phi^2}$$

with appropriate values of R, V and I_a or T_e. Both I_a and T_e will be negative

$$k_{\phi(\text{rated})} = \frac{250 - 0.13 \times 60}{2\pi \times 500/60} = \frac{242.2}{52.36} = 4.626 \text{ Nm/A or V/rad/s}$$

Substituting this value of k_ϕ will give all the answers from the general expression for the speed/torque curve.

(a) $I_a = -60\,\text{A}.\ \omega_m = \dfrac{250 - 0.13(-60)}{4.626} = 55.73\,\text{rad/s} = \underline{532\,\text{rev/min}}$

(b) I_a must be limited to 3 (−60) A, and $V = -250$ V. Assuming that speed does not change in the short time of current reversal:

$$I_a = \frac{V - E}{R} \text{ so } R = \frac{-250 - 242.2}{3(-60)} = 2.734\,\Omega,$$

the e.m.f. $k_\phi \omega_m$ being unchanged momentarily.

This is the total circuit resistance which means an external resistor of 2.6 Ω is required.

$$\text{Speed for full-load torque} = \frac{-250 - 2.734(-60)}{4.626} = -18.6\,\text{rad/s} = \underline{-177\,\text{rev/min}}$$

(c) I_a limited to 2(−60) A and $V = 0$ so: $R = \dfrac{0 - 242.2}{2(-60)} = 2.018 = \text{extra } 1.89\,\Omega.$

$$\text{Speed for full-load torque} = \frac{0 - 2.018(-60)}{4.626} = 26.18\,\text{rad/s} = \underline{250\,\text{rev/min.}}$$

(d) $I_a = -60\,\text{A};\ V = 125\,\text{V},\ \omega_m = \dfrac{125 - 0.13(-60)}{4.626} = 28.7\,\text{rad/s} = \underline{274\,\text{rev/min.}}$

(e) $I_a = -60\,\text{A};\ \omega_m = \dfrac{-52.36}{2}.\ V = k_\phi \omega_m + RI_a = 4.626 \times (-26.18) + 0.13(-60)$

$= \underline{-128.9\,\text{V.}}$

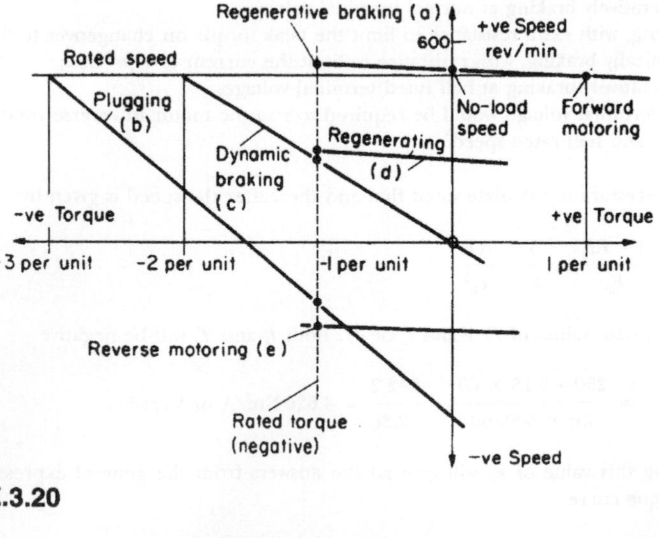

Figure E.3.20

Note that the motoring equation has been used throughout, even though most modes are generating. This simplifies the concepts but requires the insertion of a negative sign for current, torque or power if these are specified. Alternatively, if the correct signs of speed, voltage and flux are inserted the signs of I_a, T_e and P will come out naturally in the calculations.

The speed/torque curves for each setting are shown on the attached 4-quadrant diagram (Figure E.3.20) and the above answers at one particular torque (−1 *per unit*) are indicated. The dynamic changeover between quadrants and curves will be illustrated by examples in Chapter 6.

Example 3.21

A d.c. series motor drives a hoist. When lowering a load, the machine acts as a series generator, a resistor being connected directly across the terminals – dynamic braking mode. Determine:

(a) the range of resistance required so that when lowering maximum load (450 lbf ft) the speed can be restrained to 400 rev/min, and for light load (150 lbf ft), the speed can be allowed to rise to 600 rev/min.

(b) What resistance would be required if the light-load speed was maintained instead at 400 rev/min and what would then be the saving in external resistance loss at this load? What total mechanical power is gravity providing under this condition? Neglect the mechanical losses. Armature-circuit resistance 0.1 Ω.

Before giving the magnetisation data, it should be pointed out that the machine is going to be operating as a self-excited series generator and a step-by-step calculation will be required. A braking condition with a series motor driving a hoist has already been encountered in Example 3.17. The mode here was 'plugging'; also with forward torque and reverse speed, but a different circuit connection.

The following magnetisation curve was taken at a speed of 400 rev/min; i.e. $\omega_{m(test)} = 41.89$.

Field current	30	50	70	90	110	A
Generated e.m.f.	114	179	218	244	254	V
$k_\phi = E/41.89$	2.72	4.27	5.2	5.83	6.06	Nm/A
Torque $T_e = k_\phi I_f$	81.6	213.5	364	524.7	666.6	Nm

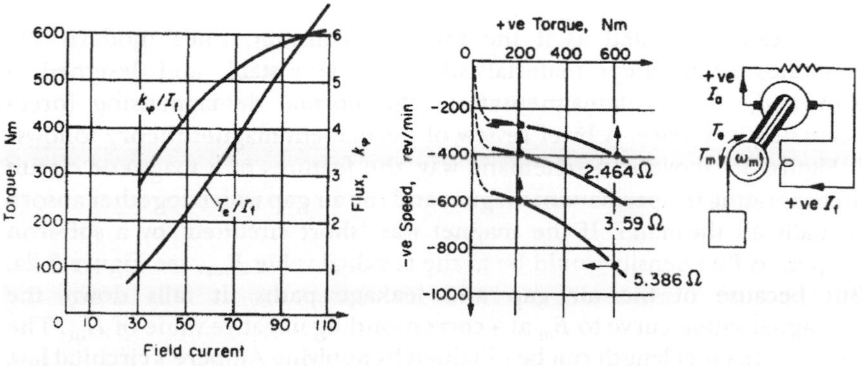

Figure E.3.21

Flux and torque are plotted against current on Figure E.3.21. The question specifies torques so the curves will be used to read k_ϕ and I_a (= I_f) at the T_e values.

$$\text{For } 450\,\text{lbf ft} = 450 \times \frac{746}{550} = 610.4\,\text{Nm};\ I_a = 102\,\text{A and } k_\phi = 6\,\text{Nm/A}$$

$$\text{For } 150\,\text{lbf ft} \quad = 203.4\,\text{Nm};\ I_a = 49\,\text{A and } k_\phi = 4.2\,\text{Nm/A}$$

$$\text{For dynamic braking:} \quad I_a = \frac{0 - k_\phi \omega_m}{R} \text{ so required resistance is } R = \frac{-k_\phi \omega_m}{I_a}$$

$$\text{(a) For } -400\,\text{rev/min and } 450\,\text{lbf ft; } R = \frac{-6 \times \frac{2\pi}{60} \times (-400)}{102} = 2.464\,\Omega;\ \text{extra } \underline{2.364\,\Omega}$$

$$\text{For } -600\,\text{rev/min and } 150\,\text{lbf ft; } R = \frac{-4.2 \times (-20\pi)}{49} = 5.386\,\Omega;\ \text{extra } \underline{5.286\,\Omega}$$

$$\text{(b) For } -400\,\text{rev/min and } 150\,\text{lbf ft; } R = \frac{-4.2 \times (-40\pi/3)}{49} = 3.59\,\Omega;\ \text{extra } \underline{3.49\,\Omega}$$

$$\text{Difference in power loss} = (5.286 - 3.49) \times 49^2 = \underline{4.31\,\text{kW}}$$

$$\text{Total mechanical power is that dissipated; } 3.59 \times 49^2 = \underline{8.62\,\text{kW}} = E \cdot I_a$$

The various features of interest are shown on the speed/torque curves which, as an exercise, could be plotted from the above data following a few additional calculations of speed, from

$$\omega_m = -RT_e/k_\phi^2.$$

As a practical point, the reversal of e.m.f. with rotation maintains the direction of I_f, thus permitting self excitation on changeover from motoring. Changeover to +ve speed and – ve torque would require an armature-connection reversal to give $I_a = -I_f$.

3.6 Permanent-magnet machines

These can be treated as if the flux was constant, since modern PM machines use magnetic materials which are very stable and designed to withstand, without demagnetisation, the normal demagnetising forces occurring in service. A brief review of permanent-magnet theory follows.

Figure 3.1 shows in a schematic way, the features of a magnetic circuit incorporating the permanent magnet and the air gap which together absorb virtually all the m.m.f. If the magnet was 'short circuited' by a soft-iron keeper, its flux density would be at the residual value B_{res}, see Figure 3.2a. But because of the air gap and leakage paths, it falls down the demagnetisation curve to B_m at a corresponding negative value of H_m. The required magnet length can be obtained by applying Ampere's circuital law, equating the current enclosed to zero, i.e. $0 = H_m l_m + H_g l_g$ from which:

$$l_m = -\frac{H_g}{H_m} \cdot l_g = -\frac{B_g}{\mu_0} \cdot \frac{l_g}{H_m}. \ (H_m \text{ is -ve})$$

The required area of magnet is obtained by equating the magnet flux to the sum of air-gap and leakage fluxes, i.e. leakage factor (LF) × $B_g A_g = B_m A_m$ from which the magnet area

$$A_m = \text{LF} \cdot \frac{B_g}{B_m} \cdot A_g$$

Hence the magnet volume required, using the modulus of H_m, is:

$$l_m \cdot A_m = \frac{B_g}{\mu_0} \cdot \frac{l_g}{H_m} \cdot \text{LF} \cdot \frac{B_g}{B_m} \cdot A_g = \frac{B_g^{\,2}}{\mu_0} \cdot \text{LF} \cdot \frac{\text{Air-gap volume}}{B_m H_m}. \tag{3.11}$$

The product $B_m H_m$ has the dimensions of energy density per unit volume and can exceed 250 kJ/m^3 for some of the newer materials. It is clearly desirable to operate near the maximum *BH* product which occurs at a point a little higher than $B_{res}/2$.

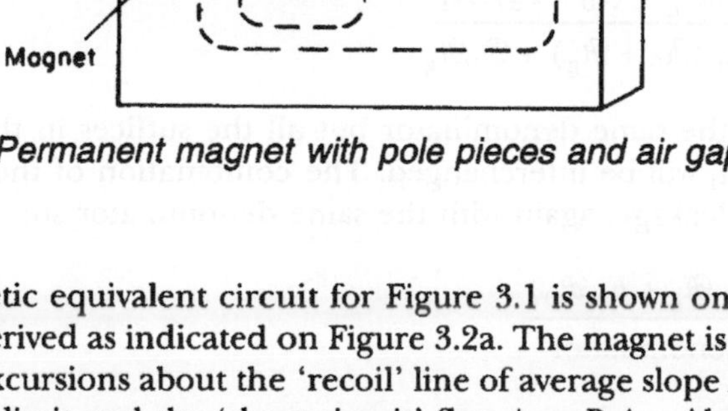

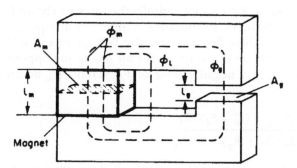

Figure 3.1 *Permanent magnet with pole pieces and air gap.*

The magnetic equivalent circuit for Figure 3.1 is shown on Figure 3.2b and can be derived as indicated on Figure 3.2a. The magnet is stabilised by minor-loop excursions about the 'recoil' line of average slope Λ_0, between the stabilised limit and the 'short-circuit' flux $\phi_r = B_r A_m$. Along this line, the magnet behaves as if it were a source of m.m.f., viz

$$F_0 = H_0 l_m, \text{ behind a reluctance } \mathcal{R}_0 \ (= 1/\Lambda_0) = \frac{\text{m.m.f.}}{\text{flux}} = \frac{H_0 l_m}{B_r A_m}.$$

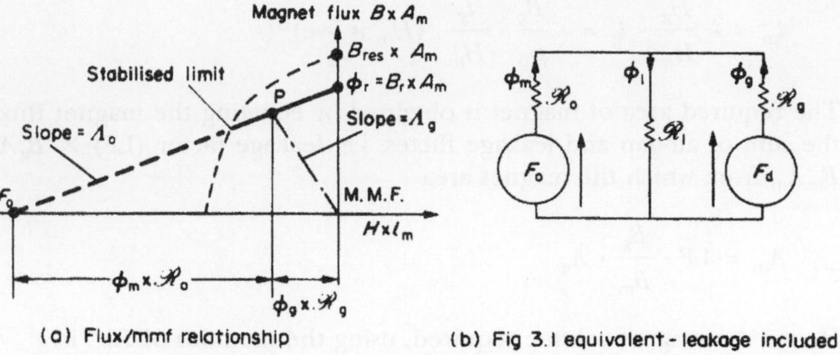

Figure 3.2 *Equivalent circuit development.*

The presence of the air gap reduces the flux below ϕ_r since it absorbs an m.m.f. $\mathcal{R}_g\phi_g$, the remainder of F_0 being absorbed by the magnet. Certain of the newer materials do in fact have a straight-line operating region down to zero flux but on Figure 3.2a, F_0 is a mathematical abstraction defining the intersection of the 'recoil' line with zero flux. On the equivalent circuit, $\mathcal{R}_l$ represents the reluctance of the leakage path and F_d the demagnetising effect of any external m.m.f., e.g. due to armature reaction. It may be noticed that Figure 3.2b is topologically the same as Figure E.2.11 for two paralleled transformers, so the same mathematical equations apply, but with different variables. For example, the leakage flux is analagous to the load current I and the flux through the air gap in the opposite sense to F_d will be indicated by a negative value of ϕ_g. Making ϕ_m analagous to I_A its value will be:

$$\phi_m = \frac{F_0\mathcal{R}_g + (F_0 - F_d)\,\mathcal{R}_l}{\mathcal{R}_l\,(\mathcal{R}_0 + \mathcal{R}_g) + \mathcal{R}_0\mathcal{R}_g} \tag{3.12}$$

ϕ_g will have the same denominator but all the suffices in the numerator apart from $\mathcal{R}_l$ will be interchanged. The combination of these two fluxes will give the leakage, again with the same denominator so:

$$\phi_l = \frac{(F_0\,\mathcal{R}_g + F_d\,\mathcal{R}_0)}{\text{denominator}}$$

Example 3.22

A PM machine uses a material with a straight-line demagnetisation characteristic cutting the axes at $B_r = 0.95\,\text{T}$ and $H_0 = -720 \times 10^3\,\text{At/m}$. The cross-sectional area of the magnet is to be the same as that of the air gap in which a flux density of 0.6 T is required. Take the leakage factor as 1.2. Calculate the required length of magnet if the air-gap length is 2 mm.

To allow for design errors, calculate how much the air-gap flux will fall:

(a) if the leakage factor is in fact 1.4;
(b) if in addition there is an armature demagnetising m.m.f. of 500 At/pole?

What would be the required magnet length if operating at the optimum point on the curve and what would then be the air-gap flux density if the leakage factor is 1.2?

From eqn (3.11), the magnet volume

$$l_m A_m = \frac{B_g^2}{\mu_0} \text{LF} \frac{\text{Air-gap volume } (l_g A_g)}{B_m H_m}$$

and since $A_m = A_g$, B_m will have to be $1.2B_g$ to sustain a leakage factor of 1.2, so $B_m = 1.2 \times 0.6 = 0.72$ T. The corresponding value of H_m follows from Figure E3.22:

$$\frac{0.72}{0.95} = \frac{720 - H_m}{720} \text{ from which } |H_m| = 174.3 \text{ kAt/m}$$

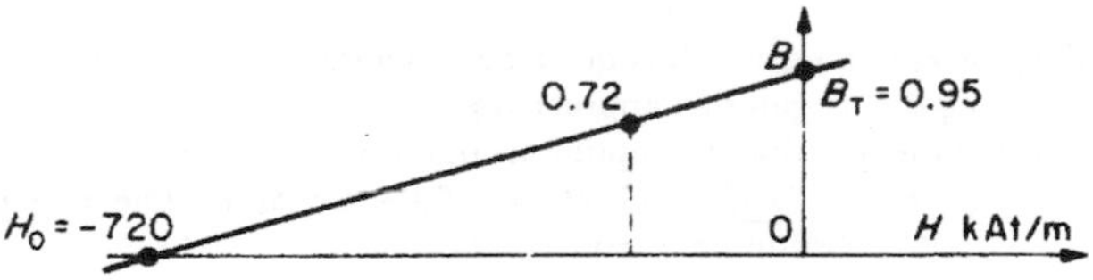

Figure E.3.22

Substituting in eqn (3.11)

$$l_m A_m = \frac{0.6^2}{4\pi/10^7} \times 1.2 \times \frac{l_g A_g}{0.72 \times 174\,300} = 2.74 \times l_g A_g$$

so the magnet length must be 2.74 × 2 = 5.48 mm, since $A_m = A_g$.
The equivalent circuit can be used for parts (a) and (b) for which the components are:

$$\mathcal{R}_g = \frac{l_g}{\mu_0 A_g}; \mathcal{R}_0 = \frac{H_0 l_m}{B_r A_m} = \frac{720\,000 \times 2.74 l_g}{0.95 \times A_g} \times \frac{\mu_0}{\mu_0} = 2.61\,\mathcal{R}_g$$

Now $\phi_l = 0.2\,\phi_g$ from LF = 1.2, so $\mathcal{R}_l$ must be 5 times $\mathcal{R}_g$ and for LF = 1.4 $\mathcal{R}_l = 2.5\,\mathcal{R}_g$. $F_0 = 720\,000 \times 2.74 \times 2 \times 10^{-3} = 3945.6$ At and $F_a = 0$ or 500 At. The denominator of eqn (3.12) for the two cases, $\mathcal{R}_l = 5\,\mathcal{R}_g$ and $\mathcal{R}_l = 2.5\,\mathcal{R}_g$:

$$\text{Denom} = \mathcal{R}_l(\mathcal{R}_0 + \mathcal{R}_g) + \mathcal{R}_0\mathcal{R}_g = 5\,\mathcal{R}_g(2.61 + 1)\mathcal{R}_g + 2.61\,\mathcal{R}_g^2 = 20.66\,\mathcal{R}_g^2$$

$$\text{or for (a) and (b):} \quad = 2.5\,\mathcal{R}_g(3.61\,\mathcal{R}_g) + 2.61\,\mathcal{R}_g^2 = 11.635\,\mathcal{R}_g^2$$

$$\text{Hence, with LF} = 1.2\ \phi_g = \frac{-F_0\,\mathcal{R}_l}{denom.} = \frac{-3945.6 \times 5\,\mathcal{R}_g}{20.66\,\mathcal{R}_g^2} = \frac{-954.88\mu_0 A_g}{2 \times 10^{-3}} = -0.5999\,A_g$$

which checks with the specified air-gap density. Note –ve sign, following Figure 3.2b.

$$\text{(a)}\ \phi_g = \frac{-3945.6 \times 2.5\, \mathcal{R}_g}{11.635\, \mathcal{R}_g^{\,2}} = \frac{-847.8\mu_0\, A_g}{2 \times 10^{-3}} = \underline{-0.533\, A_g}$$

$$\phi_m = \frac{3945.6 \times \mathcal{R}_g + 3945.6 \times 2.5\, \mathcal{R}_g}{11.635\, \mathcal{R}_g^{\,2}} = 0.745\, A_g$$

and the leakage flux $\phi_l = \phi_m + \phi_g = 0.212\ A_g$ which checks as $0.4 \times \phi_g$. The air-gap flux has fallen by more than 10% but the magnet flux has increased slightly to accommodate the new leakage. Eqn (3.11) could be solved for B_g with this new value of B_m to give 0.533 T as above.

$$\text{(b)}\ \phi_g = \frac{500 \times 2.61\, \mathcal{R}_g + (500 - 3945.6) \times 2.5\, \mathcal{R}_g}{11.635\, \mathcal{R}_g^{\,2}} = \frac{-628.2\mu_0\, A_g}{2 \times 10^{-3}} = \underline{-0.395\, A_g}$$

$$\phi_m = \frac{3945.6 \times \mathcal{R}_g + (3945.6 - 500) \times 2.5\, \mathcal{R}_g}{11.635\, \mathcal{R}_g^{\,2}} = 0.678\, A_g$$

the leakage flux is now $(0.678 - 0.395)\ A_g = 0.283\ A_g$ so LF is $\dfrac{0.678}{0.395} = 1.72.$

The effect of the demagnetising m.m.f. is quite considerable and drives more flux into the leakage paths reducing the air-gap flux appreciably.

For the final part of the question, the optimum point of the straight-line demagnetisation characteristic is at $B_m = B_r/2 = \underline{0.475T}$, with $H_m = -360 \times 10^3$ At/m. The air-gap flux density will be $0.475/1.2 = 0.396$. Substituting in eqn (3.11):

$$l_m A_m = \frac{0.396^2}{4\pi/10^7} \times 1.2 \times \frac{l_g A_g}{0.475 \times 360\,000} = \underline{0.875 l_g A_g}$$

Thus the magnet is much smaller than the previous figure of 2.74 ×, since the maximum $B_m H_m$ point has been used. However, for the same condition, the flux density is only about 2/3 so there would have to be a 50% increase in machine size to give the same total air-gap flux as before. This situation has highlighted the overall problem of design optimisation.

4 Induction machines

For drives, the important characteristic is that relating speed and torque. Using the same axes as for the d.c. machine, with $\omega_m = f(T_e)$ as the dependent variable, the basic speed/torque curve on a 4-quadrant diagram is shown on Figure 4.1a. It can be compared with that shown for the d.c. machine in Example 3.20. Basic operation as a motor is at speeds near to synchronous, $n_s = f/p$ rev/sec, with small values of slip $s = (n_s - n)/n_s$. There are other significant operational modes however, e.g. starting, generating and braking. Adopting a motoring convention, for which P_{elec} and P_{mech} are both positive, the various modes are shown, covering slip variations from small negative values (generating) to larger values $s \rightarrow 2$, where braking occurs. Changing the ABC supply sequence to the primary – usually the stator winding – will reverse the rotation of the magnetic field and give a mirror-image characteristic as indicated. Note also the typical mechanical characteristic; $T_m = f(\omega_m)$, its intersection with the $\omega_m = f(T_e)$ characteristic which determines the steady-state speed, and its reversal, as a passive load, if rotation reverses. Although the natural induction-machine characteristic as shown is quite typical, it is possible to change it by various means, for example to cause change of speed or improve the starting torque. The later questions in this chapter and in Section 7.4 are much concerned with such changes.

4.1 Revision of equations

Figure 4.1b shows a power-flow diagram for the induction machine. Reference back to Section 1.3, Figures 1.9(b) and (e), leads directly to Figure 4.1c, the 'exact' (Tee) equivalent circuit, per phase, the approximate circuit being indicated by the transfer of the magnetising branch to the terminals. These are not the only ways of presenting the equivalent circuit but they have the advantage of preserving the identity of important

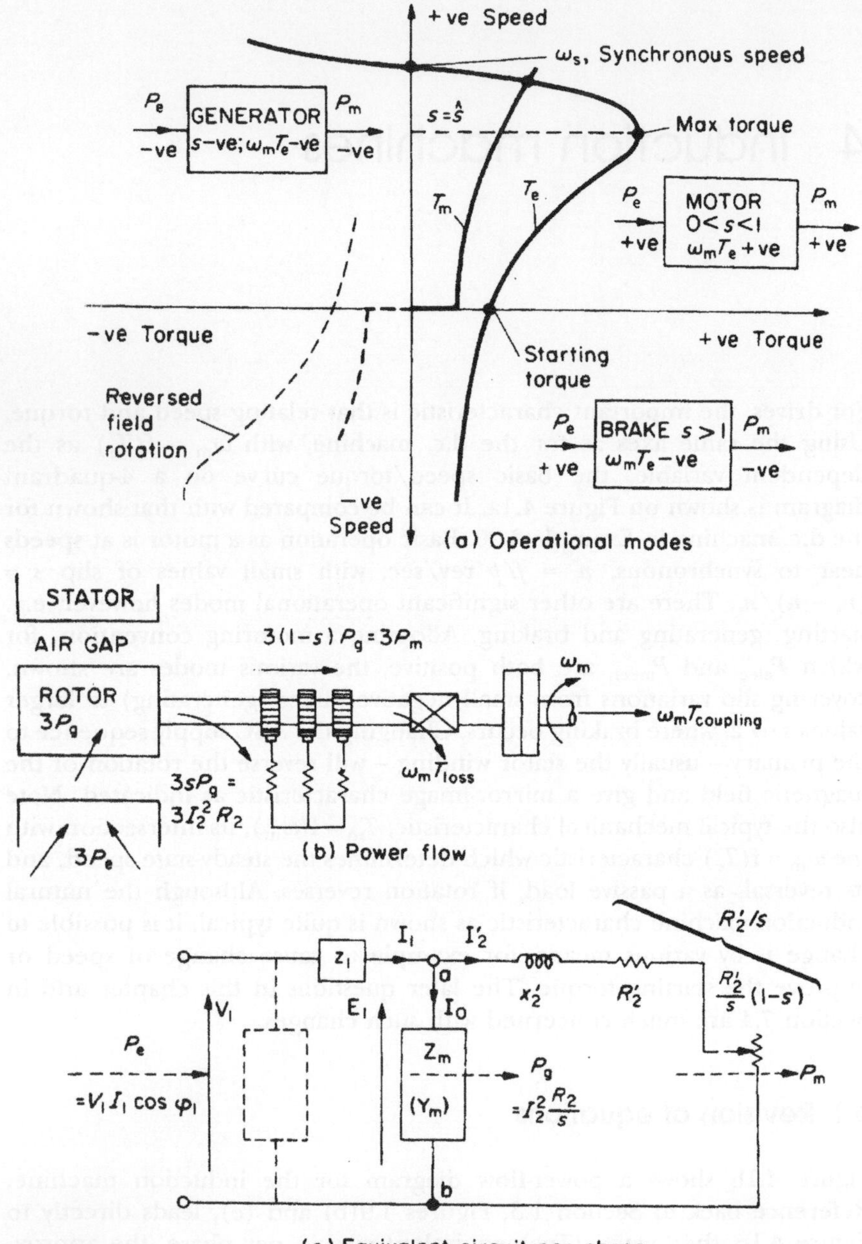

Figure 4.1 *3-phase induction machine.*

physical features like the magnetising current I_m, and the winding impedances. $z_1 = R_1 + jx_1$ and $Z_2' = R_2'/s + jx_2'$. The magnetising branch admittance Y_m is $1/R_m + 1/jX_m$. The approximate circuit makes calculations very easy and is justified if a general idea of performance is required, having an accuracy within about 10%. The worked examples in this chapter, apart from those in Section 4.4, refer to 3-phase machines under balanced conditions, but the phase calculations would apply to any polyphase machine. Section 4.4 deals with unbalanced and single-phase operation.

Referring to Figure 4.1c, the power-balance equation yields the important relationships:

Electrical terminal power per phase

$$= P_e = V_1 I_1 \cos\varphi_1 = E_1 I_2' \cos\varphi_2 + \text{Stator loss}$$

Air-gap power per phase

$$= P_g = E_1 I_2' \cos\varphi_2 = I_2'^2 R_2'/s \tag{4.1}$$

Hence: rotor-circuit power per phase

$$= I_2'^2 R_2' = I_2^2 R_2 = sP_g \tag{4.2}$$

and: mechanical power per phase

$$= P_m = (1-s)P_g = \frac{(1-s)}{s} I_2^2 R_2 \tag{4.3}$$

from which: electromagnetic torque

$$= T_e = \frac{3P_m}{\omega_m} = \frac{3P_g(1-s)}{\omega_s(1-s)} = \frac{3}{\omega_s} P_g = \frac{3}{\omega_s}\frac{I_2^2 R_2}{s} \tag{4.4}$$

From the expressions for P_e, P_g and P_m it can be seen that P_e and P_m are both positive (motoring) when $0 < s < 1$. When s is negative, P_g, P_e and P_m are negative (generating), and when $s > 1$, P_g and P_e are positive but P_m is negative, i.e. power flow is inwards at both sets of 'terminals', which is a braking condition.

The mechanical 'coupling' power is:

$$P_{coupling} = 3I_2^2 R_2 \frac{(1-s)}{s} - \omega_m T_{loss}$$

Note that the rotor-circuit power-loss can be expressed either as $3I_2^2R_2$ or $3I_2'^2R_2'$, and the rotor is assumed to carry the secondary winding, the usual arrangement. If the approximate circuit is used, eqn (4.4) becomes:

$$T_e = \frac{3}{2\pi \times f/p} \times \frac{V_1^2}{(R_1 + R_2'/s)^2 + (x_1 + x_2')^2} \times \frac{R_2'}{s} \text{ Nm} \tag{4.5}$$

where:

$$\mathbf{I}_2' = \frac{V_1 + \mathrm{j}0}{(R_1 + R_2'/s) + \mathrm{j}(x_1 + x_2')} = (Real) - \mathrm{j}(Imag.) \tag{4.6}$$

$$\mathbf{I}_0 = \frac{V_1}{R_m} - \mathrm{j}\frac{V_1}{X_m} = I_p - \mathrm{j}I_m \tag{4.7}$$

and $\mathbf{I}_1 = \mathbf{I}_2' + \mathbf{I}_0 = [(Real) + I_p] - \mathrm{j}[(Imag.) + I_m] = I_1 \cos\varphi_1 + \mathrm{j}I_1 \sin\varphi_1$. Note that φ_1 and $\sin\varphi_1$ are taken as –ve for lagging p.f. For a generator with s negative, the expression for $\mathbf{I}_2'$ is of the form:

$$\frac{\mathbf{V}_1}{-A + \mathrm{j}B} \times \frac{-A - \mathrm{j}B}{-A - \mathrm{j}B} \text{ which becomes of the form: } \mathbf{V}_1\ (-a - \mathrm{j}b),$$

the real part of the current being negative. The machine is not a 'positive' motor as the motoring-convention equations have assumed, but a 'negative' motor indicating reverse power flow. If we reverse the convention, changing the signs, the real part becomes positive, and also the imaginary part, showing that as a generator, the induction machine operates at leading power factor.

For calculations using the 'exact' circuit, the following arrangement preserves the connection with the equivalent circuit as a useful reference:

$$\begin{aligned}
\mathbf{Z}_{input} &= \mathbf{z}_1 + \mathbf{Z}_{ab} \\
&= R_1 + \mathrm{j}x_1 + \frac{1}{\dfrac{1}{R_m} + \dfrac{1}{\mathrm{j}X_m} + \dfrac{1}{R_2'/s + \mathrm{j}x_2'}} \qquad (4.8) \\
&= R_1 + \mathrm{j}x_1 + R_{ab} + \mathrm{j}X_{ab} \\
&= R_{input} + \mathrm{j}X_{input}
\end{aligned}$$

Hence:

$$\mathbf{I}_1 = \frac{V_1}{Z_{\text{input}}}\left(\frac{R_{\text{in.}}}{Z_{\text{in.}}} - \mathrm{j}\frac{X_{\text{in.}}}{Z_{\text{in.}}}\right) = I_1 \cos \varphi_1 + \mathrm{j}I_1 \sin \varphi_1$$

$$\mathbf{E}_1 = \mathbf{V}_1 \times \frac{\mathbf{Z}_{\text{ab}}}{\mathbf{Z}_{\text{input}}} \text{ and } \mathbf{I}_0 = \frac{E_1}{R_{\text{m}}} - \frac{\mathrm{j}E_1}{X_{\text{m}}} \qquad (4.9), (4.10)$$

$$I_2'^2 = \frac{E_1^2}{(R_2')/s)^2 + x_2'^2} \qquad (4.11)$$

From among these equations, the majority of the Chapter 4 examples are solved, but other special equations are developed later as required. For example, an important quantity is the maximum torque. This can be obtained from the approximate expression eqn (4.5), either by differentiating or considering the condition for maximum power transfer, taking the load as the power consumed ($I_2'^2R_2'/s$) in the apparent rotor resistance. From these considerations, $R_2'/\hat{s}$ must be equal to:

$$R_2'/\hat{s} = \sqrt{R_1^2 + (x_1 + x_2')^2}, \qquad (4.12)$$

giving a maximum torque on substitution of the corresponding slip $\hat{s}$ as:

$$\text{Max}\, T_e = \frac{3}{\omega_s} \times \frac{V_1^2}{2[\pm\sqrt{(R_1^2 + (x_1 + x_2')^2)} + R_1]} \qquad (4.13)$$

which is seen to be independent of the value of R_2'. It will also be noticed, from the equations, that if all parameters are fixed apart from speed, a particular value of R_2'/s gives a particular current and torque. If R_2 is controllable, the speed for any particular torque can be obtained from the corresponding value of R_2'/s.

4.2 Solution of equations

As for the d.c. machine, the flow diagram on p. 82 has been prepared to act as a guide to the kind of problems which might be posed, and to indicate in a general way the approach to solutions. The reader must also be prepared to refer back to the equations just developed and to Sections 1.3, 1.4 and Chapter 2 if necessary, when trying to understand the solutions in the following examples. N.B. A hand calculator has been used and the subsection and final answers rounded off.

Induction-Machine Solution Programme

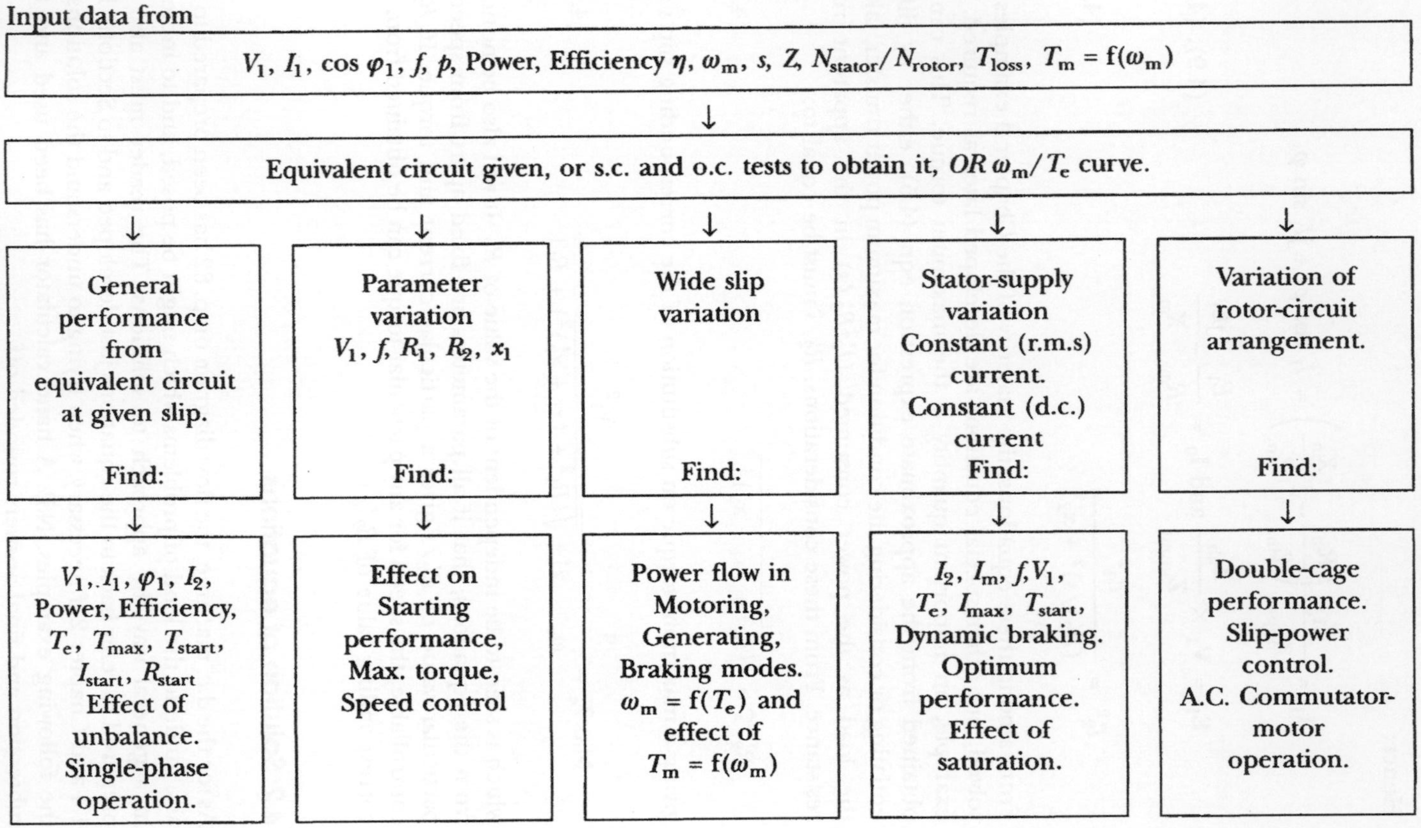

Input data from
V_1, I_1, cos φ_1, f, p, Power, Efficiency η, ω_m, s, Z, N_{stator}/N_{rotor}, T_{loss}, $T_m = f(\omega_m)$
Equivalent circuit given, or s.c. and o.c. tests to obtain it, *OR* ω_m/T_e curve.
General performance from equivalent circuit at given slip.
Find:
V_1, I_1, φ_1, I_2, Power, Efficiency, T_e, T_{max}, T_{start}, I_{start}, R_{start}
Effect of unbalance.
Single-phase operation.
Parameter variation
V_1, f, R_1, R_2, x_1
Find:
Effect on Starting performance, Max. torque, Speed control
Wide slip variation
Find:
Power flow in Motoring, Generating, Braking modes.
$\omega_m = f(T_e)$ and effect of $T_m = f(\omega_m)$
Stator-supply variation
Constant (r.m.s) current.
Constant (d.c.) current
Find:
I_2, I_m, f, V_1, T_e, I_{max}, T_{start},
Dynamic braking.
Optimum performance.
Effect of saturation.
Variation of rotor-circuit arrangement.
Find:
Double-cage performance.
Slip-power control.
A.C. Commutator-motor operation.

Example 4.1

The equivalent circuit of a 440-V, 3-phase, 8-pole, 50-Hz star-connected induction motor is given on the figure. The short-circuit test is conducted with a locked rotor and a line current of 80 A. The 'open-circuit' test is conducted by supplying the primary winding at rated voltage and at the same time driving the rotor in the same direction as the rotating field, at synchronous speed; $s = 0$. Determine:

(a) the line voltage and power factor on the short-circuit test;
(b) the line current and power factor on the 'open-circuit' test;
(c) the equivalent circuit per-phase if these tests were analysed on an approximate basis; i.e. neglecting the magnetising branch when analysing the s.c. test and neglecting the series leakage impedances for analysis of the o.c. test.

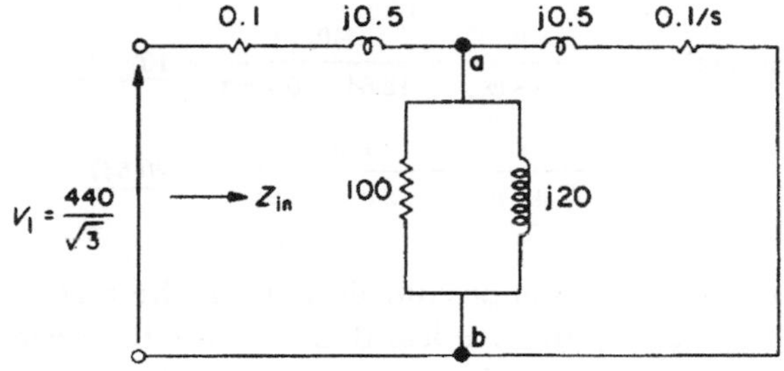

Figure E.4.1

(a) Input impedance $\mathbf{Z}_{in} = z_1 \qquad + \mathbf{Z}_{ab}(1/\mathbf{Y}_{ab})$

$$= 0.1 + j0.5 + \cfrac{1}{\cfrac{1}{100} + \cfrac{1}{j20} + \cfrac{1}{0.1 + j0.5}}$$

$$\cfrac{1}{0.01 - j0.05 + \cfrac{0.1 - j0.5}{0.1^2 + 0.5^2}}$$

$$\frac{1}{0.395 - j1.973}$$

$$= 0.1 + j0.5 + 0.097 + j0.487$$

$$= 0.197 + j0.987 = 1.006\angle 78°.7\,\Omega \text{ per phase}$$

Input voltage on s.c. $= \sqrt{3} \times 1.006 \times 80 =$ 140 Line V at 0.1957 power factor

(b) Input impedance ($R_2/s = \infty$) $= 0.1 + j0.5 + \dfrac{1}{0.01 - j0.05}$ $(1/\mathbf{Y}_m)$

$$= 0.1 + j0.5 + 3.85 + j19.23$$

$$= 3.95 + j19.73$$

$$= 20.1\underline{/78^\circ.7}\ \Omega \text{ per phase}$$

$$\text{Input current } I_0 = \frac{440/\sqrt{3}}{20.1} = \underline{12.64\,\text{A at } 0.1963\text{ p.f.}}$$

(c) $$\mathbf{Z}_{sc} = \frac{140/\sqrt{3}}{80}\underline{/\cos^{-1} 0.1957} = 1.006\,(0.196 + j0.98) = \underline{0.197 + j0.987\,\Omega}$$

Dividing this equally between stator and rotor: $\mathbf{z}_1 = \mathbf{z}_2' = \underline{0.0985 + j0.494\,\Omega}$

$$\text{From 'o.c. test': } R_m = \frac{V}{I_0 \cos\varphi_{oc}} = \frac{440/\sqrt{3}}{12.64 \times 0.1963} = \underline{102.4\,\Omega}$$

$$X_m = \frac{V}{I_0 \sin\varphi_{oc}} = \frac{440/\sqrt{3}}{12.64 \times 0.98} = \underline{20.5\,\Omega}$$

The largest errors with this approximation are in the magnetising-circuit parameters, about 2.5%, and are less than this for the series-impedance elements, which are more decisive on overall performance. The errors depend on the relative magnitudes of the parameters and are not always so small, unlike the transformer case where X_m is relatively high. Note that test information would usually be given in terms of line voltage and current and total power. The power factor would have to be calculated from:

$$\cos\varphi = \frac{\text{Total power}}{\sqrt{3} \times V_{\text{line}} \times I_{\text{line}}}$$

In this example, the o.c. and s.c. power factors are about the same, but this is not usually so. An 'exact' analysis of the o.c. and s.c. tests to derive the Tee-equivalent circuit, is described in Reference 4. Note also that it is not always possible to drive the machine for the 'o.c. test'. Running on no load, where the slip is very small, is a close approximation and allowance can be made for the error if assuming $\mathbf{z}_2' = \infty$, as described in Reference 1.

Example 4.2

For the machine and equivalent circuit given in Example 4.1, calculate, at a slip of 3%, the input stator-current and power factor; the rotor current referred to the stator; the electromagnetic torque; the mechanical output power and the efficiency. Also calculate the starting torque. Take the mechanical loss as 1 kW. Consider:

(a) the equivalent circuit neglecting stator impedance altogether;
(b) the approximate circuit;
(c) the 'exact' circuit.
(d) Repeat (c) working in *per unit* and taking the calculated output, rated voltage and synchronous speed as base quantities.

Synchronous speed $\omega_s = 2\pi \times f/p = 2\pi \times 50/4 = 78.54$ rad/s. $\omega_m = \omega_s(1 - 0.03) = 76.18$ rad/s.

	(a) Stator impedance neglected	(b) Approximate equivalent circuit
$I'_2 =$	$\frac{254}{0.1/0.03 + j0.5} = 74.5 + j11.15$	$\frac{254}{(0.1 + 0.1/0.03) + j1} = 68.2 - j19.9$
$I_0 =$	$\frac{254}{100} - \frac{j254}{20} = 2.54 - j12.7$	$= 2.54 - j12.7$
$I_1 =$	$I'_2 + I_0 = 77.04 - j23.85$	$70.74 - j32.6$
$=$	$80.6\text{A at } \cos\varphi_1 = 0.955$	$77.9\text{A at } \cos\varphi_1 = 0.908$
$I'_2 =$	$\sqrt{74.5^2 + 11.15^2} = 75.3\text{A}$	$\sqrt{68.2^2 + 19.9^2} = 71\text{A}$
$T_e =$	$\frac{3}{78.54} \times 75.32^2 \times \frac{0.1}{0.03} = 721.9\text{Nm}$	$\frac{3}{78.54} \times 71^2 \times \frac{0.1}{0.03} = 641.8\text{Nm}$
$P_{mech} =$	$76.18 \times 722.2 - 1000 = 54\text{kW}$	$76.18 \times 641.8 - 1000 = 47.9\text{kW}$
$P_{elec} =$	$\sqrt{3} \times 440 \times 77.04 = 58.7\text{kW}$	$\sqrt{3} \times 440 \times 70.74 = 53.9\text{kW}$
Effy. $\eta =$	$54/58.7 = 92\%$	$47.9/53.9 = 88.8\%$
$T_{start} =$	$\frac{3}{78.54} \times \frac{254^2}{0.1^2 + 0.5^2} \times \frac{0.1}{1} = 947.8\text{Nm}$	$\frac{3}{78.54} \times \frac{254^2}{0.2^2 + 1^2} \times \frac{0.1}{1} = 236.9\text{Nm}$

It can be seen that neglecting stator impedance gives appreciable differences in the answers and at starting they are utterly erroneous. From the next, exact-circuit calculation, it will be found that the approximate circuit gives answers well within 10% of these correct results.

(c) Impedance across points 'ab' $= \mathbf{Z}_{ab} = \dfrac{1}{\dfrac{1}{100} + \dfrac{1}{j20} + \dfrac{1}{\dfrac{0.1}{0.03} + j0.5}}$

$$= \frac{1}{0.01 - j0.05 + 0.293 - j0.044} = 3.01 + j0.934 = 3.15\underline{/17°.2}$$

Adding stator impedance $\mathbf{z}_1$: $\mathbf{Z}_{in} = 3.11 + j1.434 = 3.42\underline{/24°.7}$

$$\mathbf{I}_1 = \frac{254}{3.42\underline{/24°.7}} = \underline{74.3\text{ A at } 0.908 \text{ power-factor lagging}}$$

$$E_1 = \frac{|Z_{ab}|}{|Z_{in}|} \times V_1 = \frac{3.15}{3.42} \times 254 = 234\,V. \therefore \mathbf{I}_2' = \frac{234}{\sqrt{3.33^2 + 0.5^2}} = \underline{69.4\,\text{A}}$$

$$\mathbf{I}_0 = E_1/R_m - \text{j}\,E_1/X_m = \underline{2.34 - \text{j}11.7\,\text{A}}$$

$$\text{Torque } T_e = \frac{3}{78.54} \times 69.4^2 \times 3.33 = \underline{612.4\,\text{Nm}}$$

$$P_{mech} = 76.18 \times 612.4 - 1000 = \underline{45.66\,\text{kW}}$$

$$P_{elec} = \sqrt{3} \times 440 \times 74.3 \times 0.908 = \underline{51.4\,\text{kW}}$$

$$\text{Efficiency } \eta = 45.66/51.4 = \underline{88.8\%}$$

For starting torque, $s = 1$, some data are already available from Example 4.1:

$$Z_{ab} = \sqrt{0.097^2 + 0.487^2} = 0.496\,\Omega \text{ and } Z_{in} = 1.006\,\Omega$$

$$\text{Hence } T_{start} = \frac{3}{78.54} \times \left(\frac{0.496}{1.006} \times 254\right)^2 \times \frac{1}{0.5^2 + 0.1^2} \times \frac{0.1}{1} = \underline{230.4\,\text{Nm}}$$

(d) For the *per-unit* notation, we must first establish the base quantities:

Rated output = 45.66 kW = P_{base} (total) or 15.22 kW (phase value)
Rated voltage = 440 V = V_{base} (line) or 254 V (phase value)
Synchronous speed = $2\pi \times f/p$ = 78.54 rad/s = Speed $\omega_{m(base)} = \omega_s$.

These are the usual base quantities chosen for induction motors and the rest follow as below:

$$\mathbf{I}_{base} \text{ (per phase)} = \frac{P_{rated}/3}{V_{rated}/\text{phase}} \; (= 59.9\,\text{A}).$$

$$\mathbf{I}_{rated} = \frac{P_{rated}/3}{V_{rated} \times \cos\varphi_{rated} \times \eta_{rated}} \; (= 74.3\,\text{A})$$

$$\mathbf{I}_{rated} \text{ in } \textit{per unit} = \frac{I_{rated}}{I_{base}} = \frac{1}{\cos\varphi_{rated} \times \eta_{rated}} \; (= 1.24 \textit{ per unit})$$

$$\omega_{m(rated)} = \omega_s\,(1 - s_{rated});\ \text{so } \omega_{m(rated)} \text{ in } \textit{per unit} = 1 - s_{rated}$$

$$\text{Torque}_{base} = \frac{P_{base}}{\omega_{m(base)}} \text{ and Torque}_{rated} = \frac{P_{rated}/\omega_{m(rated)}}{P_{rated}/\omega_{m(base)}} = \frac{1}{1 - S_{rated}} \text{ in } \textit{per unit.}$$

$$Z_{base} = \frac{V_{base}\text{(per phase)}}{I_{base}\text{(per phase)}} = \frac{V_{base}}{P_{base}\text{(per phase)}/V_{base}} = \frac{V^2_{base}}{P_{base}} \text{ ohms}$$

Applying these relationships to the question: $Z_{base} = \dfrac{(440/\sqrt{3})^2}{45\,660/3} = 4.24\,\Omega$

$$\text{Hence } R_1 = R_2' = 0.1/4.24 = 0.02358 \textit{ per unit}$$

$$x_1 = x_2' = 0.5/4.24 = 0.11792 \textit{ per unit}$$

$$R_m = 100/4.24 = 23.58 \textit{ per unit}$$

$$X_m = 20/4.24 = 4.717 \textit{ per unit}$$

$$\text{Mechanical loss} = 1000/45\,660 = 0.0219 \textit{ per unit}$$

Calculations now proceed as in part (c) but $V = 1$; $\omega_s = 1$ and all other quantities are in *per unit*.

$$\mathbf{Z}_{ab} = \frac{1}{\dfrac{1}{23.58} + \dfrac{1}{j4.717} + \dfrac{1}{0.02358/0.03 + j0.11792}}$$

$$= \frac{1}{0.0424 - j0.212 + 1.2443 - j0.18677}$$

$$= 0.7091 + j0.2197 = 0.74235 \text{ (modulus)}$$

$$\text{add } \mathbf{z}_1 \qquad \underline{0.02358 + j0.11792}$$

$$\mathbf{Z}_{in} = \underline{0.7327 + j0.3376} = 0.80674 \text{ (modulus)}$$

$$\text{Hence} \quad I_1 = \frac{1}{0.80674} = 1.24 \textit{ per unit} \text{ at } \cos\varphi_1 = 0.7327/0.80674 = 0.908 \text{ lagging}$$

$$E_1 = \frac{0.74235}{0.80674} \times 1 = 0.92$$

$$I_2' = \frac{0.92}{\sqrt{0.786^2 + 0.1179^2}}$$

$$= 1.157 \text{ (Check from actual values } 69.4/59.9 = 1.158)$$

$$\mathbf{I}_0 = \frac{0.92}{23.58} - \frac{j0.92}{4.717} = 0.039 - j0.195.$$

$$\text{Coupling torque} = \frac{1}{1} \times 1.157^2 \times \frac{0.02358}{0.03} - 0.0219 \text{ (mech. loss)}$$

$$= 1.03 \; [= 1/(1 - 0.03)]$$

$$P_{elec} \; (V \times I \times \cos\varphi_1) = 1 \times 1.24 \times 0.908 = 1.126 \textit{ per unit} \text{ so } \eta = 1/1.126 = 0.888$$

All the *per-unit* values check with part (c). The final calculations are somewhat neater and the method has advantages when many repetitive calculations are required, comparisons are being made or large systems are being studied. For computer simulations, especially for transient analyses, the scaling problem is eased considerably.

Example 4.3

A 3-phase, 440-V, delta-connected, 4-pole 50-Hz induction motor runs at a speed of 1447 rev/min when operating at its rated load. The equivalent circuit has the following per-phase parameters:

$$R_1 = 0.2\,\Omega,\ R_2' = 0.4\,\Omega;\ x_1 = x_2' = 2\,\Omega;\ R_m = 200\,\Omega;\ X_m = 40\,\Omega$$

(a) Using the approximate circuit, determine, for rated load, the values of line current and power factor, torque, output power and efficiency. The mechanical loss is 1000 watts.
(b) Determine the same quantities, if the machine is run as a generator with the same numerical value of slip.

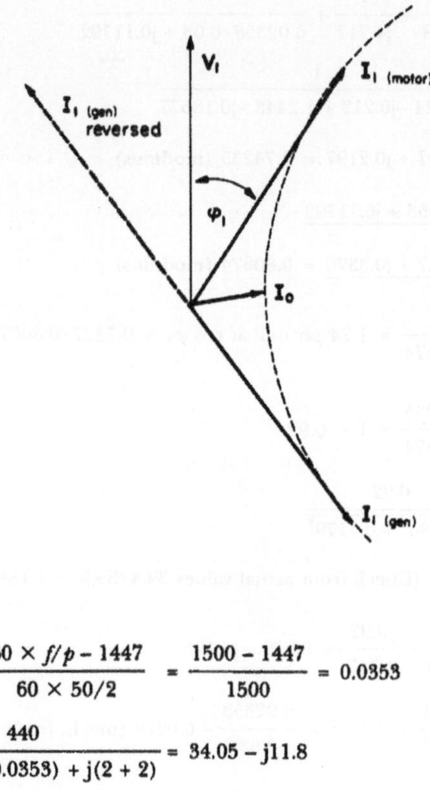

Figure E.4.3

(a) Full-load slip $= \dfrac{60 \times f/p - 1447}{60 \times 50/2} = \dfrac{1500 - 1447}{1500} = 0.0353$

$$I_2' = \frac{440}{(0.2 + 0.4/0.0353) + j(2+2)} = 34.05 - j11.8$$

$$I_0 = \frac{440}{200} + \frac{440}{j40} = \underline{2.2 - j11}$$

$$I_1 = \qquad I_1 = \underline{36.25 - j22.8}$$

$$\text{Line current} = \sqrt{3 \times (36.25^2 + 22.8^2)} = \underline{74.2\,\text{A}} \text{ at } 36.25/42.8 = \underline{0.847\,\text{p.f.}}$$

$$T_e = \frac{3}{2\pi \times 50/2} \times (34.05^2 + 11.8^2) \times \frac{0.4}{0.0353} = \frac{3}{157.1} \times 36.04^2 \times 11.33 = \underline{281\,\text{Nm}}$$

$$P_{mech} = (1 - 0.0353) \times 157.1 \times 281 - 1000 \text{ (mech. loss)} = \underline{41.59\,\text{kW}} = 55.75\,\text{h.p.}$$

$$P_{elec} = 3 \times 440 \times 36.25 = 47.85\,\text{kW. Efficiency} = 41.59/47.85 = \underline{86.9\%}$$

(b) As a generator, slip = −0.0353; hence:

$$\mathbf{I}_2' = \frac{440}{(0.2 - 0.4/0.0353) + j4} = \frac{440}{-11.13 + j4} = -35 - j12.6$$

$$\mathbf{I}_0 = \underline{2.2 - j11}$$

$$\mathbf{I}_1 = \underline{-32.8 - j23.6}$$

$$\text{Line current} = \sqrt{3 \times (32.8^2 + 23.6^2)} = \underline{70\,\text{A}} \text{ at } -32.8/40.4 = \underline{0.81 \text{ p.f.}}\ (-144°)$$

Note, that since a motoring equation has been used, the real part of $\mathbf{I}_1$ is negative. If the convention is reversed it will be positive and can be seen that I_1 is at leading power factor. Phasor diagrams are helpful here and are shown alongside. The current locus is circular and circle diagrams can be used for rapid solutions. They are less important nowadays but the technique, with numerical illustration, is dealt with in Reference 1.

$$T_e = \frac{3}{157.1} \times (35^2 + 12.6^2) \times (-11.33) = \underline{-299.4\,\text{Nm}}$$

$$P_{mech} = (1 + 0.0353) \times 157.1 \times (-299.4) - 1000 = \underline{-49.7\,\text{kW}}$$

$$P_{elec} = 3 \times 440 \times (-32.8) = -43.3\,\text{kW. Efficiency} = 43.3/49.7 = \underline{87.1\%}$$

NOTE AGAIN THAT THE MOTORING CONVENTION RESULTS IN NEGATIVE SIGNS IN THE ANSWERS.

Example 4.4

In a certain 3-phase induction motor, the leakage reactance is five times the resistance for both primary and secondary windings. The primary impedance is identical with the referred secondary impedance. The slip at full load is 3%. It is desired to limit the starting current to three times the full-load current. By how much must:

(a) R_1 be increased?
(b) R_2 be increased?
(c) x_1 be increased?

How would the maximum torque be affected if the extra impedance was left in circuit?

This question gives minimum information but can be solved most conveniently by taking ratios and thus cancelling the common constants. Let the new impedances be expressed as $k_a R$, $k_b R_2'$ and $k_c x_1$ respectively, and $R_1 = R_2' = R$; $x_1 = x_2' = 5R$.

$$\text{Full-load current} = \frac{V}{\sqrt{(R + R/0.03)^2 + (5R + 5R)^2}} = \frac{V/R}{\sqrt{1278.8}}$$

$$\text{Starting current} = \frac{V}{\sqrt{(k_a R + k_b R)^2 + (5k_c R + 5R)^2}}$$

$$= \frac{V/R}{\sqrt{(k_a + k_b)^2 + 25(k_c + 1)^2}}$$

$$\text{Current ratio } \frac{I_s}{I_{fl}} = \sqrt{\frac{1278.8}{(k_a + k_b)^2 + 25(k_c + 1)^2}}; \text{ – required to be 3.}$$

Using this ratio and for (a) and (b), if $k_c = 1$ and either k_a or $k_b = 1$:

$$1278.8 = 9 \times [(k+1)^2 + 100]$$

so k (= k_a or k_b) = 5.49 ($5.49R_1$ or $5.49R_2$)

(c) If $k_a = k_b = 1$: $1278.8 = 9 \times [2^2 + 25(k_c + 1)^2]$ from which $k_c = 1.35$ ($1.35x_1$)

Expression for full-load torque [eqn (4.5)]

$$= \frac{3V_1^2}{\omega_s \times [(R + R/0.03)^2 + (10R)^2]} \times \frac{R}{0.03}$$

$$= \frac{3V_1^2}{\omega_s R} \times \frac{1}{38.36}$$

Expression for starting torque [eqn (4.5), $s = 1$]

$$= \frac{3V_1^2 \times R}{\omega_s \times [(k_a R + k_b R)^2 + 25(k_c R + R)^2]}$$

$$= \frac{3V_1^2}{\omega_s R} \times \frac{1}{(k_a + k_b)^2 + 25(k_c + 1)^2}$$

Expression for maximum torque [eqn (4.13)]

$$= \frac{3V_1^2}{\omega_s \times 2[\sqrt{(k_a R)^2 + 25(k_c R + R)^2} + k_a R]}$$

$$= \frac{3V_1^2}{\omega_s R} \times \frac{1}{2[\sqrt{k_a^2 + 25(k_c + 1)^2} + k_a]}$$

$$\text{Ratio } \frac{\text{Starting torque}}{\text{Full-load torque}} = \frac{T_s}{T_{fl}} = \frac{38.36}{(k_a + k_b)^2 + 25(k_c + 1)^2}$$

$$\text{Ratio } \frac{\text{Maximum torque}}{\text{Full-load torque}} = \frac{T_{max}}{T_{fl}} = \frac{38.36}{2[\sqrt{k_a^2 + 25(k_c + 1)^2} + k_a]}$$

Results	$k_c = 1.35$; $k_a = k_b = 1$ x_1 increased	$k_b = 5.49$; $k_a = k_c = 1$ R_2 increased	$k_a = 5.49$; $k_b = k_c = 1$ R_1 increased	$k_a = k_b = k_c = 1$ No change
T_s/T_{fl}	0.27	0.27	0.27	0.369
T_{max}/T_{fl}	1.5	1.735	1.135	1.735
I_s/I_{fl}	3	3	3	3.507

The table shows that there is no loss of maximum torque if the starting current is limited by rotor resistance. Based on considerations of maximum-torque loss alone, x_1 is a preferable alternative to R_1 as a limiting impedance, but it must be remembered that this question is meant to illustrate the use of the equations rather than draw profound conclusions from a much restricted investigation into starting methods. A reference to Figure 8.4, p. 310, would be instructive. It shows a more comprehensive picture of the overall starting process.

Example 4.5

A 3-phase, 200-hp, 3300-V, star-connected induction motor has the following equivalent-circuit parameters per phase:

$$R_1 = R_2' = 0.8\,\Omega \quad x_1 = x_2' = 3.5\,\Omega$$

Calculate the slip at full load if the friction and windage loss is 3 kW. How much extra rotor resistance would be necessary to increase the slip to three times this value with the full-load torque maintained? How much extra stator resistance would be necessary to achieve the same object and what loss of peak torque would result?

$$\text{Mechanical power} = 3P_g(1 - s) - \text{mechanical loss}$$

$$200 \times 746 = \frac{3 \times V_1^2}{(R_1 + R_2'/s)^2 + (x_1 + x_2')^2} \times \frac{R_2'}{s}(1 - s) - 3000 \text{ (mech. loss)}$$

$$146\,200 = \frac{3 \times (3300/\sqrt{3})^2}{(0.8 + 0.8/s)^2 + 7^2} \times \frac{0.8(1 - s)}{s}$$

$$0.01747\left(0.64 + \frac{1.28}{s} + \frac{0.64}{s^2} + 49\right) = \frac{1}{s} - 1$$

$$167.26 - \frac{87.445}{s} + \frac{1}{s^2} = 0$$

$$\text{solving the quadratic: } \frac{1}{s} = \frac{87.445 \pm \sqrt{87.445^2 - 4 \times 167.26}}{2}$$

from which the lower value of s comes from $\frac{1}{s} = 85.5$; $\underline{s = 0.0117}$

From the torque equation (4.5), it can be seen that torque will be a fixed value, with all other parameters constant, if the quantity R_2'/s is unchanged; i.e. if R_2' changes in proportion to any slip change, the torque will be unaltered. In this case, the slip is to be 3 ×, so the extra rotor resistance, referred to the primary, will be $2 \times 0.8 = 1.6\,\Omega$. Maximum torque is unaffected by change of R_2', as shown in Example 4.4.

For increase of stator resistance, $\frac{3}{\omega_s} \times I_2'^2 \times \frac{R_2'}{s}$ to be unchanged; i.e. $\frac{V_1^2}{Z^2} \times \frac{R_2'}{s}$ must be the same so, since V_1 and R_2' are unchanged then: $1/(Z^2 \times s)$ must be the same for the same torque. Equating:

$$\frac{1}{\{[R_1 + 0.8/(3 \times 0.01117)]^2 + 49\} \times (3 \times 0.0117)} = \frac{1}{\{(0.8 + 0.8/0.0117)^2 + 49\} \times 0.0117}$$

$$\frac{1}{519.5 + 45.584R_1 + R_1^2 + 49} = \frac{3}{4675.2 + 49}$$

from which: $R_1^2 + 45.584R_1 + 1043 = 0$ and the lowest value of $R_1 = 16.6\,\Omega$. Extra stator resistance required is therefore $\underline{15.8\,\Omega}$

$$\text{Now maximum torque [eqn (4.13)]} = \frac{3V_1^2}{2\omega_s[\sqrt{R_1^2 + (x_1 + x_2')^2} + R_1]}$$

$$\text{proportional to: } \frac{1}{\sqrt{R_1^2 + (x_1 + x_2')^2} + R_1}$$

$$\text{so torque ratio} = \frac{\sqrt{0.8^2 + 7^2} + 0.8}{\sqrt{16.6^2 + 7^2} + 16.6} = \frac{7.845}{34.6} = 0.227 = \underline{77.3\%\text{ reduction}}$$

Example 4.6

A 3-phase, 4-pole, 3300-V, 50-Hz, star-connected induction motor has identical primary and referred secondary impedances of value $3 + j9\,\Omega$ per phase. The turns-ratio per phase is 3/1 (stator/rotor), and the rotor winding is connected in delta and brought out to slip rings. Calculate:

(a) the full-load torque at rated slip of 5%;
(b) the maximum torque at normal voltage and frequency;
(c) the supply voltage reduction which can be withstood without the motor stalling;
(d) the maximum torque if the supply voltage and frequency both fall to half normal value;
(e) the increase in rotor-circuit resistance which, at normal voltage and frequency, will permit maximum torque to be developed at starting. Express this: (i) as a fraction of normal R_2 and (ii) as (3) ohmic values to be placed in series with each of the slip-ring terminals and star connected.

The approximate circuit may be used and the magnetising branch neglected.

$$\text{(a) Full-load torque} = \frac{3 \times (3300/\sqrt{3})^2}{2\pi \times 50/2} \times \frac{1}{(3 + 3/0.05)^2 + 18^2} \times \frac{3}{0.05} = \underline{969\text{ Nm}}$$

$$\text{(b) Maximum torque, eqn (4.13)} = 69\,328 \times \frac{1}{2 \times (\sqrt{3^2 + 18^2} + 3)} = \underline{1631\text{ Nm}}$$

(c) With voltage reduced, the torque must not fall below 969 Nm, and since $T_e \propto V^2$ [eqn (4.5)]

$$\frac{969}{1631} = \left(\frac{\text{Reduced } V}{\text{Normal } V}\right)^2 \quad \text{so: Reduced } V = \sqrt{\frac{969}{1631}} = 0.77 \textit{ per unit} = \underline{23\%\text{ reduction}}$$

(d) This situation could arise if the supply-generator speed was to fall without change of its excitation; both voltage and frequency would fall together. Correcting all affected parameters in the appropriate equations:

$$\hat{s} = \frac{3}{\sqrt{3^2 + (18 \times \frac{1}{2})^2}} = 0.316; \text{ eqn (4.12), allowing for the reduced frequency.}$$

Substituting in the first equation:

$$\text{Max. } T_e = \frac{3 \times (\frac{1}{2} \times 3300/\sqrt{3})^2}{2\pi \times (\frac{1}{2} \times 50)/2} \times \frac{1}{(3 + 3/0.316)^2 + (18/2)^2} \times \frac{3}{0.316}$$

$$= \underline{1388 \text{ Nm}}$$

The answer could have been obtained directly by substituting the reduced parameters into the second equation used, (4.13), in part (b).

(e) Normally, maximum torque occurs at a slip of $\hat{s} = \dfrac{3}{\sqrt{3^2 + 18^2}} = 0.1644$

The required value of R_2 could be obtained by substituting $\hat{s} = 1$ in eqn (4.12), or alternatively, since R_2/s is a constant for any given torque:

$$\frac{\text{New } R_2'}{1} = \frac{3}{0.1644} \text{ Hence } \frac{\text{New } R_2'}{\text{Old } R_2'} = \frac{1}{0.1644} = 6.082$$

Hence, additional R_2 required = 5.082 times original R_2. Since the turns ratio is 3/1, the actual additional resistance per rotor phase must be $5.082 \times 3\,\Omega/3^2 = 1.694\,\Omega$. However, this would carry the phase current and either by considering the delta/star transformation or the fact that the line current of a star-connected load across the slip rings would carry $\sqrt{3}$ times the phase current, three external line resistors of value $1.694/3 = \underline{0.565\,\Omega}$ would dissipate the same power and avoid the necessity of bringing out expensive additional connections and slip rings, if inserting resistance in each phase.

Example 4.7

It is required to specify external rotor resistors for a 3-phase induction motor used for laboratory demonstration purposes. Examine the possible requirements and derive suitable criteria for making approximate estimates which neglect the machine impedance. Compare results using a 4-pole, delta-connected, 50-Hz machine with $R_1' = R_2 = 2\,\Omega$ and $x_1' = x_2 = 5\,\Omega$. Rated speed is 1455 rev/min. The rotor o.c. voltage is 240 V per phase.

Consider an induction motor with a normal o.c. rotor voltage of E_2 per phase, a rated rotor current of I_R per phase and a rated torque of T_R with rated slip s_R. A rotor external resistor can be used:

(a) to limit the starting current, with full voltage and frequency applied;
(b) to give speed control down to speed $(1 - s)$ *per unit* at a particular torque;
(c) to get maximum torque at starting when it normally occurs at slip $\hat{s}$;
(d) to get a particular starting torque.

The resistor rating must allow for short-term overload at starting and its continuous rating should reasonably be expected to correspond with I_R.

In answering this question, certain basic equations, which can be checked in Section 4.1, should be noted. At normal voltage and frequency:

A particular torque corresponds to a particular value of R_2/s and I_2.
Maximum torque occurs when $R_2/\hat{s} = \sqrt{R_1'^2 + (x_1' + x_2)^2}$ so if $\hat{s}$ is known, R follows. Starting currents may be 4–7 *per unit* without external limiting impedance.

The various points can now be dealt with, working in per-phase values. Allowance for delta-connected rotor would proceed as in Example 4.6.

(a) Suppose the starting current is to be limited to kI_R (k *per unit*) then required $R = E_2/kI_R$, e.g. if k = 2, $R = \frac{1}{2}E_2/I_R$.

(b) For any known torque T_x, with corresponding current I_x and a requirement for speed control down to s_x, then required $R = s_x E_2/I_x$.

(c) At starting, $\frac{R}{1} = \frac{R_2}{\hat{s}}$ so required total R follows. Starting currents with this external resistance would still be greater than I_R since $R_2/\hat{s} \ll R_2/s_R$. k would be E_2/RI_R.

(d) For the known current corresponding to the specified torque, T', say I', $R = E_2/I'$.
Certain machine features will be calculated first for reference purposes.

$$\text{Rated rotor current – all quantities are referred to the rotor} = \frac{240}{\sqrt{(2 + 2/0.03)^2 + 10^2}}$$

$$I_R = 3.46\,\text{A}$$

$$\text{Rated torque} = \frac{3}{2\pi \times 50/2} \times 3.46^2 \times \frac{2}{0.03} = 15.23\,\text{Nm}$$

$$\text{Slip for maximum torque } \hat{s} = 2/(\sqrt{2^2 + 10^2}) = 0.196$$

$$\text{Current at maximum torque} = \frac{240}{\sqrt{(2 + 2/0.196)^2 + 10^2}} = 15.2\,\text{A} = 4.4 \textit{ per unit}$$

$$\text{Normal starting current} = \frac{240}{\sqrt{(2 + 2)^2 + 10^2}} = 22.3\,\text{A} = 6.44 \textit{ per unit}$$

For (a), the required values of R will be approximated using the equations given at three different values of k; 1, 2 and 3 *per unit.*

For (b), rated torque will be considered and hence $s_x = RI_R/E_2 = 1/\text{k}$ so the speed range for the values of R calculated in (a) will be given by $1 - s_x = 1 - 1/\text{k}$.

The approximate values in (a) and (b) above are shown in the table below and the actual values of starting current, together with the actual speed range, are worked out with the machine impedance included.

$$\text{The actual starting current} = \frac{240}{\sqrt{(2 + 2 + R)^2 + 10^2}}$$

The actual speed range follows by solving for slip s in the following equation:

$$\text{Rated torque} = 15.23 = \frac{3}{2\pi \times 50/2} \times \frac{240^2}{\sqrt{(2 + (2 + R)/s)^2 + 10^2}} \times \frac{2 + R}{s}$$

k values (starting current in *per unit*)	1	2	3
(a) R required, ohms.	69.4	34.7	23.1
(b) Speed range $1 - s = 1 - 1/k$ (min. speed *p.u.*)	0	0.5	0.67
Actual starting current (*per unit*)	3.24(0.94)	6(1.74)	8.3(2.4)
Actual minimum speed at rated torque (*p.u.*)	–0.07	0.45	0.62

It can be seen that the estimates are the more accurate for low values of k and the consequent higher, dominating values of external resistance.

(c) R for maximum torque is $R_2/\hat{s} = 2/0.196 = 10.2\,\Omega$. This is the total circuit resistance so the extra per phase is $8.2\,\Omega$ which is much less than for the maximum k value above. Actually, k will be 4.4, the *per-unit* current at maximum torque.

(d) R has been worked out for three current values including rated torque and current in (a) above.

Example 4.8

Using the approximate circuit for the motor of Example 4.2, calculate the mechanical coupling power at speeds of 0, 720, 780 and –720 rev/min; positive speed being taken as in the direction of the rotating field. For the last case show, on a power-flow diagram, all the individual power components, to prove that the total input power is absorbed in internal machine losses. Take the mechanical loss as constant at all speeds other than zero, where it too is zero.

Synchronous speed $= N_s = 60 \times f/p = 60 \times 50/4 = 750$ rev/min

Series-circuit impedance $= \sqrt{(0.1 + 0.1/s)^2 + 1^2} = Z$

Mechanical coupling power $= 3P_m - \text{mech. loss} = 3\dfrac{I_2'^2 R_2'}{s}(1-s) - 1000 = P_{\text{coupling}}$

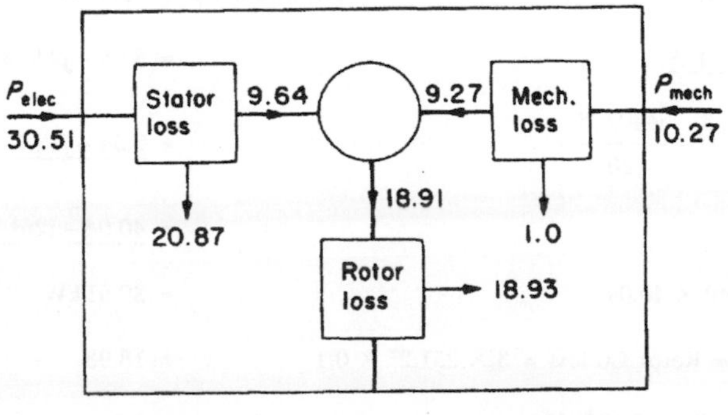

Figure E.4.8

Speed, rev/min.	0	720	780	−720
$\text{Slip} = \dfrac{750 - \text{Speed}}{750}$	1	0.04	−0.04	1.96
$R'_2/s = 0.1/s$	0.1	2.5	−2.5	0.051
Z	1.02	2.786	2.6	1.011
$I'_2 = \dfrac{440/\sqrt{3}}{Z}$ A	249	91.2	97.7	251.2
$3P_m$ = kW	0	59.86	−74.45	−9.268
$P_{coupling} = P_{mech}$	0	58.86	−75.45	−10.268
$T_e = \dfrac{3}{2\pi \times 50/4} \times \dfrac{I'^2_2 R'_2}{s}$ Nm	236.8	790.8	−911.5	122.9

These four sets of readings correspond to four significant points on the speed/torque curve; starting, motoring at full load, generating at the same, but negative slip, and reverse-current braking (plugging) see Figures 4.1(a) and E.4.17. In this last case, the values are those which would occur momentarily if the motor, running at full speed in the reverse sense, suddenly had its phase sequence and rotating field reversed. The values of currents, powers and torques should be studied to gain better understanding of induction machine operation.

For −720 rev/min, $\cos\varphi = (R_1 + R'_2/s)/Z = 0.1493$ and $\sin\varphi = -0.9888$.

$\therefore \mathbf{I}'_2 = 251.2\underline{/-84^\circ.3}$ $= 37.5 - j248.4$

and $\mathbf{I}_0 = \dfrac{440/\sqrt{3}}{100} - \dfrac{j440/\sqrt{3}}{20}$ $= \underline{2.54 - j12.7}$

$\therefore \mathbf{I}_1$ $= \underline{40.04 - j261.1}$

$P_{elec} = \sqrt{3} \times 440 \times 40.04$ $= 30.51$ kW

Stator Cu loss = Rotor Cu loss $= 3 \times 251.2^2 \times 0.1$ $= 18.93$

Stator Fe loss $= 3 \times (254)^2/100$ $= \underline{1.94}$

Total stator loss $= 20.87$

Mechanical loss = 1 kW; with rotor Cu loss (18.93 kW) = 19.93

Total machine loss = 40.8 kW

Total machine input = $P_{elec} + P_{mech}$ = 30.51 + 10.27 = 40.78 kW

The slight differences, e.g. between the input and the loss, are due to rounding-off errors. The figure shows the power distribution for this braking condition.

Example 4.9

A 3-phase, 6-pole, 50-Hz induction motor has a peak torque of 6 Nm and a starting torque of 3 Nm when operating at full voltage. Maximum torque occurs at a slip of 25%. When started at 1/3 of normal voltage the current is 2 A.

(a) What is the mechanical power, at peak torque when operating at normal voltage?
(b) What maximum torque would the machine produce at 1/3 of normal voltage?
(c) What starting current would the machine take when supplied with normal voltage?
(d) What extra rotor-circuit resistance, as a percentage, would be required to give maximum torque at starting and what would then be the current, in terms of that at peak torque without external resistance?

This is basically a simple problem to bring out certain elementary relationships. The curves sketched on Figure E.4.9 indicate the main points for the solution. No additional equations from the ones used previously are involved.

(a) Power at maximum torque = $\omega_m T_e = 2\pi \times \frac{50}{3} \times (1 - 0.25) \times 6$ = 0.471 kW

(b) Torque $\propto V^2$ hence, reduced maximum torque = $(1/3)^2 \times 6$ = 2/3 Nm
(c) Current $\propto V$, hence $I_{start} = 3 \times 2$ = 6 A
(d) A given torque requires a particular value of R'_2/s. Since s changes from 0.25 to 1, then the total rotor circuit resistance must change in the same ratio; i.e. by 4 times. Hence extra rotor resistance = 300% R_2.

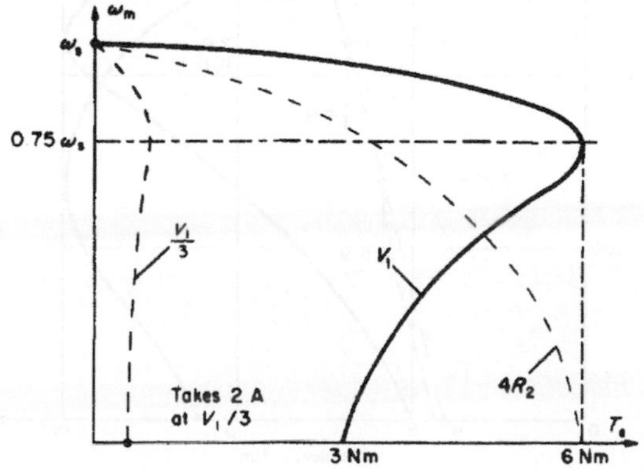

Figure E.4.9

Since R_2'/s is constant and since this is the only equivalent-circuit impedance which could vary with speed, the total impedance presented to the terminals is unchanged and so the current is the same as at $s = 0.25$.

Example 4.10

An induction motor has the following speed/torque characteristic:

Speed	1470	1440	1410	1300	1100	900	750	350	0 rev/min
Torque	3	6	9	13	15	13	11	7	5 Nm

It drives a load requiring a torque, including loss, of 4 Nm at starting and which increases linearly with speed to be 8 Nm at 1500 rev/min.

(a) Determine the range of speed control obtainable, without stalling, by providing supply-voltage reduction.

(b) If the rotor was replaced by one having the same leakage reactance but with a doubled resistance, what would then be the possible range of speed variation with voltage control?

For each case, give the range of voltage variation required.

This is a 'drives' problem and must be solved graphically from the data given. The solution depends on the simple relationships that $T_e \propto V^2$, and for any given torque R_2'/s is constant, if all other parameters are constant; eqn (4.5).

(a) Plotting the speed = $f(T_e)$ and $T_m = f(\omega_m)$ characteristics from the data gives the normal steady-state speed at their intersection. The T_e characteristic is reduced proportionally until maximum T_e intersects the T_m characteristic. This occurs at a torque of 6.9 Nm and

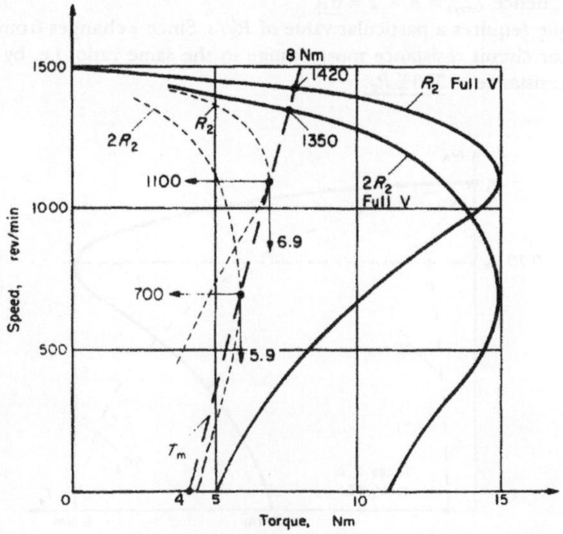

Figure E.4.10

speed of 1100 rev/min. Hence voltage reduction is $\sqrt{15/6.9} = 1.47/1$ or 100% to 68% volts, giving a speed reduction from 1420 to 1100 rev/min.

(b) At various values of T_e, the slip, which is proportional to the speed difference from the synchronous value, is noted and a new speed plotted for this same T_e but with this speed difference doubled, since R_2 is doubled. Again, from the intersections of the T_m characteristic with the two new curves, the speed range is seen to be 1350 to 700 rev/min. The torque value on the reduced curve for $2R_2$ is 5.9, so the ratio of peak torques gives the appropriate voltage reduction as $\sqrt{15/5.9} = 1.59/1$ or 100% to 63% volts. This greater speed range, for a similar voltage reduction, is obtained at the penalty of additional rotor-circuit losses, but nevertheless such schemes are sometimes economically suitable because of their simplicity, for certain types of load where torque falls off appreciably with speed.

The curves also show the speed range obtainable with resistance control; in this case, doubling the rotor resistance reduces the speed from 1420 to 1350 rev/min.

Example 4.11

A 440-V, 3-phase, 6-pole, 50-Hz, delta-connected induction motor has the following equivalent-circuit parameters at normal frequency:

$R_1 = 0.2\,\Omega$; $R_2' = 0.18\,\Omega$; $x_1 = x_2' = 0.58\,\Omega$ – all per phase values.

(a) The machine is subjected in service to an occasional fall of 40% in both voltage and frequency. What total mechanical load torque is it safe to drive so that the machine just does not stall under these conditions?

(b) When operating at normal voltage and frequency, calculate the speed when delivering this torque and the power developed. Calculate also the speed at which maximum torque occurs.

(c) If V and f were both halved, what would be the increase in starting torque from the normal direct-on-line start at rated voltage and frequency?

(d) If now the machine is run up to speed from a variable-voltage, variable-frequency supply, calculate the required terminal voltage and frequency to give the 'safe' torque calculated above: (i) at starting and (ii) at 500 rev/min.

(e) Repeat (d) for the machine to develop a torque equal to the maximum value occurring at rated voltage and frequency. In both (d) and (e), the criteria is that the air-gap flux per pole is maintained constant for any particular torque.

(a) The question asks effectively for the maximum torque with voltage and frequency reduced to 0.6 of rated values. Substituting in the maximum-torque eqn. (4.13) with appropriate correction of parameters:

$$\text{Max. } T_e = \frac{3}{2\pi \times 0.6 \times 50/3} \times \frac{(0.6 \times 440)^2}{2[\sqrt{0.2^2 + (0.6 \times 1.16)^2} + 0.2]} = \underline{\underline{1800\,\text{Nm}}}$$

(b) From the general torque expression eqn. (4.5), with normal supply, the required slip to produce this torque is obtained by equating:

$$1800 = \frac{3}{2\pi \times 50/3} \times \frac{440^2}{(0.2 + 0.18/s)^2 + 1.16^2} \times \frac{0.18}{s}$$

$$\text{giving: } 1.803\left(0.04 + \frac{0.072}{s} + \frac{0.0324}{s^2} + 1.3456\right) = \frac{1}{s}$$

$$\text{from which: } \frac{1}{s^2} - \frac{214.9}{s} + 42.77 = 0$$

and the smaller value of s on solution is 0.0907 corresponding to a speed of:

$$1000\,(1 - s) = \underline{909\text{ rev/min.}}$$

$$\text{Power developed} = \frac{2\pi}{60} \times 909 \times 1800 = \underline{171.3\,\text{kW}} = \underline{230\,\text{hp}}$$

$$\text{Speed for maximum torque from eqn. (4.12) } \hat{s} = \frac{0.18}{\sqrt{0.2^2 + 1.16^2}} = 0.1529$$

so speed = 847 rev/min.

(c) The expression for starting torque is:

$$\frac{3}{2\pi \times f/p} \times \frac{V^2}{(R_1 + R_2')^2 + (x_1 + x_2')^2} \times \frac{R_2'}{1}$$

Using the ratios to cancel the constants in the expression:

$$\frac{\text{Starting torque at } \tfrac{1}{2}V \text{ and } f}{\text{Normal starting torque}} = \frac{1}{(f/2)/f} \times \frac{[(V/2)/V]^2}{(0.38^2 + (1.16/2)^2)/(0.38^2 + 1.16^2)}$$

$$= 2 \times \frac{0.25}{0.4808/1.49} = \underline{1.55 \text{ times normal.}}$$

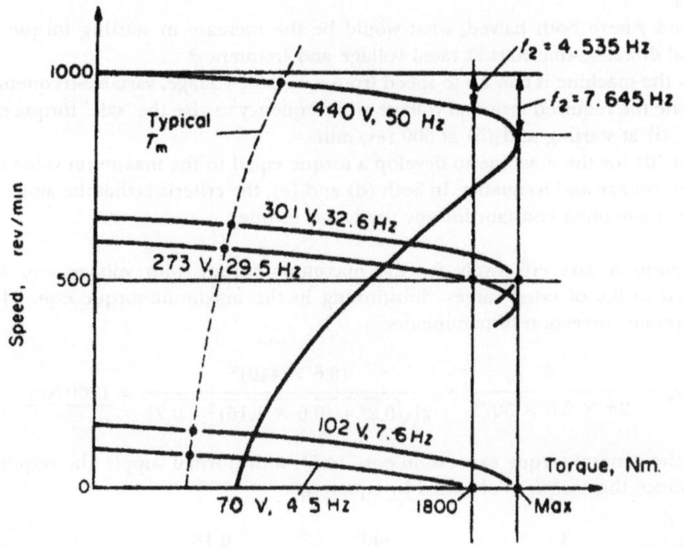

Figure E.4.11

(d) and (e) As will be shown in the next example, if the flux per pole ($E_1/f \propto \hat{\phi}$, see eqn 2.1) is maintained constant by adjustment of voltage and frequency, any particular torque occurs at a unique value of slip frequency f_2 and rotor current I'_2. Further, since: $n_s = n + sn_s$

$$pn_s = pn + p \times \frac{f_2}{f_{supply}} \times n_s$$

$$f_{supply} = pn + f_2 \text{ since } pn_s \text{ is the supply frequency.}$$

The questions ask for the supply frequency and voltage to produce the maximum torque and a torque of 1800 Nm, at two different speeds, 500 rev/min and zero. The required values of f_2 and I_2 can be deduced from those occurring at normal voltage and frequency for these particular torques. The supply voltage required will be $I'_2 Z$ where the reactive elements in the impedance will be corrected for f_{supply} as calculated. Parts (d) and (e) are worked out in the following table.

	Starting $n = 0$		Speed n = 500/60 rev/min	
	1800 Nm	Max. torque	1800 Nm	Max. torque
Slip at 50 Hz	0.0907	0.1529	0.0907	0.1529
Slip frequency $f_2 = s \times 50$	4.535	7.645	4.535	7.645
Supply frequency $= pn + f_2$	4.535	7.645	29.535	32.645
$I'_2 = \dfrac{440}{\sqrt{(0.2 + 0.18/s)^2 + 1.16^2}}$	177.9	244.4	177.9	244.4
s at new frequency $= f_2/f_{supply}$	1	1	0.1536	0.234
R'_2/s (new slip)	0.18	0.18	1.172	0.7692
$X = 1.16 \times f_{supply}/50$	0.1052	0.177	0.6852	0.757
$Z = \sqrt{(0.2 + R'_2/s)^2 + X^2}$	0.3943	0.4192	1.534	1.23
$V_{supply} = I'_2 Z$	70.1	102.5	272.8	300.6
Max. torque $= \dfrac{3}{2\pi \times 7.645/3} \times 244.4^2 \times \dfrac{0.18}{1} = 2014$ Nm				

The various speed/torque curves are sketched on the figure, for the criteria of constant flux per pole. It can be seen that they are very suitable for speed-controlled applications, with maximum torque being available over the whole range. For part (a) (264 V 30 Hz), the curve would be close to the one at 29.5 Hz but peaking at 1800 Nm, because the flux per pole is only approximately constant; $V_1/f \approx E_1/f$.

4.3 Constant- (primary) current operation: improved starting performance

Constant-current operation requires a source which has sufficient voltage available to regulate the phase current rapidly and accurately. This mode is of particular interest for field-oriented control of cage-rotor induction motors (Section 7.4), and also applies during rheostatic braking where the constant current is usually of zero frequency. The value of primary impedance is only required for the calculation of supply voltage and does not influence the electromechanical performance. Consequently, for such calculations, the equivalent circuit can omit the primary impedance. The magnetising resistance will also be omitted, without significant loss of accuracy. It must be emphasised that the induction machine equivalent circuit can be used at any frequency over a wide range (including approximate allowance for the time harmonics by superposition), providing *all* frequency-sensitive parameters are given their appropriate values. However, it is sometimes useful to define, and after modification, work with the parameters specified at a particular frequency f_{base}, which will usually be the rated value.

Example 4.12

The values of E_1, X_m and x_2' for a particular induction motor are known at a frequency f_{base}.

(a) Develop the expressions which show that the rotor current and torque are independent of the supply frequency but depend on the slip frequency, f_2; providing that the flux per pole (E_1/f) is constant.
(b) Show also, independently of the above, that for any given primary current I_1, the rotor current I_2' is governed by f_2 and explain how this is related to the constant-flux condition.
(c) Finally, derive the expression for the slip $\hat{s} = \hat{f}_2/f$ at which maximum torque occurs for a constant-current drive and hence show that the maximum torque *capability* is independent of both supply and slip frequencies.

(a) At any frequency f, and slip f_2/f, then sx_2' becomes: $(f_2/f)(x_2' \times f/f_{base}) = x_2' \times f_2/f_{base}$. *With constant flux per pole,* the referred secondary e.m.f. E_1 at standstill is proportional to the supply frequency f, so sE_1 becomes: $(f_2/f)(E_1 \times f/f_{base}) = E_1 \times f_2/f_{base}$. Hence rotor current:

$$\mathbf{I}_2' = \frac{\mathbf{E}_1 \times f_2/f_{\text{base}}}{R_2' + \mathrm{j}(x_2' \times f_2/f_{\text{base}})} = \frac{\mathbf{E}_1}{\dfrac{R_2'}{f_2/f_{\text{base}}} + \mathrm{j}x_2'} \tag{4.14}$$

To get this expression for I_2', which applies for any frequency, the equivalent circuit has to be modified slightly to Figure E.4.12. Eqn (4.4) was derived without reference to a particular frequency so at any frequency f, the general torque expression is:

$$\frac{3}{2\pi \times f/p} \times I_2'^2 \times \frac{R_2'}{f_2/f} \times \frac{(f_{\text{base}})}{(f_{\text{base}})}$$

By incorporating f_{base} in a unity multiplier, cancelling f and rearranging:

$$T_e = \frac{3}{2\pi \times f_{\text{base}}/p} \times I_2'^2 \times \frac{R_2'}{f_2/f_{\text{base}}} \tag{4.15}$$

showing that since I_2' is independent of f, eqn (4.14), then so is T_e – *for the constant-flux condition.*

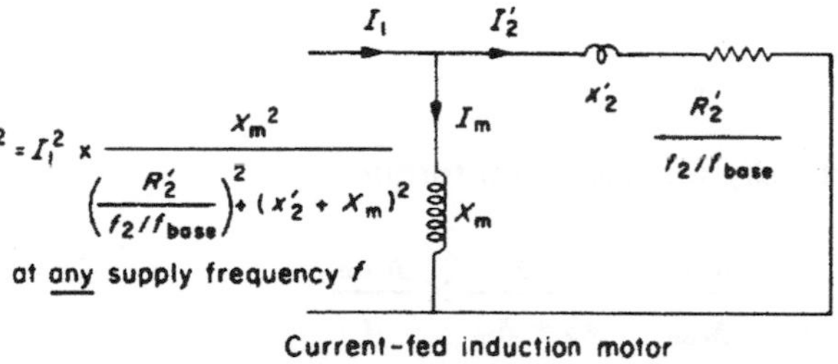

Figure E.4.12

(b) From the rules for parallel circuits and correcting reactance parameters for frequency, the equivalent circuit of Figure E.4.12 yields the following equation:

$$\mathbf{I}_2' = \mathbf{I}_1 \times \frac{\mathrm{j}X_m(f/f_{\text{base}})}{R_2'/(f_2/f) + \mathrm{j}(x_2' + X_m)(f/f_{\text{base}})}$$

Dividing throughout by f/f_{base} and squaring:

$$I_2'^2 = I_1^2 \times \frac{X_m^2}{\left(\dfrac{R_2'}{f_2/f_{\text{base}}}\right)^2 + (x_2' + X_m)^2} \tag{4.16}$$

A similar expression, with $X_m{}^2$ in the numerator replaced by

$$\left[\left(\frac{R_2'}{f_2/f_{\text{base}}}\right)^2 + x_2'^2\right]\text{, will give } I_m{}^2.$$

The expression for I_2' suggests that its value depends on f_2, not on supply frequency, whether or not the flux is constant. However, at a particular slip frequency f_2, eqn (4.16) modified to give $I_m{}^2$ shows that I_m has a unique value related to I_1 and hence corresponds to a particular flux. *Hence, for a given I_1 and f_2, there is a unique value of I_2' and of I_m, flux $\hat{\phi}$ and torque T_e from eqns (4.15) and (4.16). Alternatively if voltage and frequency are controlled to give constant flux (E_1/f), any particular f_2 will define the currents and the torque.* This gives an easy method of deriving the variable-frequency ω_m/T_e characteristic from that at a particular frequency (see Reference 1).

(c) Inserting the expression for $I_2'^2$ into the torque expression, T_e will be found to have a form similar to eqn (4.5) for the approximate circuit with $I_1 X_m$ as the coefficient replacing V_1. By comparing expressions or by differentiating, the maximum value of T_e, see eqn (4.12), occurs when $R_2'/(f_2/f_{\text{base}}) = x_2' + X_m$; so for maximum torque:

$$\frac{\hat{f}_2}{f_{\text{base}}} = \frac{R_2'}{x_2' + X_m}$$

hence, the slip for maximum torque

$$\hat{s} = \frac{\hat{f}_2}{f} \times \frac{f_{\text{base}}}{f_{\text{base}}} = \frac{R_2'}{x_2' + X_m} \times \frac{f_{\text{base}}}{f} \tag{4.17}$$

Substituting the value of $\hat{f}_2/f_{\text{base}}$ in the torque equation (4.15) gives

$$\text{Maximum } T_e = \frac{3I_1{}^2}{2\pi f_{\text{base}}/p} \times \frac{X_m{}^2}{(x_2' + X_m)^2 + (x_2' + X_m)^2} \times (x_2' + X_m)$$

$$= \frac{3I_1{}^2}{2\pi \times f_{\text{base}}/p} \times \frac{X_m{}^2}{2(x_2' + X_m)} \tag{4.18}$$

which is independent of both f and f_2. However, f_2 must have the value given above, $\hat{f}_2$, to achieve this inherent capability and it does depend on the value of $I_1{}^2$, which itself has a maximum limit.

The purpose of deriving the equations for this constant-current mode is primarily concerned with voltage/frequency control, used to maintain maximum torque as these quantities are increased. Usually constant flux per pole is aimed for, at a value corresponding to rated voltage and frequency, although moderate increases could be considered, during starting operations for example.

Note that in the relevant expressions derived, the variable f_2/f_{base} replaces $s = f_2/f$ used in the previous constant-frequency expressions. This quantity will be given the symbol S and is particularly useful when the constant primary-current is d.c. The machine is then operating as a synchronous generator at a rotor frequency $f_2 = pn$. It is convenient to take base synchronous speed as a reference, i.e. $n_s = n_{s(base)}$, and therefore:

$$S = pn/f_{base} = pn/pn_{s(base)} = n/n_s = n_s(1 - s)/n_s = 1 - s.$$

In this dynamic braking mode, the current I_1 is the equivalent r.m.s. primary current giving the same m.m.f. as I_{dc}. For star connection, usually arranged with just two line terminals taken to the d.c. supply, the equivalent $I_1 = \sqrt{(2/3)}I_{dc} = 0.816I_{dc}$. The speed/torque curves for dynamic braking are only in the 2nd and 4th quadrants with negative speed × torque product, and a shape similarity with the motoring curves, see Reference 1 and Example 4.17. A torque maximum occurs at a value of $\hat{S} = R_2'/(x_2' + X_m)$ usually at a very low speed because $S = 1 - s$ and $X_m \gg R_2'$. The relative motion for this condition is given by the expression: $Sn_s = n$ rev/s.

Example 4.13

A 3-phase, 8-pole, 50-Hz, star-connected, 500-V induction motor has the equivalent-circuit per phase shown. Calculate the torques produced at slips of 0.005, 0.025, 0.05, 1 and the maximum torque, for the following two conditions, both at 50 Hz:

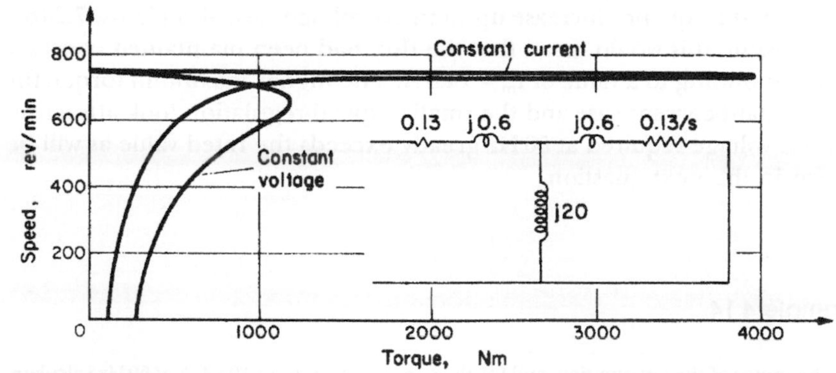

Figure E.4.13

(a) Constant (r.m.s.)-voltage drive at 500 line V;
(b) Constant (r.m.s.)-current drive at the same primary current occurring for slip = 0.05 in (a).

As distinct from the constant-voltage equations, for which I_m is relatively small so that the approximate circuit may be used, those for constant current include X_m and hence the effect of the magnetising current I_m on all the currents and torque. At very low slips, R_2'/s becomes so high that I_m becomes greater than I_2' as the latter decreases. The flux therefore increases and the very large maximum torque occurs when the two currents are about equal – at very low slips. This is brought out by the present example and the accompanying table – which does not include all the detail of the calculations.

$$\text{For const. } V\hat{s} = \frac{R_2'}{\sqrt{R_1'^2 + (x_1 + x_2')^2}} = \frac{0.13}{\sqrt{0.13^2 + 1.2^2}} = 0.1077$$

$$\text{For const. } I_1\hat{s} = \frac{R_2'}{x_2' + X_m} = \frac{0.13}{0.6 + 20} = 0.00631$$

$$T_e = \frac{3}{2\pi \times 50/4} \times I_2'^2 \times \frac{R_2'}{s} = \frac{I_2'^2}{26.2} \times \frac{R_2'}{s}$$

The two speed/torque curves are sketched from the tabulated results. They have the same torque approximately at $s = 0.05$, since the primary currents are the same. The difference is due to the approximations in the constant voltage case. For this curve, if the full-load torque is taken to correspond with a slip of 0.025, i.e. 553 Nm, the machine could be said to have an overload capacity of about 2 *per unit* (1188/553). For this, the primary current has to increase from 57.7 A to 170.3 A which is about 3 *per unit*, for a 2 *per unit* increase in torque. During starting procedure, this would be the ideal figure to be held constant by suitable matching of voltage and frequency increase up to rated voltage (see also Figure 7.24b). If successful, this would mean that the flux had been maintained constant – corresponding to a value of $I_m = 14.4$ A. The higher maximum torque for the constant-current case and the smaller speed-regulation look attractive, but the voltage required at 50 Hz greatly exceeds the rated value as will be shown in the next question.

Example 4.14

For the motor of the last question, and for the constant current of 103.4 A at 50 Hz calculate, using the available results where appropriate:

		$s =$	0.005	0.00631	0.025	0.05	0.1077	1
(a)		rev/min =	746	745	731	712.5	669	0
		R'_2/s	26	20.6	5.2	2.6	1.207	0.13
		$R_1 + R'_2/s = R =$			5.33	2.73	1.337	0.26
		$x_1 + x'_2 = X =$			1.2	1.2	1.2	1.2
		$Z = \sqrt{R^2 + X^2} =$			5.46	2.98	1.8	1.228
		$I'_2 = 500/\sqrt{3}Z =$			52.8	96.8	160.6	235.1
		$\mathbf{I}'_2 =$			51.5 – j11.6	88.7 – j38.8	119 – j107.5	49.8 – j229.8
		$\mathbf{I}_0 = 288.7/j20 =$			–j14.4	–j14.4	–j14.4	–j14.4
		$\lvert\mathbf{I}\rvert_1 = \lvert\mathbf{I}_2{}^1 + \mathbf{I}_0\rvert$			57.7	103.4	170.3	249.2
		T_e at 500 V =			553	930	1188	274
(b)		$Z^2 = \frac{R'^2_2}{s^2} + (x'_2 + X_m)^2$	1100	848.7	451.4	431.1		424.4
		$X_m{}^2/Z^2 =$	0.364	0.471	0.886	0.928		0.943
		$I'_2 = 103.4\sqrt{Xm^2/Z^2}$	62.3	71	97.3	99.6	100.2	100.4
		$I_m =$	81	73.1	25.4	13.3	6.7	3.1
		T_e at 103.4 A	3859	3965	1880	985	463	50

(a) the values of $\mathbf{I}_m$ and $\mathbf{I}'_2$ for maximum torque;
(b) the required supply voltage to sustain this primary current at $s = 0.05$;
(c) the required supply voltage to sustain this primary current at $s = 0.00631$.

Estimate these voltages using $\mathbf{I}'_2$ and the series elements of the approximate circuit.

$$\text{(a) } \mathbf{I}'_2 = \mathbf{I}_1 \times \frac{\mathrm{j}X_m}{(x'_2 + X_m) + \mathrm{j}(x'_2 + X_m)} = 103.4 \times \mathrm{j}20\,\frac{(20.6 - \mathrm{j}20.6)}{20.6^2 + 20.6^2} = 50.2 + \mathrm{j}50.2$$

$$= \underline{71\,\mathrm{A}}$$

$$\mathbf{I}_m = \mathbf{I}_1 - \mathbf{I}'_2 = 103.4 + \mathrm{j}0 - 50.2 - \mathrm{j}50.2 = 53.2 - \mathrm{j}50.2 \qquad = \underline{73.1\,\mathrm{A}}$$

$$\text{(b) } V = \sqrt{3} \times Z \times I'_2 = \sqrt{3} \times 2.98 \times 99.6 \qquad = \underline{514\,\mathrm{V}}$$

$$\text{(c) } V = \sqrt{3} \times \sqrt{20.73^2 + 1.2^2} \times 71 \qquad = \underline{2553\,\mathrm{V}}$$

Clearly, with a supply frequency of 50 Hz, the maximum torque, very-low-slip condition is not a practicable possibility. Furthermore, with a magnetising current of 73.1 A, instead of the normal 14.4 A, saturation would be considerable and X_m would fall dramatically. A method of allowing for this is examined in Example 4.16 but, in practice, such large increases would not be considered. However, more moderate increases of flux could be analysed in the same way. The next example (4.15) shows that this maximum torque can be obtained at lower frequencies since then, the slip $\hat{f}_2/f$ is higher and the component of voltage drop $I'_2 R'_2/\hat{s}$ required is therefore lower.

Example 4.15

(a) Derive an expression for the required frequency to give maximum torque with a constant-current supply.
(b) Using the same motor data as for Example 4.13, calculate the required supply frequency and voltage to give this maximum torque with constant-current drive (i) at starting (ii) at 20% of normal synchronous speed.
(c) For the same motor data, but with constant flux maintained instead, at the value corresponding to normal operation, determine the required voltage and frequency to give maximum torque at starting.

(a) From Example 4.12 $\hat{s} = \dfrac{R'_2}{x'_2 + X_m} \times \dfrac{f_{base}}{f}$ [eqn. (4.17)]

and, since $s = \dfrac{n_s - n}{n_s} = \dfrac{f/p - n}{f/p} = \dfrac{f - pn}{f}$, cancelling f after equating these two

expressions gives $\hat{f} = \underline{\dfrac{R'_2 \times f_{base}}{x'_2 + X_m} + pn}$

(b)	(b) (i)	(b) (ii)
$\hat{f} =$	$\dfrac{0.13 \times 50}{0.6 + 20} = 0.3155\,\text{Hz}$	$\dfrac{0.13 \times 50}{20.6} + 4 \times \dfrac{50}{4 \times 5} = 10.3155$
$Z_2' = \dfrac{R_2'}{s} + jx_2'\dfrac{\hat{f}}{f_{base}} =$	$0.13 + j0.6 \times \dfrac{0.3155}{50}$	$\dfrac{0.13}{0.3155/10.3155} + j0.6 \times \dfrac{10.3155}{50}$
$\mathbf{Z}_{input} = \mathbf{z}_1 + \dfrac{1}{\mathbf{Y}_m + \dfrac{1}{Z_2'}} =$	$0.1913 + j0.0687 = 0.2033\,\Omega$	$2.132 + j2.247 = 3.097\,\Omega$
$V_{supply} = \sqrt{3} \times 103.4 \times Z_{input}$	36.4 V at 0.3155 Hz	555 V at 10.3155 Hz

All the detailed calculations are not shown and note that $\mathbf{Y}_m = 1/jX_m$ must also be corrected for the frequency change from 50 Hz values. The exact circuit must be used because of the high value of $I_m = 73.1$ A, from Example 4.14. Note that normal supply voltage of 500 V limits constant-current operation at 103.4 A, 50 Hz, to rather less than 20% of normal synchronous speed, so would not reach the high torques shown on the table of p. 107.

(c) From Example 4.12, any particular torque, at constant flux, is obtained at a unique slip frequency. In this case we require maximum torque, which from Example 4.13 occurs at $i = 0.1077$ and therefore $\hat{f}_2 = 0.1077 \times 50 = 5.385$ Hz.

Because this is a constant, rated-flux condition, I_m is relatively small and the approximate circuit may be used to calculate the supply voltage from $\sqrt{3} \times Z \times I_2'$ using the results of Example 4.13. From Example 4.13 part (a), the rotor current will be the same ($I_2' = 160.6$ A) and the maximum torque too will be unchanged at 1188 Nm.

$$\text{Hence, } Z_{input} = 0.13 + j(0.6 + 0.6) \times \frac{5.385}{50} + 0.13 = 0.26 + j0.1292 = 0.29\,\Omega$$

$$V_{supply} = \sqrt{3} \times 0.29 \times 160.6 = \underline{80.8\,\text{V at } 5.385\,\text{Hz}}$$

$$I_m = \frac{80.8/\sqrt{3}}{20 \times 5.385/50} = 21.6\,\text{A} \quad \text{(difference from 14.4 A is due to approximations)}$$

It will be noticed that although the maximum torque can be maintained up to 50 Hz, it is very much less than obtained with the constant-current drive. Even with the lower (constant) primary current of 103.4 A the torque is 3965 Nm, as against 1188 Nm. This is because the frequency is very much lower at 0.3155 Hz and more of the current I_1 is therefore passed through the reduced X_m. This apparent improvement is offset, since the value of X_m collapses due to saturation at the high magnetising current. The next example illustrates this point.

Example 4.16

Once again, using the motor data of the previous examples, calculate, for a constant-current drive of 103.4 A, the maximum starting torque (i) neglecting saturation and (ii) assuming the value of X_m is reduced to 1/3 of its normal value due to saturation. Make an approximate comparison of the flux levels for (i) and (ii) compared with normal operation at slip = 0.05.

In Example 4.12 it was shown that the maximum torque under constant-current drive conditions is:

$$T_e = \frac{3I_1^2}{2\pi f_{base}/p} \times \frac{X_m^2}{2(x_2' + X_m)}$$

This expression is independent of all frequencies, though x_2' and X_m must correspond to f_{base}. It is sufficient for the purpose of answering this question. By correcting the second term for the specified saturated change of X_m, the effect of saturation on maximum torque is found simply. It is a useful exercise to check this, however, by working out the value of $\hat{s}$ for the saturated condition and hence the values of f_2, Z_2', Z_{input}, I_2' and I_m. They are as follows – using the exact circuit for solutions:

	$\hat{s}$	$\hat{f}_2$	Z_2'	Z_{input}	I_2'	I_m	$T_{e(max)}$	X_m
Unsaturated	1	0.3155	0.061 + j0.065	0.203	71	73.1	3965	20
Saturated	1	0.8944	0.055 + j0.065	0.199	67.1	73.4	1250	6.667

The unsaturated values have been worked out in previous examples. For the saturated value:

X_m reduced to: $20 \times 1/3 = 6.667\,\Omega$

$$\hat{f}_2 = \frac{R_2'}{x_2' + X_m} \cdot f_{base} = \frac{0.13 \times 50}{0.6 + 6.667} = 0.8944\,\text{Hz}$$

$$\text{Maximum } T_e = \frac{3 \times 103.4^2}{2\pi \times 50/4} \times \frac{6.667^2}{2(0.6 + 6.667)} = \underline{1250\,\text{Nm}}$$

Using a lower frequency than for the constant, rated-flux condition, it can be seen that in spite of saturating the magnetic circuit, the maximum torque is still higher at $I_1 = 103.4$ A than the 1188 Nm for $I_1 = 170.3$ A, when constant voltage is the supply condition. To check that the saturation allowance is reasonable, the flux will be worked out from $E_1/f = I_m X_m/f$, noting that X_m must be corrected for frequency.

For normal rating, flux proportional to: $12.6 \times 20/50 \qquad = 5.04$

For starting (unsat.) " " $73.1 \times \dfrac{20 \times 0.3155}{50} \Big/ 0.3155 = 29.4$

For starting (sat.) " " $73.4 \times \dfrac{6.67 \times 0.8944}{50} \Big/ 0.8944 = 9.79$

The flux ratio allowing for saturation is $9.79/5.04 = \underline{1.94}$
The I_m ratio is: $73.4/12.6 = \underline{5.8}$
By the empirical formula used in Example 3.6, this flux ratio should give an I_m ratio of:

$$\frac{0.6 \times 1.94}{1 - 0.4 \times 1.94} = \underline{5.2}$$

which is close to 5.8 above and suggests that the saturation allowance is reasonable. For an exact calculation, the magnetisation characteristic ϕ/I_m would have to be available and an iterative program devised to approach the exact solution, since the value of I_1 in the maximum torque expression is not known until the value of X_m is known. A similar method to that used for the d.c. series motor can be adopted. Here, the non-linearity of the magnetisation curve was dealt with at the beginning by taking various values of I_f and k_ϕ. In the present case, a series of I_m values would define a series of corresponding X_m values. Referring to the various expressions developed in Example 4.12, each X_m will define $\hat{s}, \hat{f}_2$, E_1 $(= I_m X_m)$, I_2' (from E_1/Z_2') and I_1 from the parallel-circuit relationships. The maximum torque follows for the various values of I_1 calculated.

The last five examples have shown, through the circuit equations, the special characteristics of the induction motor when under controlled frequency and voltage. A simple way of summarising the behaviour is through study of the m.m.f. diagram represented by the $\mathbf{I}_1$, $\mathbf{I}_2'$ and $\mathbf{I}_m$ triangle. As will be shown later in Section 5.5, which compares induction and synchronous machines, the torque is proportional to the product of any two currents and the sine of the angle between them. Now consider the two phasor diagrams of Figure 4.2. The first one is for the maximum-torque condition deduced from the constant-supply-voltage equations in Examples 4.12 and 4.13. I_1 and I_2' are relatively high but I_m remains at the level corresponding to rated flux. The voltage diagram is drawn for the approximate-circuit calculation. If this maximum-torque is required over a range of frequencies down to zero speed, the m.m.f. diagram would be unchanged, with the constant rated-flux represented by I_m. This occurs,

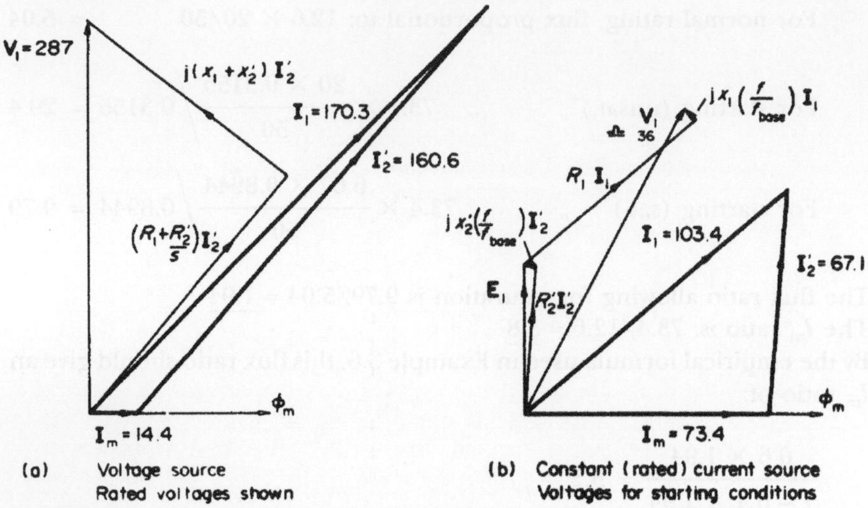

Figure 4.2 *Maximum torque with different supply conditions.*

from Example 4.15, at a constant slip-frequency of 5.385 Hz. $\mathbf{I}_2'$, $\mathbf{I}_1$ and $\mathbf{I}_m$ would be constant as supply frequency varied. Refer also to Example 4.11.

Figure 4.2b shows the condition for maximum torque deduced from the constant supply-current equations, for the rated current of I_1 = 103.4 A. Again this condition can be sustained over a range of frequencies, by suitable adjustment of V_1 and f_1. It will be noticed that I_m is very much higher, Example 4.14a, and the diagram has been constructed allowing for saturation as in Example 4.16. The voltage diagram has to be drawn this time from the exact circuit because I_m is so large. It is shown for the starting condition. V_1 was not calculated allowing for saturation but it was 36.4 line volts with saturation changes neglected in Example 4.15. The starting frequency f_2 (= f_1) was there calculated as 0.3155 Hz but in fact, allowing for saturation, f_2 should be 0.8944 Hz – Example 4.16. Even so this value is still much less than for Figure 4.2a and therefore I_m is much higher. So is the maximum torque itself, though the supply current is only 103.4 A instead of 170.3 A in Figure 4.2a. This is due to the currents being at a better displacement angle; $\mathbf{I}_2'$ and $\mathbf{I}_m$ are nearly in quadrature. Again, if it is desired to sustain the maximum torque of Figure 4.2b up to the speed where the voltage reaches its maximum – about f_1 = 10 Hz, Example 4.15 – f_2, and therefore all the currents, must be maintained constant. Thus for either constant-voltage or constant-current supplies, the condition for maximum torque is a particular constant flux, which is different for the two cases. Further discussion of this mode will be deferred till Sections 5.5 and 7.4.

Example 4.17

Using the data and the 'exact' circuit calculations for the motor of Example 4.2, investigate the following features of the performance when the machine is changed over from motoring at a slip of 3% to dynamic braking. D.C. excitation is applied to the stator, using the 2-lead connection; i.e. equivalent r.m.s. a.c. current I_1 equal to $\sqrt{2/3}\,I_{dc}$. For all cases the speed may be assumed unchanged until the switch changeover is completed. Further, the circuit is so adjusted as to retain initially the same value of rotor current as when motoring at 3% slip.

(a) Rotor-circuit resistance unchanged. Calculate the d.c. excitation voltage and current and find the initial braking torque on changeover and also the maximum torque produced during the run-down to zero speed. The d.c. excitation is maintained.
(b) Repeat the calculation for the condition where instead, the excitation is so adjusted that the air-gap flux (E_1/f) is maintained at the 3% slip value. Extra rotor-circuit resistance will now be required to keep the rotor current at the 3% value.
(c) Compare the initial braking torques and currents if instead of (a) or (b), two stator leads are reversed to cause reverse-current braking (plugging).

For (a) and (b), the magnetising resistance does not apply because the stator excitation is d.c. For part (c), the approximate circuit may be used.

(a) The equivalent circuit is shown on the figure but unlike Examples 4.13 to 4.16 the constant current is now d.c. and the relative motion is changed from $n_s - n$ to $n = Sn_s$ where $S = f_2/f_{base}$. The rotor frequency is now proportional to speed since the machine is operating as a variable-speed generator and $S = n/n_s = 1 - s$.
Before and immediately after changeover:

$$N = (1 - s)N_s = (1 - 0.03) \times 60 \times 50/4 = 727.5\,\text{rev/min}$$

$$S = 727.5/750 = 0.97$$

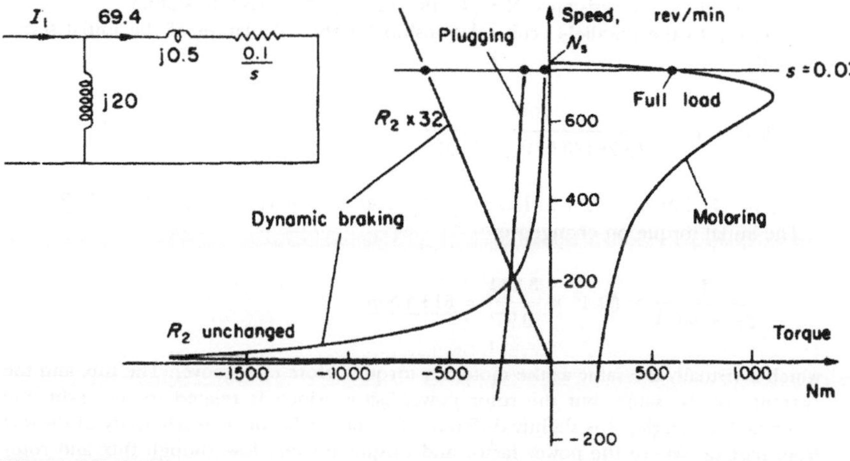

Figure E.4.17

From the parallel-circuit relationships $I_2'^2 = I_1^2 \dfrac{20^2}{(0.1/0.97)^2 + (20 + 0.5)^2}$

and since I_2', from Example 4.2 = 69.4 A, I_1 = 71.1 A.
The d.c. current to give this equivalent I_1 is 71.1 × $\sqrt{(3/2)}$ = 87.1 A
and since two stator phases are in series, the required d.c. voltage is 2 × 0.1 × 87.1 = 17.4 V

The initial torque on changeover, using the general torque expression eqn (4.15) developed in Example 4.12, is equal to:

$$\frac{3}{2\pi \times 50/4} \times 69.4^2 \times \frac{0.1}{0.97} = \underline{19\,\text{Nm}}$$

From the expression for maximum torque eqn (4.18):

$$T_e = \frac{3}{2\pi \times 50/4} \times 71.1^2 \times \frac{20^2}{2(20 + 0.5)} = \underline{1884\,\text{Nm}}$$

and this occurs at a value of $\hat{S}$ = 0.1/(20 + 0.5) = 0.0049, i.e. at speed 3.9 rev/min. The torque values should be compared with that for 3% slip motoring, i.e. 613.6 Nm. Calculate also that I_m is now 50.3 A – increased from 11.7 A.

(b) For the alternative strategy of maintaining the flux constant instead of I_1, it is more convenient to use the alternative equation for I_2' since E_1 is now known – 234 V, from Example 4.2.

$$\text{Rotor current } I_2' = \frac{SE_1}{R_2' + Sx_2'} = 69.4\,\text{A when } S = 0.97$$

$$\text{hence: } 69.4^2 = \frac{(0.97 \times 234)^2}{(R + 0.1)^2 + (0.97 \times 0.5)^2}$$

from which the extra resistance $R = \sqrt{10.46} - 0.1 = \underline{3.134\,\Omega}$ (ref. to stator).

Reverting to the parallel-circuit relationship for the calculation of the initial stator current:

$$69.4^2 = I_1^2 \times \frac{20^2}{(3.234/0.97)^2 + (20.5)^2}$$

giving I_1 = 72.1 A; I_{dc} = 72.1 × 1.225 = 88.3 A and d.c. voltage = 88.3 × 0.2 = 17.7 V.

The initial torque on changeover

$$= \frac{3}{2\pi \times 50/4} \times 69.4^2 \times \frac{3.234}{0.97} = \underline{613.4\,\text{Nm}}$$

which is virtually the same as the motoring torque before changeover. The flux and the current are the same, but the rotor power-factor, which is related to the induction machine load angle,[1] is slightly different. The power factor is nearly unity as distinct from part (a) where the power factor and torque are very low, though flux and rotor current are nearly the same.

For maximum torque, since

$$T_e = \frac{3}{2\pi \times 50/4} \times \frac{E_1^2}{(R'/S)^2 + X^2} \times \frac{R'}{S}$$

and E_1 is fixed, then $\hat{S}$ occurs when $R'/\hat{S} = X$. Hence $\hat{S} = 3.234/0.5 = 6.47$.

This means that the speed for maximum torque is impractically high above synchronous speed and therefore 613.4 Nm is the highest torque encountered in running down to zero speed.

The value of the maximum torque is:

$$\frac{3}{2\pi \times 50/4} \times \frac{234^2}{0.5^2 + 0.5^2} \times \frac{3.234}{6.47} = \underline{2091\,\text{Nm}}$$

(c) The changeover now corresponds to a reversal of n_s and slip becomes:

$$s = \frac{n_s - n}{n_s} = \frac{-750 - 727.5}{-750} = 1.97$$

Using the torque equation for constant voltage, since the only change from motoring is the phase sequence and slip:

$$T_e = \frac{3}{2\pi \times 50/4} \times \frac{(440/\sqrt{3})^2}{(0.1 + 0.1/1.97)^2 + 1^2} \times \frac{0.1}{1.97}$$

$$= \frac{1}{26.18} \times 251.2^2 \times 0.05076 = \underline{122.3\,\text{Nm}}$$

Note that although the current is nearly four times that for dynamic braking, the torque is very much less, because of the poor rotor-circuit power factor, the frequency being nearly 100 Hz initially. The various speed/torque curves are sketched on Figure E.4.17.

Example 4.18

A 3-phase, double-cage-rotor, 6-pole, 50-Hz, star-connected induction motor has the following equivalent-circuit parameters per phase:

$z_1 = 0.1 + j0.4\,\Omega$; $z_2' = 0.3 + j0.4\,\Omega$; $z_3' = 0.1 + j1.2\,\Omega$ – all at standstill.

Find, in terms of the line voltage V_1, the torque at 980 rev/min:

(a) including the outer-cage impedance;
(b) neglecting the outer-cage impedance.

What is the starting torque?

The equivalent circuit is shown on the figure, the magnetising branch being neglected. The circuit assumes that both rotor windings embrace the

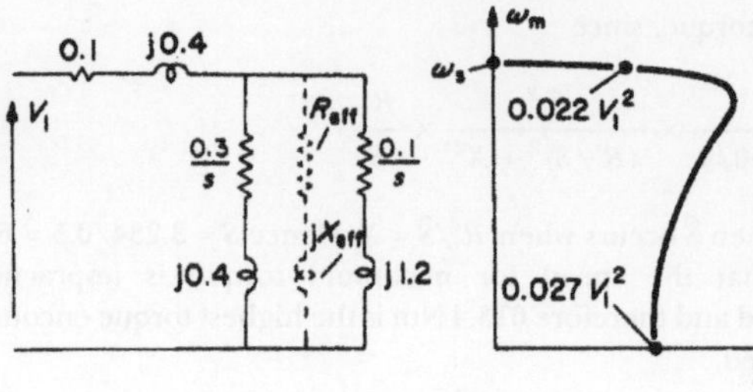

Figure E.4.18

same flux and that they only have leakage with respect to the primary winding. The calculation of torque is then virtually the same as for the single-cage rotor except that the two cages combine to an equivalent impedance $R_{eff} + jX_{eff}$. This impedance includes the effect of slip and is a function of slip. The high-resistance cage is nearer to the surface and therefore has the lower leakage reactance. It is responsible for most of the starting torque because of its lower impedance $\mathbf{z}_2'$ at standstill. $\mathbf{z}_3'$ represents the inner cage in which most of the working torque at low slip is produced, because of its lower resistance, reactances being very low at normal slip frequencies.

For a 6-pole, 50-Hz machine, N_s = 1000 rev/min so $s = (1000 - 980)/1000 = 0.02$.

(a) $$\mathbf{Z}_{rotor} = \frac{1}{\dfrac{1}{0.1/0.02 + j1.2} + \dfrac{1}{0.3/0.02 + j0.4}}$$

$$= \frac{1}{0.1891 - j0.04539 + 0.06661 - j0.00178}$$

$$R_{eff} + jX_{eff} = 3.78 + j0.698$$

Adding $\mathbf{z}_1$: $\underline{0.1 + j0.4}$

gives $\mathbf{Z}_{in} = 3.88 + j1.098 = 4.032\ \Omega$

Hence: $$T_e = \frac{3}{2\pi \times 50/3} \times \frac{(V/\sqrt{3})^2}{4.032^2} \times 3.78 = \underline{0.00222 V_1^2}$$

(b) Neglecting $\mathbf{Z}_2'$ and adding $\mathbf{z}_1 + \mathbf{Z}_3' = 5.1 + j1.6 = 5.3451\ \Omega$ then:

$$T_e = \frac{3}{2\pi \times 50/3} \times \frac{(V/\sqrt{3})^2}{5.3451^2} \times \frac{0.1}{0.02} = \underline{0.00167 V_1^2}$$

which shows that the outer cage does contribute about 25% of the rated torque.

For starting torque, $s = 1$

$$\mathbf{Z}_{\text{rotor}} = \frac{1}{\dfrac{1}{0.1 + \text{j}1.2} + \dfrac{1}{0.3 + \text{j}0.4}}$$

$$R_{\text{eff}} + \text{j}X_{\text{eff}} = 0.169 + \text{j}0.324$$

Adding $\mathbf{z}_1$: $\underline{0.1 + \text{j}0.4}$

gives $\mathbf{Z}_{\text{in}} = 0.269 + \text{j}0.724 = 0.772\,\Omega$

$$\text{Hence: } T_e = \frac{3}{2\pi \times 50/3} \times \frac{(V/\sqrt{3})^2}{0.772^2} \times 0.169 = \underline{0.00271\,V_1^{\,2}}$$

The starting torque is a little higher than the full-load torque, indicating the desired effect of the double-cage construction in improving the normally low, single-cage starting torque; see figure.

4.4 Unbalanced and single-phase operation

When the three phase-currents are unequal and/or the mutual time-phase displacement between them is not 120°, this unbalanced system can be solved by the use of symmetrical components, see Section 2.2. Deliberate unbalancing of the supply voltages is sometimes used to change the speed/torque curve and give another means of induction-motor speed-control. There is a deterioration of performance in other ways, e.g. the variation occurs partly due to a backwards torque component produced by negative-sequence currents. Positive- and negative-sequence circuits can be calculated independently, the former producing the forward torque with an apparent rotor resistance R_2/s. The negative sequence gives a reverse rotating field for which the impedance is different, since the apparent rotor resistance is $R_2/(2 - s)$.

Example 4.19

The motor for which the equivalent circuit was given in Example E.4.1, is supplied with full line voltage across terminals AB, but terminal C is reduced in potential so that both V_{BC} and V_{CA} are 330 V in accordance with the phasor diagram. Calculate the electromagnetic torque developed at a slip of 3% and compare with Example E.4.2 (b), neglecting the magnetising impedance.

$$\text{Displacement } \theta_C = 180° - \cos^{-1}\frac{330^2 + 440^2 - 330^2}{2 \times 330 \times 440} = 180° - \cos^{-1}\frac{440}{660}$$

$$= 180° - 41°.19 = 138°.81$$

$$\theta_B = 360° - 138°.81 = 221°.19$$

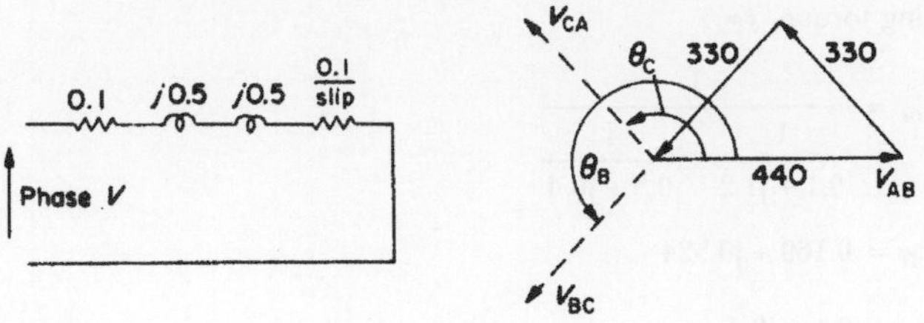

Figure E.4.19

Hence, $V_{BC} = 330\underline{/221°.19}$, $V_{CA} = 330\underline{/138°.81}$, $V_{AB} = 440\underline{/0°}$ and eqn (2.3) is used for the transformation to symmetrical components. The zero sequence is of no practical interest because such currents cannot flow in a star-connected circuit without neutral wire and even in a closed delta, where they could circulate, they could only produce a pulsating torque.

$$\begin{bmatrix} V_{A(+)} \\ V_{A(-)} \\ V_{A(0)} \end{bmatrix} = \frac{1}{3}\begin{bmatrix} 1 & \underline{/120°} & \underline{/240°} \\ 1 & \underline{/240°} & \underline{/120°} \\ & & \end{bmatrix}\begin{bmatrix} 440\underline{/0°} \\ 330\underline{/221°.19} \\ 330\underline{/138°.81} \end{bmatrix} = \frac{110}{3}\begin{bmatrix} 4 + 3\underline{/341°.19} + 3\underline{/378°.81} \\ 4 + 3\underline{/461°.19} + 3\underline{/258°.81} \\ \end{bmatrix}$$

$$= \frac{110}{3}\begin{bmatrix} 4 + 3(0.947 - j0.322) + 3(0.947 + j0.322) \\ 4 + 3(-0.194 + j0.981) + 3(-0.194 - j0.981) \end{bmatrix} \begin{bmatrix} 355\,\text{V} \\ 104\,\text{V} \end{bmatrix}$$

These are line voltages so the positive- and negative-sequence phase-voltages are $355/\sqrt{3} = 205\,\text{V}$ and $104/\sqrt{3} = 60\,\text{V}$ respectively.

The positive-sequence torque at full voltage has already been calculated in Example E.4.2 as 641.8 Nm at full voltage. At 355 line volts, it will be reduced to:

$$641.8 \times \frac{355^2}{440^2} = 417.8\,\text{Nm}$$

For the negative sequence, the slip is $2 - 0.03 = 1.97$ and $R_2/(2 - s) = 0.0508$. The negative-sequence torque is calculated in the same way as the positive sequence with the appropriate change to apparent rotor resistance and the applied voltage. Its value is thus, using eqn (4.5):

$$\frac{3}{2\pi \times \frac{50}{4}} \cdot \frac{60^2}{0.1508^2 + 1^2} \cdot 0.0508 = 6.8\,\text{Nm}$$

This is a torque in the reverse sense so the total machine torque is reduced from 641.8 to:

$$417.8 - 6.8 = \underline{411\,\text{Nm}}$$

A reduction in torque would occur over the whole slip range and the effect is similar to normal voltage reduction – Examples 4.9 and 4.10. The contribution of the negative sequence is small but there are simple circuits,[3] which vary the unbalance over such a range that positive and negative speeds are both covered. As for balanced-voltage control, high-resistance rotors improve the characteristic though the torque capability is still reduced due to the lowered sequence voltages and the two sequence components opposing one another. The method of calculation used above is of general application, e.g. to determine the effect of unbalance as a possible operational hazard, see Tutorial Example T4.19.

Single-phase operation

The most common form of unbalanced operation occurs with motors energised from single-phase supplies, representing in fact the majority of motor types and almost entirely in the small-power (< 1 kW) range. Many special designs of very small machines have been developed for which the theory and calculating methods will not be discussed here. Single-phase commutator and permanent-magnet machines have been reviewed briefly in Sections 3.4 and 3.6, the latter being sometimes energised from rectified single-phase supplies. For an induction motor, single-phase excitation produces only a pulsating field and zero starting torque since positive- and negative-sequence components are equal. Using phase-splitting or static-phase-converter circuits will give an approximation to a 2-phase supply, which together with a machine having two windings in space quadrature can be analysed using 2-phase symmetrical components. One of the windings may just be used for starting and then cut out, or, by adjustment of its external-circuit components, approximate to balanced 2-phase operation. However, even a single winding will produce a net forward torque once motion is initiated.

A 3-phase induction motor can be started and run up from a single-phase supply with a suitable capacitor connected. Reference 5 suggests external-circuit connections for both star and delta arrangements, for each of which, two machine line terminals are connected to the single-phase supply and the remaining machine terminal is taken through a static phase-converter to one supply terminal. The analysis shows that for maximum starting torque, the admittance of the capacitor required for a star-connected machine is:

$$2\pi fC = \frac{2}{3} \times \frac{10^6}{Z}\,\mu F \quad \text{where } Z \simeq \sqrt{(R_1 + R_2')^2 + (x_1 + x_2')^2}$$

Applying this formula to the machine of Example 4.2 gives the capacitor value as:

$$C = \frac{1}{2\pi \times 50} \times \frac{2}{3} \times \frac{10^6}{\sqrt{0.2^2 + 1^2}} = 2081\,\mu F$$

This is a very high value but the machine is large, about 50 kW and therefore has a low *per-unit* impedance. For a more typical power rating where such starting methods might be used, viz. $\not>$ about 1–2 kW, the impedance might be 10–50 times higher, depending on the rating, with corresponding reduction in capacitor size.

Example 4.20

To gain some idea of the change of capacitance required between starting and running, assume that the same machine has two phase-windings in space quadrature, each with the same equivalent circuit as for Example 4.2. Neglect the magnetising impedance and determine the external components in series with one phase so that (a) at starting and (b) when operating at 3% slip, the currents in the two phases are in time quadrature and of equal magnitude when the supply available is single phase. The torque can also be calculated from the rotor copper loss as for the 3-phase machine.

(a) For the phase connected directly to the supply, $Z = 0.2 + j1\,\Omega = 1.02\underline{/78°.69}$

For the other phase, the impedance presented to the supply to give 90° shift and same magnitude: $= 1 - j0.2\,\Omega = \underline{1.02/-11°.31}$

Hence required external impedance must be: $0.8 - j1.2\,\Omega$

The capacitative component of this would be $\dfrac{10^6}{2\pi \times 50 \times 1.2} = \underline{2626\,\mu F}$

which is of similar order to the previous value obtained.

The starting torque would be 2/3 of the value obtained from part (b) of Example 4.2, since the impedance per phase is the same but there are only two phases. This applies to the running condition also, so the torques are 158 Nm and 428 Nm respectively.

(b) For a slip of 3%, $Z = 0.1 + 0.1/0.03 + j1 = 3.433 + j1\,\Omega$

For the second phase to give 90° shift, impedance $= \underline{1 - j3.433\,\Omega}$

Hence required external impedance must be: $-2.433 - j4.433\,\Omega$

The capacitative component of this would be: $\dfrac{10^6}{2\pi \times 50 \times 4.433} = \underline{718\,\mu F}$

This last answer indicates the big difference between starting and running requirements. It also indicates that perfect balance would require a negative resistance of 2.433 Ω! so some unbalance has to be tolerated with such simple phase-conversion arrangements. As for the previous illustration, the capacitor values are very high because the machine in the example is untypically large.

An understanding of single-phase operation with one winding alone can be approached through consideration of the 3-phase motor having one supply lead opened. The motor would then be subjected to single-phase excitation and with star connection say, $I_A = 0$ and $I_B = -I_C$. The 1-phase pulsating m.m.f. F can be resolved into equal synchronously rotating forward and reverse m.m.f.s of half magnitude $F/2$, as will be understood if two oppositely rotating space-phasors are combined.(1) Each component m.m.f. is considered to act separately on the rotor with the appropriate slip. At standstill, positive- and negative-sequence impedances, connected in series, are each equal to half of the short-circuit impedance between two terminals, i.e. approximately the same as the per-phase equivalent circuit in the 3-phase mode, for each sequence, Figure 4.3. The

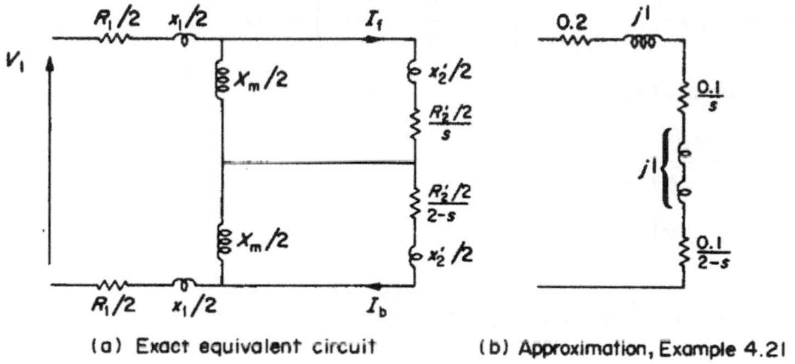

Figure 4.3 *Single-phase induction motor.*

magnetising reactance will be much smaller however due to interference with the flux by the backwards rotating field component. When the machine is stationary, the two equal and opposite torque components cancel so there is no starting torque. But with rotation, the impedances change due to the different slip values and though I_f, the forward-sequence current is only a little greater than I_b, the voltages across the two sections are very different. The positive-sequence torque $I_f^2 R_2'/2s\omega_s$, is much larger than the negative-sequence torque, $I_b^2 R_2'/2(2-s)\omega_s$.

The next example uses the approximate circuit for simplicity. Neglecting the magnetising impedance is a more drastic step than for the balanced polyphase machine but a comparison will be made through Tutorial Example T4.20 which calls for an exact solution and shows all the various components and their interactions. On Figure 4.3, the magnetising resistance is omitted, the iron loss being treated separately since superposition is not a satisfactory method of allowing for this.

Example 4.21

The 3-phase machine having the equivalent circuit used in Examples 4.1 and 4.2 is operating on no load when a fuse in one line blows to give a single-phasing condition. Determine the torque when the machine is now loaded and the speed falls to give the rated slip of 3%. The approximate equivalent circuit can be used. In order to make a general comparison of single-phase and 3-phase performance, even though this machine is of much higher rating than the maximum for single phase, not usually greater than 1 kW, calculate also the maximum torque as a 3-phase machine and for single-phase operation calculate the torque at slips of 7% and 10% which will cover the maximum torque in this mode.

If the approximate circuit is used, Z_m need not be included in calculations of torque. For the 3-phase machine, the impedance required to calculate performance is:

$$0.1 + j0.5 + j0.5 + 0.1/s = (0.1 + 0.1/s) + j1\,\Omega \text{ per phase.}$$

For the single-phase machine the impedance required is:

$$0.2 + j1 + j1 + \frac{0.1}{s} + \frac{0.1}{2-s} = \left(0.2 + \frac{0.1}{s} + \frac{0.1}{2-s}\right) + j2\,\Omega$$

For the 3-phase machine

$$I_2' = \frac{440/\sqrt{3}}{\sqrt{(0.1 + 0.1/s)^2 + 1}}$$

For the 1-phase machine

$$I_f = I_b = \frac{440}{\sqrt{[0.2 + 0.1/s + 0.1/(2-s)]^2 + 2^2}}$$

For the maximum torque on the 3-phase machine,

$$\hat{s} = \frac{R_2'}{\sqrt{R_1^2 + (x_1 + x_2')^2}} = \frac{0.2}{\sqrt{0.01 + 1}} = 0.0995$$

Hence

$$T_{max} = \frac{3}{78.54} \times \frac{(440/\sqrt{3})^2}{(0.1 + 0.1/0.0995)^2 + 1^2} \times \frac{0.1}{0.0995} = \underline{1115\,\text{Nm}}$$

$$T_e \text{ at } s = 0.03 \text{ from Example 4.2} = \underline{642\,\text{Nm}}$$

For the single-phase machine:

Slip	=	0.03	0.07	0.1
$0.1/s$	=	3.3333	1.4285	1
$0.1/(2-s)$	=	0.0508	0.0518	0.0526
$R = 0.2 +$ rotor resis.	=	3.584	1.68	1.253
$Z = \sqrt{R^2 + 2^2}$	=	4.104	2.612	2.36
$I_f = I_b$	=	107.2	168.5	186.4
$I_f^2\left(\frac{0.1}{s} - \frac{0.1}{2-s}\right)$	=	37 722	39 088	32 883
T_e – divide by ω_s	=	480	498	419

The maximum torque will be just over 500 Nm which is less than rated torque as a 3-phase machine for which the full-load current is less than 80 A, from Example 4.2. The deterioration for 1-phase operation is thus considerable, the maximum torque being rather less than 50% and the currents being much higher, when compared with the 3-phase motor. There is also a double-frequency torque pulsation with consequent vibration and noise. 'Exact'-circuit calculations required in Tutorial Examples T4.20 and T7.7 will further illustrate the above points.

The equivalent-circuit parameters can be obtained from o.c. (no-load), and s.c. tests, though approximations are usually necessary, e.g. the transfer of the magnetising reactance to the terminals and noting that at low slips, the lower $X_m/2$ branch is almost short circuited by the $R_2'/2(2-s)$ term in series with $[R_1 + j(x_1 + x_2')]/2$ so that $Z_{n.1} \simeq X_m/2$.

4.5 Speed control by slip-power recovery

This method of speed control involves the application to the secondary terminals, of a voltage V_3 from an active source, which may provide (or accept) power, increasing (or decreasing) the speed. It must automatically adjust itself to slip frequency and this can be done with commutator machines or by power-electronic switching circuits. An external resistor, carrying the slip-frequency current, though only a passive load, does give slip frequency automatically since $V_3 = R_3 I_2$. For an active source, the slip power, $P_3 = V_3 I_2$, $\cos\varphi_3$, can be fed back to the supply so that power changes with speed giving approximate constant-torque characteristics. If the slip-power is fed to a suitable machine on the main shaft, a 'constant'-power drive is formed. Independently of the method, the speed variation can be calculated by applying $\mathbf{V}_3'/s$ to the rotor terminals on the equivalent circuit, $\mathbf{V}_3$ having been transformed for turns ratio and for frequency, as for all the other rotor-circuit parameters.

The rotor current becomes:

$$I_2' = \frac{|\mathbf{V}_1 - \mathbf{V}_3'/s|}{\sqrt{(R_1 + R_2'/s)^2 + (x_1 + x_2')}} \tag{4.19}$$

and the rotor-circuit power per phase:

$$sP_g = sE_1 I_2' \cos\varphi_2 = I_2^2 R_2 + P_3.$$

The torque

$$T_e = \frac{3P_g}{\omega_s} = \frac{3}{\omega_s} \times \frac{(I_2'^2 R_2' + V_3' I_2' \cos\varphi_3)}{s}. \tag{4.20}$$

V_3 can be used to change the slip and/or the power factor.

Example 4.22

A 3-phase, 440-V, star-connected (stator and rotor), wound-rotor induction motor has the following equivalent-circuit parameters per phase:

$$R_1 = 0.2\,\Omega; \quad x_1 = 0.8\,\Omega; \quad R_2 = 0.06\,\Omega; \quad x_2 = 0.25\,\Omega.$$

The magnetising branch may be neglected and the stator/rotor turns ratio is 2:1. With the slip rings short circuited, the motor develops full load torque at 3% slip.

(a) Determine the voltage, in phase with supply, which applied to the slip rings will cause the motor to develop full-load torque at 25% slip.
(b) What voltage in lagging quadrature with the supply is required to give unity power-factor at the full-load slip? What effect will this have on the torque?

The approximate equivalent circuit is shown on the figure. The rotor impedance has been referred to the stator winding; i.e. $R_2' = 2^2 \times 0.06 = 0.24\,\Omega$; $x_2' = 4 \times 0.25 = 1\,\Omega$.

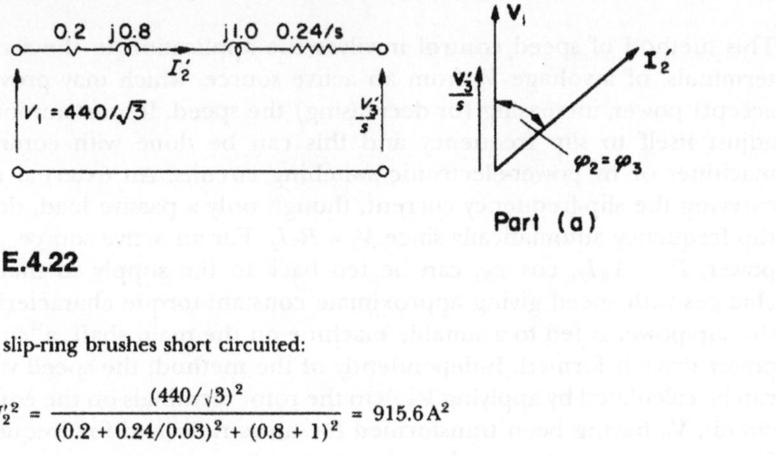

Figure E.4.22

(a) With slip-ring brushes short circuited:

$$I_2'^2 = \frac{(440/\sqrt{3})^2}{(0.2 + 0.24/0.03)^2 + (0.8 + 1)^2} = 915.6\,\text{A}^2$$

With applied voltage V_3 at 25% slip:

$$I_2'^2 = \frac{(440/\sqrt{3} - V_3'/0.25)^2}{(0.2 + 0.24/0.25)^2 + 1.8^2} = (118.63 - 1.868\,V_3')^2$$

The equivalent-circuit impedance is $\sqrt{1.16^2 + 1.8^2} = 2.1414\,\Omega$

and since $\mathbf{V}_3$ is in phase with $\mathbf{V}_1$, power factor $\cos\varphi_2 = \dfrac{1.16}{2.414} = \cos\varphi_3$.

These last expressions for $I_2'^2$ and for $\cos\varphi_3$ are now substituted in the torque equation (4.20) and equated to the full-load torque for short-circuited brushes.

With V_3 applied: $I_2'^2 R_2 \quad = (14\,073 - 443.2\,V_3' + 3.489\,V_3'^2) \times 0.24$

and $V_3' I_2' \cos\varphi_3 \quad = \underline{V_3'(118.63 - 1.868\,V_3') \times 1.16/2.1414}$

$$\frac{\text{Total rotor-circuit power}}{0.25} = -0.698V_3'^2 - 168.5V_3' + 13\,510 \text{ after some simplification.}$$

This is equated to:

$$\frac{\text{Normal rotor-circuit power}}{s} = \frac{915.6 \times 0.24}{0.03} = 7324.8 \text{ watts/phase}$$

Again, after some simplification, this yields the expression $V_3'^2 + 241.4V_3' - 8861.3 = 0$ and the quadratic solution gives, as the only positive value, $V_3' = 32.4\,\text{V}$.

Allowing for the 2/1 turns ratio: $\underline{V_3 = 16.2\,\text{V}}$, in phase with V_1, will reduce speed to 0.75 N_s while delivering full-load torque.

(b) With $\mathbf{V}_3'$ lagging $\mathbf{V}_1$ by 90° and at full-load slip of 3%:

$$\mathbf{I}_2' = \frac{440/\sqrt{3} - (-jV_3'/0.03)}{(0.2 + 0.24/0.03) + j1.8} = \frac{254 + j33.3V_3'}{8.2 + j1.8}$$

For unity power factor, this expression must be 'real'. This in turn means that the 'real' and 'imaginary' parts of numerator and denominator must be in the same ratio.

Hence:

$$\frac{254}{33.3V_3'} = \frac{8.2}{1.8} \text{ from which } V_3' = 1.673\,\text{V and } \underline{V_3 = 0.836\,\text{V}}$$

Although this voltage is very low, it must be remembered that the rotor e.m.f. at this small slip is also low.

Substituting the value of V_3':

$$\mathbf{I}_2' = \frac{254 + j33.3 \times 1.673}{8.2 + j1.8} = 33.1\underline{/+11^\circ.35}$$

and since $\mathbf{V}_3'$ is lagging $\mathbf{V}_1$, the total angle of lead $\varphi_3 = 11^\circ.35 + 90^\circ = 101^\circ.35$

The power P_3 is therefore $V_3'I_2' \cos\varphi_3 = 1.673 \times 33.1 \times \cos 101^\circ.35 = -21.8$ watts/phase

$$\text{Hence: } T_e = \frac{3P_s}{\omega_s} = \frac{3}{\omega_s} \times \frac{(33.1^2 \times 0.24 - 21.8)}{0.03} = \underline{\frac{3}{\omega_s} \times 8038\,\text{Nm}}$$

This compares with the full-load torque of $\underline{(3/\omega_s) \times 7324.8\,\text{Nm}}$ so there would be a tendency for speed to rise with the additional torque and V_3 would have to be modified slightly to give a speed-reducing component. This tendency can also be understood from the sign of P_3 which is negative, indicating a power input to the rotor, increasing the speed. The phase angle of $\mathbf{I}_2'$ with respect to $\mathbf{V}_3$ is greater than 90°. The exact calculations of $\mathbf{V}_3$ for a given speed and power factor are more complex than the above.[(1)]

Example 4.23

A 6-pole, 50-Hz, wound-rotor induction motor drives a load requiring a torque of 2000 Nm at synchronous speed. It is required to have speed variation down to 50% of synchronous speed by slip-power control. Determine the maximum kW rating of the injected power source, (a) assuming $T_m \propto \omega_m^2$, and (b) assuming $T_m \propto \omega_m$. Neglect all the machine losses.

In this case, a standard induction motor is provided with an external slip-power source, which could be a commutator machine, or more usually nowadays, a power-electronic circuit to give frequency conversion from supply frequency to slip frequency.[1]

Since rotor copper loss is being neglected, the question is asking for the maximum rotor-circuit power $3sP_g = 3P_3$, as speed is reduced by slip-power control down to half synchronous-speed. The mechanical output $3(1-s)P_g$, see Figure 4.1b, also varies with speed and the nature of this variation governs the magnitude of P_3.

At synchronous speed, the mechanical power $3\hat{P}_m = 2000 \times 2\pi \times 50/3 = 209.4\,\text{kW}$

At any other speed $\omega_m = \omega_s(1-s)$; i.e. the *per-unit* speed is $(1-s)$ so:

$$P_m/\hat{P}_m = (1-s)^x$$

Since mechanical power is $\omega_m T_m$, the index 'x' is either 3 for (a) or 2 for (b).

Equating expressions for power $3(1-s)\,P_g = 3\hat{P}_m(1-s)^x$

from which: $$P_g = \hat{P}_m(1-s)^{(x-1)}$$

and $$P_3 = sP_g = s\hat{P}_m(1-s)^{(x-1)}$$

differentiating: $$\frac{dP_3}{ds} = \hat{P}_m[(1-s)^{(x-1)} + (-1) \times s \times (x-1)(1-s)^{(x-2)}]$$

This is zero when: $(1-s)^{(x-1)} = s(x-1)(1-s)^{(x-2)}$

i.e. when $$s = 1/x$$

Substituting this value of s gives $P_3 = \dfrac{1}{x} \times P_m \times \left[\dfrac{x-1}{x}\right]^{(x-1)} = \hat{P}_m \dfrac{(x-1)^{(x-1)}}{x^x}$

Hence for $T_m \propto \omega_m^2$, x = 3 and $3P_3 = 209.4 \times \dfrac{2^2}{3^3} = \underline{31\,\text{kW}}$

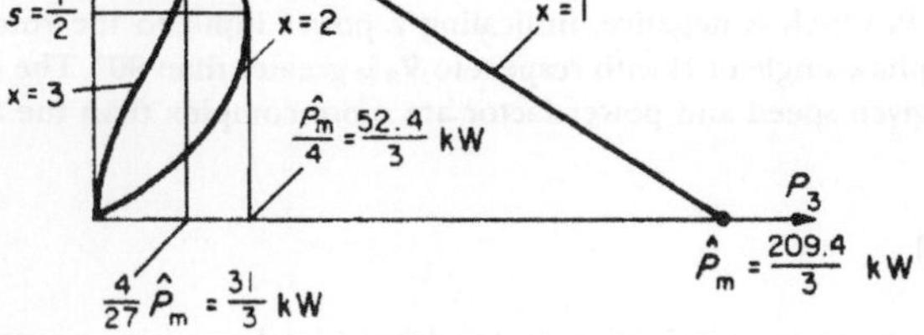

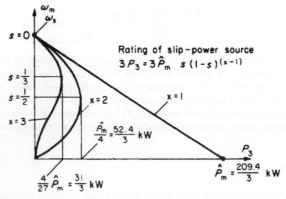

Figure E.4.23

and for $T_m \propto \omega_m$, x = 2 and $3P_3 = 209.4 \times \frac{1^1}{2^2} = \underline{52.4\,\text{kW}}$

The variations of P_3 with speed, for different values of x, see Figure E.4.24, show that the above figures are maximum values occurring over the speed range. It can be seen too, that the higher the value of x, the more attractive is this method of speed control because the P_3 rating is reduced. Hence this slip-power, (Scherbius or Kramer) system (see Reference 1 and p. 285), finds use in fan and pump drives, where T_m falls considerably as speed reduces.

This chapter has surveyed the more usual methods of controlling induction motors. But there are other ways of doing this for which Reference 3 could be consulted. Discussion of power-electronic control will be deferred till Chapter 7.

5 Synchronous machines

Electrical power is generated almost entirely by synchronous machines, of individual ratings up to and beyond a million kW (1 GW). Consequently, problems of power-system generation, transmission, distribution, fault calculations and protection figure very largely in synchronous-machine studies and these receive more attention for example in Reference 6. The purpose of this present text is to place more emphasis on electrical drives, though some generator problems are given, e.g. determination of equivalent circuit from generating tests, calculation of excitation, simple multi-machine circuits and operating charts. Synchronous-motor drives are the fewest in actual numbers but they are used up to the highest ratings. The facility for power-factor control is an important decisive element if constant speed is suitable and, in addition, synchronous machines have the highest efficiencies. With the advent of static variable-frequency supplies, variable-speed synchronous motors are gaining wider application, having a set speed as accurate as the frequency control. For steady-state operation, the equivalent circuit is simpler than for the induction machine and there is the additional, straightforward control facility – the excitation. Since the dominating magnetising reactance carries the total armature current, its variation with different air-gap flux and saturation levels should be allowed for. Although most synchronous motors are of salient-pole construction for which the equations are less simple than for the round-rotor equivalent circuit of Figure 1.9c, this latter circuit still gives a fairly accurate answer to the general operating principles for steady state. A few examples with salient-pole equations are worked out at the end of this chapter.

5.1 Summary of equations

Figure 1.9c shows all the parameters on the equivalent circuit. R_a is usually much less than the synchronous reactance $X_s = x_{al} + X_m$ and is often neglected in circuit calculations. It becomes important in efficiency calculations of course and also when operating at fairly low frequencies when reactances have fallen appreciably. The earlier problems in this chapter will include the resistance, the equations being quoted directly from Reference 1 when they are not developed in the text. The equivalent circuit is normally derived from tests at very low power-factor and this is not difficult to achieve, except for very small machines where R_a becomes relatively high.[4] For such tests, the reactance voltage drops are virtually in phase with all the other voltages and can be combined therewith algebraically, to derive and separate the leakage and magnetising components of the synchronous reactance. Examples 5.1–5.3 will help in understanding the following terms and equations which relate to both the time-phasors for voltages and the interconnection with the space-phasors of m.m.f. The conversion is made through the magnetising-curve sensitivity in volts, per unit of m.m.f. Note that the m.m.f.s are all expressed in terms of field turns through which they are measured. The average sensitivity varies from the unsaturated value on the air-gap line (k_f volts/At) to k_{fs}, that of the mean slope through the operating point determined by the air-gap e.m.f. E. The equations use motor conventions but can be used as they stand for a generator if this is considered as a 'negative' motor, the $\mathbf{I}_a$ phasor being at an angle greater than 90° from the reference terminal voltage $\mathbf{V}$ (Figure 5.1).

Circuit equations:
$$\mathbf{E} = \mathbf{V} - R_a\mathbf{I}_a - jx_{al}\mathbf{I}_a \tag{5.1}$$

$$\mathbf{E}_f = \mathbf{E} - jX_m\mathbf{I}_a \tag{5.2}$$

M.M.F. equation:
$$\mathbf{F}_r = \mathbf{F}_f + \mathbf{F}_a \tag{5.3}$$

multiplied by k_{fs} volts/At:
$$k_{fs}\mathbf{F}_r = k_{fs}\mathbf{F}_f + k_{fs}\mathbf{F}_a \tag{5.4}$$

gives voltage components:
$$\mathbf{E} = \mathbf{E}_f + jX_{ms}\mathbf{I}_a \tag{5.5}$$

corresponding to flux component equation:

$$\mathbf{\Phi}_m = \mathbf{\Phi}_f + \mathbf{\Phi}_a \tag{5.6}$$

Mag. curve gives general relationship:

$$E = f(F_r) \tag{5.7}$$

At zero power factor, eqns (5.1)–(5.6) become algebraic. For zero leading as a motor, or zero lagging as a generator; V, I_a, F_{f1}, F_{f2} and k_f having been measured and noting that $\mathbf{F}_a$ is then completely demagnetising:

	Short-circuit test ($V = 0$)	Zero p.f. test ($V \neq 0$; $I_a = I_{asc}$)
Circuit equations are:	$E_1 = x_{al} I_{asc}$ (5.1a)	$E_2 = V + x_{al} I_a$ (5.1b)
M.M.F. equations are:	$F_{f1} = F_a + F_{r1}$ (5.3a)	$F_{f2} = F_a + F_{r2}$ (5.3b)
Magnetisation curve gives:	$E_1 = k_f F_{r1}$ (5.7a)	$E_2 = f(F_{r2})$ (5.7b)

From these two tests and the six equations, a solution for one unknown will yield all the other five. It is only really necessary to solve for x_{al} though sometimes the armature m.m.f. F_a is required if the complete phasor diagram is to be drawn. Eliminating all unknowns apart from E_2 and F_{r2} leaves eqn (5.7b) and:

$$E_2 = V - k_f(F_{f2} - F_{f1}) + k_f F_{r2} \tag{5.8}$$

The intersection of the straight line (5.8) with the curve (5.7b) yields E_2 and F_{r2}; hence x_{al} from (5.1b) and F_a from (5.3b); see Example 5.1.

The short-circuit test and the o.c. curve (5.7) also yield the unsaturated synchronous reactance;

since on o.c.; $I_a = 0 \quad \therefore V = E = E_f$;

and on s.c.; $V = 0 \quad \therefore E = x_{al} I_a$ and $E_f = (x_{al} + X_{mu}) I_a = X_s I_a$.

The unsaturated value of the synchronous reactance X_{su} and of magnetising reactance X_{mu} are thus derived from the air-gap line, x_{al} having been found previously.

In using this equivalent-circuit information to determine the excitation for any specified terminal voltage, current and power factor, the air-gap e.m.f. E is first found from eqn (5.1). This gives the operating point and the appropriate value of k_{fs}. Hence the correct saturated value of magnetising reactance is $X_{mu} k_{fs}/k_f$ and completing the eqn (5.2) gives the value of E_f, the e.m.f. behind synchronous reactance. The required excitation is E_f/k_{fs}: see Examples 5.1–5.3 and Reference 4.

Electromechanical equations

Considerable insight is gained into the essential aspects of synchronous-machine control and behaviour if the machine losses are neglected. The

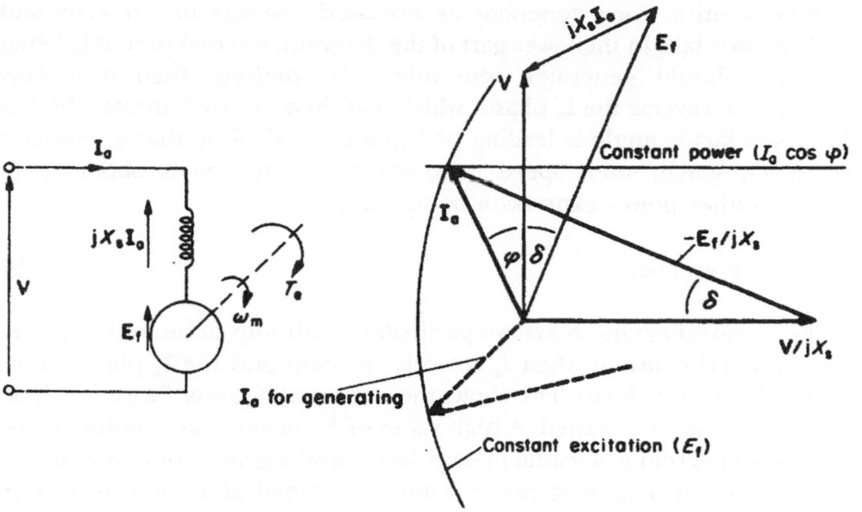

Figure 5.1 *Approximate equivalent circuit per phase and phasor diagram (motor conventions).*

approximate circuit and the phasor diagram for this condition are shown on Figure 5.1. We have the equation:

$$V\angle 0 = E_f\angle\delta + jX_sI_a\angle\varphi \quad (5.9)$$

$$\text{and } I_a\angle\varphi = \frac{V\angle 0}{jX_s} - \frac{E_f\angle\delta}{jX_s} \quad (5.9a)$$

Equation (5.9) shows the terminal voltage with its two components. Equation (5.9a) which is a rearrangement of eqn (5.9) shows the terminal current as the sum of two components, each lagging 90° behind its corresponding voltage. Both equations are shown on the phasor diagram and power (which is the same for input and output since losses are being neglected) can be obtained from either equation, $\mathbf{I}_a$ being resolved along $\mathbf{V}$ directly, or in its two components. Note that $\mathbf{V}/jX_s$ resolves to zero.

Hence, for a 3-phase machine:

$$\text{Power} = 3VI_a\cos\varphi \quad (5.10)$$

$$\text{or} \quad = -3V\frac{E_f}{X_s}\sin\delta \quad (5.10a)$$

The negative sign is explained by the choice of motor conventions. The load angle δ is negative (rotor falling back) when motoring, and power will then be positive. For a generator, as indicated, the sign of δ reverses and the $\mathbf{I}_a$ phasor falls in the lower part of the diagram, the real part of $\mathbf{I}_a$ being negative. Should generator conventions be desired, then it is only necessary to reverse the $\mathbf{I}_a$ phasor which will then clearly indicate whether the power-factor angle is leading or lagging on **V**. Note that at constant frequency, synchronous speed is constant, and torque is obtained on dividing either power expression by $\omega_s = 2\pi f/p$;

$$T_e = \text{Power}/\omega_s \qquad (5.10b)$$

The phasor diagram shows two particular conditions of interest. If power is maintained constant, then $I_a \cos \varphi$ is constant and the $\mathbf{I}_a$ phasor must follow a horizontal locus. This shows the variation of power factor and load angle as excitation is varied. A high value of E_f means that a motor *receives* power with current at a leading power-factor and a generator *delivers* power with current at a lagging power-factor; δ is small and there is a large overload capacity before sin δ reaches unity. A low excitation leads to the opposite behaviour. If instead the power is allowed to vary but the excitation is constant; then $\mathbf{I}_a$ must follow a circular locus determined by the end of the $\mathbf{E}_f/\mathrm{j}X_s$ phasor. Now, power, power-factor, load angle and function can change. There is another important condition where the current phasor follows a horizontal locus along the zero power axis; i.e. on no load as a 'motor', $\mathbf{I}_a$ is completely leading **V** (at high values of E_f) or completely lagging **V** (at low values of excitation). This is operation as a synchronous compensator: similar to a capacitor when overexcited and to a lagging reactor when underexcited. Example 5.4 illustrates all these modes.

Generated-e.m.f. equation

This is little different from the average voltage for the d.c. machine eqn (3.1), except that it is the r.m.s. value which is required so must be multiplied by 1.11. Further, the number of conductors in series z_s must refer to one phase of the winding. The distribution of the coils round the machine periphery is carried out in many different ways and there is inevitably a loss of total phase e.m.f. because the individual conductor voltages are slightly out of phase with one another. So the whole equation must be multiplied by a winding factor, typically about 0.9 for the fundamental voltage. It is different for the harmonic voltages and the winding is deliberately designed to suppress these. The overall effect is that the expression for the fundamental r.m.s. voltage per phase is almost the same as for the average voltage given by eqn (3.1). Alternatively, the transformer e.m.f. equation can be used for a.c. machines; i.e. induced r.m.s. e.m.f. = $4.44 \times \hat{\phi} \times f \times N_s \times k_w$, where $\hat{\phi}$ is the maximum fundamental flux per pole, N_s the turns in series per phase and k_w the winding factor. No worked examples will be provided to illustrate this since a more detailed study of windings is really required, see Reference 1. It is sufficient to note that in the machine equations, the e.m.f. is proportional to the speed (or the frequency), and the flux component being considered.

5.2 Solution of equations

A general plan for guidance on synchronous-machine problems is given on p. 134.

Example 5.1

The test results on a 5-MVA, 6.6-kV, 3-phase, star-connected synchronous generator are as follows:

Open-circuit test

Generated (line) e.m.f.	3	5	6	7	7.5	7.9	8.4	8.6	8.8 kV
Field current	25	42	57	78	94	117	145	162	181 A

Short-circuit test, at rated armature current, required 62 field amperes.
Zero power-factor lagging test, at 6.6 kV and rated current, required 210 field amperes.

If the field resistance is 1.2 Ω cold, 1.47 Ω hot, calculate for normal machine voltage the range of exciter voltage and current required to provide the excitation from no load up to full load at 0.8 p.f. lagging. The armature resistance is 0.25 Ω per phase.

$$\text{Rated armature current } I_{aR} = \frac{5000}{\sqrt{3} \times 6.6} = 437.4\,\text{A}$$

The o.c. magnetisation curve is plotted on Figure E.5.1 and the z.p.f. data (F_{f1} = 62 A and F_{f2} = 210 A) permit the line $E_2 = V - k_f(F_{f2} - F_{f1}) + k_f F_{r2}$, eqn (5.8) to be plotted. k_f, the unsaturated slope of the magnetisation curve, is 6000/50 = 120 line V/A. Hence:

$$E_2 = 6.6 - 0.12(210 - 62) + 0.12F_{r2} = -11.16 + 0.12F_{r2}\,\text{kV (line)}$$

Note, although the equivalent circuit parameters are per-phase values, it is merely a matter of convenience to use the given kV line voltages, but the scaling factor involved must be allowed for as below. Phase volts could of course be used to avoid the faintest possibility of error.

The intersection of the above line with the $E = f(F)$ curve gives simultaneous solution of the two equations at E_2 = 8.6 kV. F_{r2} is not usually required. Hence

$$\sqrt{3}x_{al}I_a = E_2 - V\,\text{(line)} = 8.6 - 6.6 = 2\,\text{kV} \quad \therefore x_{al} = \frac{2000}{\sqrt{3} \times 437.4} = \underline{2.64\,\Omega.}$$

Synchronous-Machine Solution Programme

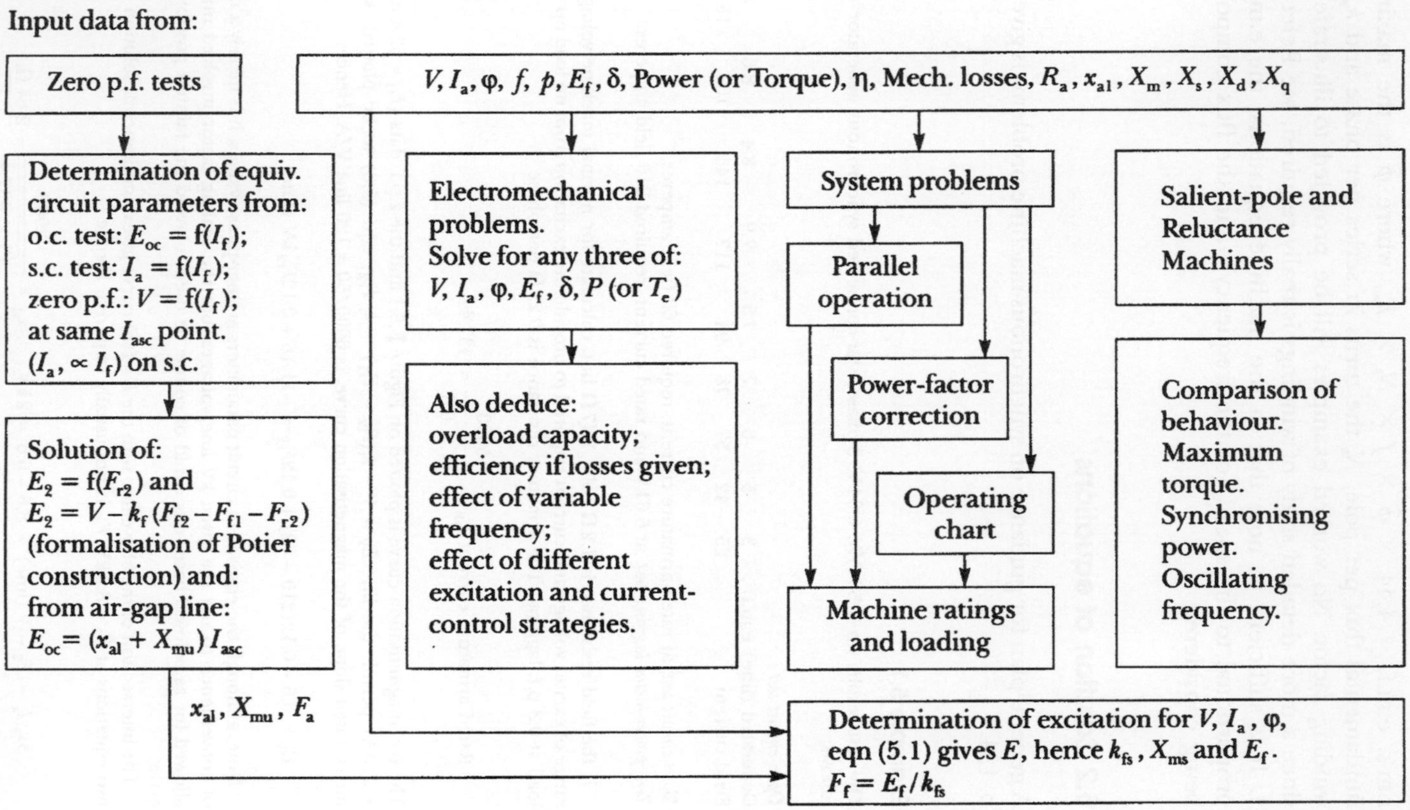
Input data from:
Zero p.f. tests
$V, I_a, \varphi, f, p, E_f, \delta$, Power (or Torque), η, Mech. losses, $R_a, x_{al}, X_m, X_s, X_d, X_q$
Determination of equiv. circuit parameters from:
o.c. test: $E_{oc} = f(I_f)$;
s.c. test: $I_a = f(I_f)$;
zero p.f.: $V = f(I_f)$;
at same I_{asc} point.
$(I_a, \propto I_f)$ on s.c.
Solution of:
$E_2 = f(F_{r2})$ and
$E_2 = V - k_f(F_{f2} - F_{f1} - F_{r2})$
(formalisation of Potier construction) and:
from air-gap line:
$E_{oc} = (x_{al} + X_{mu}) I_{asc}$
x_{al}, X_{mu}, F_a
Electromechanical problems.
Solve for any three of:
$V, I_a, \varphi, E_f, \delta, P$ (or T_e)
Also deduce:
overload capacity;
efficiency if losses given;
effect of variable frequency;
effect of different excitation and current-control strategies.
System problems
Parallel operation
Power-factor correction
Operating chart
Machine ratings and loading
Salient-pole and Reluctance Machines
Comparison of behaviour.
Maximum torque.
Synchronising power.
Oscillating frequency.
Determination of excitation for V, I_a, φ,
eqn (5.1) gives E, hence k_{fs}, X_{ms} and E_f.
$F_f = E_f / k_{fs}$

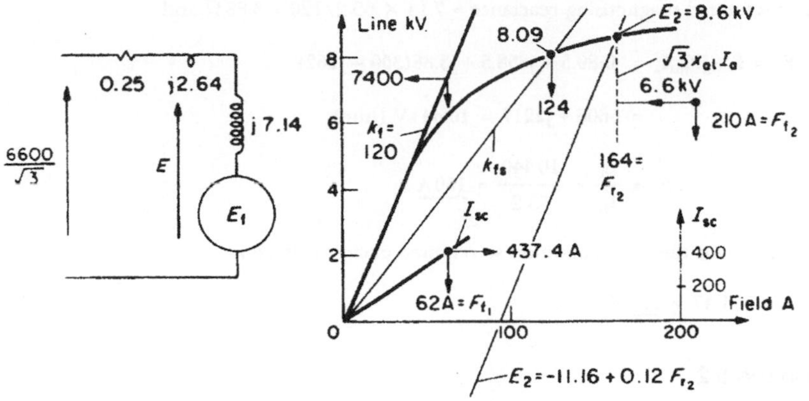

Figure E.5.1

For total (unsaturated) synchronous reactance consider point on o.c. and s.c. at $F = 62\,\text{A}$

$$X_{su} = \frac{7400/\sqrt{3}}{437.4} = 9.78\,\Omega \text{ giving}$$

$$X_{mu} = 9.78 - 2.64 = 7.14\,\Omega = \text{unsaturated magnetising reactance}$$

The equivalent circuit per phase is also shown on the figure.

The construction carried out is virtually the same as the Potier construction also described in Reference 1, but put into a more direct mathematical form. The armature m.m.f. F_a at rated current from the Potier triangle is equivalent to 210 – 164 = 46 A times field turns.

Excitation calculations

On no load, when the machine is cold, and the terminal voltage is the same as the air-gap e.m.f., the required field current from the o.c. curve is 69 A, requiring 69 × 1.2 = 83 V from the exciter.

On load, with $\mathbf{I}_a = 437.4(0.8 - j0.6) = 350 - j262\,\text{A}$

the air-gap e.m.f. from the generator equation $\mathbf{E} = \mathbf{V} + \mathbf{zI}$ is:

$$\mathbf{E}\text{ (phase)} = \frac{6600}{\sqrt{3}} + (0.25 + j2.64)(350 - j262)$$

$$= 3810.5 + 779 + j858.5$$

$$= 4589.5 + j858.5 = 8.087\,\text{kV (line)}$$

From the o.c. curve this gives a saturated volts/field A of:

$$k_{fs} = \frac{8087}{124} = 65.2 \text{ compared with } k_f = 120 \text{ line V/field A}$$

Hence saturated magnetising reactance = 7.14 × 65.2/120 = 3.88 Ω and

$$\mathbf{E}_f = \mathbf{E} + jX_{ms}\mathbf{I}_a = 4589.5 + j858.5 + j3.88(350 - j262)$$

$$= 5606 + j2217 = 10.44\,\text{kV (line)}$$

$$\therefore F_f = \frac{E_f}{k_{fs}} = \frac{10\,440}{65.2} = \underline{160\,\text{A}}$$

This will be for the 'hot' condition so required exciter voltage is:

$$160 \times 1.47 = \underline{235\,\text{V}}$$

Example 5.2

A 3-phase, 500-kVA, 3.3-kV, star-connected synchronous generator has a resistance per phase of 0.3 Ω and a leakage reactance per phase of 2.5 Ω. When running at full load, 0.8 p.f. lagging, the field excitation is 72 A. The o.c. curve at normal speed is:

Line voltage	2080	3100	3730	4090	4310 V
Field current	25	40	55	70	90 A

Estimate the value of the full-load armature ampere-turns per pole in terms of the field turns and hence calculate the range of field current required if the machine has to operate as a synchronous motor at full kVA from 0.2 leading to 0.8 lagging p.f.

In this example, as an alternative to the previous circuit approach, the phasor diagrams will be drawn for both time-phasors and m.m.f. space-phasors. On full load, F_f is given and by constructing $\mathbf{V} + \mathbf{R}_a\mathbf{I}_a + jx_{al}\mathbf{I}_a$ to get the gap e.m.f. E, the value of F_r can be read from the o.c. curve. It is shown as a space-phasor lagging $\mathbf{E}$ by 90°. The phasor $\mathbf{F}_a$ is known in direction, being in antiphase with $\mathbf{I}_a$ for a generator,[(1)] and must intersect with an arc drawn for F_f = 72 A as shown. Hence its length is determined and the angle between $\mathbf{F}_f$ and $\mathbf{F}_r$, which is also the angle between $\mathbf{E}$ and $\mathbf{E}_f$, is available if it is desired to draw in the $\mathbf{E}_f$ phasor and the closing $jX_{ms}\mathbf{I}_a$ phasor. An analytical solution based on the voltage equations is left as an

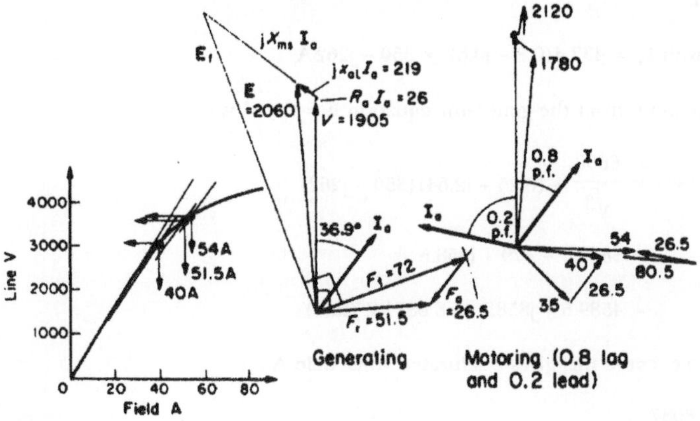

Figure E.5.2

exercise; it leads to a quadratic in X_{ms} for which the value on load is 12.3 Ω. The unsaturated value on allowance for k_f and k_{fs} is 14.3 Ω.

$$\text{Rated armature current} = \frac{500}{\sqrt{3} \times 3.3} = 87.47\,\text{A}$$

$$\mathbf{E} \text{ phasor constructed from } V, R_aI_a, x_{al}I_a = \frac{3300}{\sqrt{3}},\ 0.3 \times 87.47,\ 2.5 \times 87.47$$

$$= 1905, \quad 26, \quad 219$$

The problem will now be solved entirely by the use of phasor diagrams. Accuracy will not of course be as good as by analysis. The first phasor diagram is for the loaded, generating condition giving first the value of E as 2060 volts/phase from which a resultant m.m.f. F_r = 51.5 A is read off from the o.c. curve. The m.m.f. diagram follows as explained above and F_a by measurement is 26.5 A. The $\mathbf{E}_f$ phasor is sketched in lightly at right angles to $\mathbf{F}_f$ and the closing vector is $\mathrm{j}X_{ms}\mathbf{I}_a$.

The procedure for finding the motoring excitation for the two specified power factors is not very different. The armature current and m.m.f. are the same throughout, but unless treated as a 'negative' generator, the motor equation must be used. This gives **E** lagging instead of leading **V** as in the generator case. $\mathbf{F}_a$ is now drawn in phase with $\mathbf{I}_a$ for the motor[1] and F_f is therefore found in each case from $\mathbf{F}_f = \mathbf{F}_r - \mathbf{F}_a$. The values for the two different power factors are, by measurement:

E = 1780 V/phase (0.8 p.f.) E = 2120 V/phase (0.2 p.f.)

F_r from o.c. curve = 40 A (0.8 p.f.) F_r = 54 A (0.2 p.f.)

F_f = 35 A (0.8 p.f.) F_f = 80.5 A (0.2 p.f.)

The final answers could also be obtained by the same analytical method as in Example 5.1.

Example 5.3

A 3-phase, 5-kVA, 1000-V, star-connected synchronous machine has R_a = 4 Ω and x_{al} = 12 Ω per phase. The o.c. curve is as follows:

Field current	4	6	8	10	12	14	16 A
Armature line voltage	490	735	900	990	1070	1115	1160 V at rated speed.

On a short-circuit test, 7 A was required in the field to circulate rated armature current.

Determine the field current and voltage required for operation at constant terminal voltage of 1000 V and (a) no-load current; (b) rated current as a generator at 0.8 p.f. lagging; (c) rated current as a motor at unity power factor.

(d) The machine is going to be considered for operation as a synchronous capacitor. Plot a curve of reactive VAr in this mode against the required field current, up to about rated current, and hence determine the permissible rating of the machine, if the field current can be increased up to the generating value of part (c) above. Resistance may be neglected for this part (d).

$$\text{Rated armature current } I_{aR} = \frac{5000}{\sqrt{3} \times 1000} = 2.89\,\text{A}.$$

From o.c. curve at $I_f = 7\,\text{A}$:

$$X_{su} = \frac{\text{e.m.f. on air-gap line}}{I_{sc}(2.89\,\text{A})} = \frac{870/\sqrt{3}}{2.89} = 174\,\Omega \text{ per phase}$$

Hence

$$X_{mu} = 174 - 12 = 162\,\Omega \text{ per phase}$$

$$k_f = 870/7 \quad = 124.3 \text{ line V/fld A}$$

Mode	(a) No load	(b) Generating	(c) Motoring
$\mathbf{I}_a$	0	2.89(0.8 – j0.6) 2.31 – j1.73	2.89(1 + j0) 2.89 + j0
$\mathbf{E} = \mathbf{V} \pm z\mathbf{I}_a$	$1000/\sqrt{3}$ 577.4	577.4 + (4 + j12) × (2.31 – j1.73) 607.3 + j20.8	577.4 – (4 + j12) × (2.89) 565.7 – j34.7
E r.m.s. line V	1000	1052.5	981.7
k_{fs} line V/fld A	99	1052.5/11.5 = 91.5	981.7/9.7 = 101.2
$X_{ms} = 162 \times \dfrac{k_{fs}}{k_f}$		119.25	131.9
$\mathrm{j}X_{ms}\mathbf{I}_a$		206.3 + j275.5	j381.1
$\mathbf{E}_f = \mathbf{E} \pm \mathrm{j}X_{ms}\mathbf{I}_a$ E_f line V	577.4 1000	813.6 + j296.27 1500	565.7 – j415.8 1216
$F_f = E_f/k_{fs}$	10.1	16.4	12

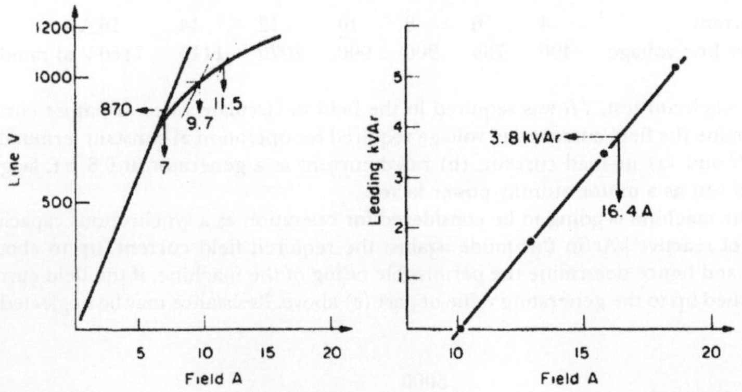

Figure E.5.3

(d) Calculations for excitation are similar, but for various currents, all of them leading voltage by 90°, up to j3 A and with R_a neglected, $jx_{al}\mathbf{I}_a = -12I_a$.

$\mathbf{I}_a$	0	j1	j2	j3
$\mathbf{E} = \mathbf{V} - jx_{al}\mathbf{I}_a$ E r.m.s. line V	577.4 1000	577.4 + 12 1021	577.4 + 24 1042	577.4 + 36 1062
k_{fs}	99	96.2	92.6	89.6
X_{ms}		125.4	120.7	116.8
$jX_{ms}\mathbf{I}_a$		−125.4	−241.4	−350.4
$\mathbf{E}_f = \mathbf{E} - jX_{ms}\mathbf{I}_a$ E_f line V	577.4 1000	589.4 + 125.4 1239	601.4 + 241.4 1460	614.4 + 350.4 1669
$F_f = E_f/k_{fs}$	10.1	12.88	15.8	18.6
$kVA_r = \sqrt{3} \times 1 \times I_a$	0	1.73	3.46	5.2

The curve of kVAr against field current is plotted and although saturation has been allowed for, it is virtually a straight line. When the $\mathbf{E}_f$ equation is divided by k_{fs}, since $X_{ms} = k_{fs}X_{mu}/k_f$, the last term is unaffected by saturation, so the major change of F_f due to armature current is a proportional one. At the generating field current of 16.4 A, the kVA_r is 3.8, which is therefore the permissible rating as a synchronous capacitor. As a synchronous reactor, with negative jI_a, the required excitation falls and the permissible rating is then limited from stability considerations.

5.3 *Per-unit* notation

For synchronous machines this notation is very commonly used and is perhaps the most straightforward in application, especially for the normal, constant-voltage and constant-frequency arrangement. Rated voltage, current and therefore rated kVA are taken as base quantities, plus the constant, synchronous speed. Rated electrical terminal power will therefore be cos φ_R in *per unit.* 1 *per-unit* impedance will be rated-voltage-per-phase/rated-current-per-phase and the synchronous reactance is typically of this order, approaching 2 *per unit* for large turbo generators and rather less than 1 *per unit* for motors with large overload capacity. The next example uses *per-unit* notation for convenience and is meant to illustrate in a simple manner, the essential aspects of synchronous machine behaviour; conventions, load angle, motoring, generating and synchronous-compensator operation. This will be achieved by calculating the current from specified voltage, e.m.f., load angle and reactance. A motor convention will be used and the synchronous reactance X_s assumed to be 1 *per unit*.

From eqn (5.9):

$$\mathbf{V} = \mathbf{E}_f + jX_s\mathbf{I}_a \text{ neglecting resistance;}$$

at rated voltage: $1 = E_f\underline{/\delta} + jI_a\underline{/\varphi}$

from which: $I_a\underline{/\varphi} = \dfrac{1 - E_f\underline{/\delta}}{j} = -j(1 - E_f\underline{/\delta})$ *per unit.*

The solution for $\mathbf{I}_a$ will yield the mode; real-part positive means motoring, real-part negative means generating and real-part zero means operation as a synchronous compensator. The real part also represents the *per-unit* power and the *per-unit* torque, since losses are being neglected. The solution also gives the power factor, though for a generator this is best detected for lagging or leading by reversing the current phasor, as shown on the phasor diagrams in the next example.

Example 5.4

A synchronous machine has $X_s = 1$ *per unit* and operates at rated voltage $V = 1$ *per unit*. Determine the *per-unit* values of current, power, torque, and the power factor and state the machine function when the e.m.f. due to field current (E_f in *per unit*) and the load angle δ have the following values: (a) $0.5\underline{/0°}$; (b) $1\underline{/0°}$; (c) $1.5\underline{/0°}$; (d) $0.5\underline{/-30°}$; (e) $1\underline{/-30°}$; (f) $1.5\underline{/-30°}$; (g) $0.5\underline{/+30°}$; (h) $1\underline{/+30°}$; (i) $1.5\underline{/+30°}$.

The expression for calculating $\mathbf{I}_a$ has just been developed as: $-j(1 - E_f\underline{/\delta})$ and it is only a matter of substituting the values above as in the table on p. 141.

The above example is a very simple application of the *per-unit* system but if it still causes a difficulty, take V as, say, 1000 V per phase and E_f will then be 500, 1000, 1500 V. If X_s is taken as 10 Ω for 1 *per unit*, then I_{aR} must be 100 A. All the currents in the above table, which are in *per unit*, will be multiplied by 100.

5.4 Electromechanical problems

In the circuit of Figure 1.9c, it will be assumed that the synchronous impedance $R_a + jX_s$ is given, though R_a may be neglected as for Figure 5.1. The excitation will be left in terms of E_f, knowing that the calculation of field current will require the techniques used in Examples 5.1–5.3. Apart from frequency and the related synchronous speed, there are six quantities of special interest; V, I_a, φ, E_f, δ, and total power P (or torque). The circuit equation, using motor conventions, is:

Synchronous-Machine Operating Modes (Example 5.4).
Illustrates the effect of excitation for various particular load angles.

E_f	$\delta°$	I_a		Mode	Phasor diagrams
(a) 0.5	0	–j(1 – 0.5) = –j0.5	= 0.5	Synchronous reactor	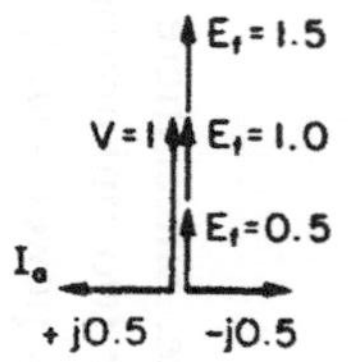
(b) 1.0	0	–j(1 – 1.0) = 0	= 0	Zero current	
(c) 1.5	0	–j(1 – 1.5) = + j0.5	= 0.5	Synchronous capacitor	
(d) 0.5	–30	–j[1 – 0.5(cos – 30 + j sin – 30)] = 0.25 – j0.567	= 0.62	Motoring 0.4 lagging	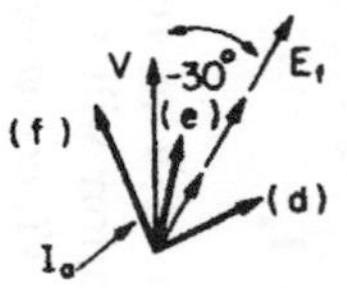
(e) 1.0	–30	–j[1 – 1.0(0.866 – j0.5)] = 0.5 – j0.134	= 0.52	Motoring 0.96 lagging	
(f) 1.5	–30	–j[1 – 1.5(0.866 – j0.5)] = 0.75 + j0.3	= 0.81	Motoring 0.93 leading	
(g) 0.5	+30	–j[1 – 0.5(cos + 30 + j sin + 30)] = –0.25 – j0.567	= 0.62	Generating 0.4 leading	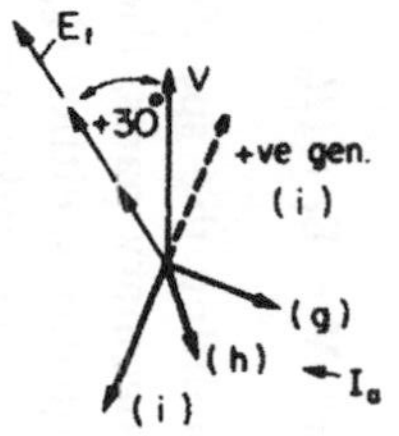
(h) 1.0	+30	–j[1 – 1.0(0.866 + j0.5)] = –0.5 – j0.134	= 0.52	Generating 0.96 leading	
(i) 1.5	+30	–j[1 – 1.5(0.866 + j0.5)] = –0.75 + j0.3	= 0.81	Generating 0.93 lagging	

$$V + j0 = E_f(\cos\delta + j\sin\delta) + (R_a + jX_s) \times I_a \times (\cos\varphi + j\sin\varphi)$$

or: $\mathbf{V} = \mathbf{E}_f + (R_a + jX_s) \times \mathbf{I}_a$;

omitting the specification of **V** as reference. Since there are real and imaginary parts, the equation will solve for two unknowns. The power equations, either $P_{elec} = 3\,VI_a\cos\varphi$ or Mech. power $= \omega_s T_e$, will solve for only one unknown. Both of these powers can be expressed in terms of the load angle δ, which is advanced (positive for a generator) or retarded (negative for a motor). They are[1]:

$$P_e \text{ (per phase)} = VI_a\cos\varphi = \frac{-VE_f}{Z_s}\sin(\delta+\alpha) + \frac{V^2 R_a}{Z_s^2} \qquad (5.11)$$

$$\text{and: } P_g \text{ (per phase)} = P_m \text{ for synchronous machine}$$

$$= -\frac{VE_f}{Z_s}\sin(\delta-\alpha) - \frac{E_f^2 R_a}{Z_s^2} \qquad (5.12)$$

where $\alpha = \sin^{-1} R_a/Z_s$.

If the electrical loss is neglected, both of these equations reduce to:

$$P_e = P_g = P_m = \frac{-VE_f}{X_s}\sin\delta$$

as also deduced from the phasor diagram of Figure 5.1 as eqn (5.10a). With resistance neglected, the approximate-circuit equations become as (5.9) and (5.9a):

$$\mathbf{V} = \mathbf{E}_f + jX_s\mathbf{I}_a \text{ and } \mathbf{I}_a = \frac{\mathbf{V}}{jX_s} - \frac{\mathbf{E}_f}{jX_s}.$$

There is enough information in the equations above to solve for any three of the six quantities, given the other three. Although V is commonly constant, the equations are not restricted to this condition. There are other possibilities which are briefly touched on in some of the later examples and Section 5.5 discusses constant-current operation. Typical problem types are outlined below.

(A) Given V, P, E_f	Vary P with constant E_f to find I_a, φ, δ variation up to pull-out torque at $(\delta - \alpha) = \pm 1$. Or vary E_f with constant P to find the same quantities.
(B) Given V, φ, P (or I_a)	Power-factor control. Excitation required for specified power factor and for a given power (or current). Compensator.

(C) Given V, I_{aR}, Pull-out torque	Pull-out torque (or power) is really specifying the rated load angle δ_R, to permit this overload at maximum angle. Solution yields required excitation, power at full load and power factor.
(D) Given V, I_a, φ (or P)	Find required excitation and power developed (or power factor), load angle and hence overload capacity.
(E) Given I_a, P, φ	Find V required $[P/(I_a \cos \varphi)]$ and excitation. Constant-current drive.
(F) Given I_a, P, δ	Find V required and excitation. Constant current with δ control.
(G) Given V, P, φ	Fixed power factor (by excitation control) changes Power/δ characteristic. Data for required excitation-control yielded.

Items (E), (F) and (G) would require special power-electronic control, possibly with microprocessor supervision. They are included here to show that the fundamental performance is still governed by the normal machine equations. Examples will be given later and there will be a more detailed treatment of power-electronic control in Chapter 7.

In the examples illustrating this section, to help understanding, the label from the nearest problem-type is quoted from the list just described; see below.

Example 5.5

A 3-phase, 8-pole, 50-Hz, 6600-V, star-connected synchronous motor has a synchronous impedance of 0.66 + j6.6 Ω per phase. When excited to give a generated e.m.f. of 4500 V per phase, it takes an input of 2500 kW.

(a) Calculate the electromagnetic torque, the input current, power factor and load angle.
(b) If the motor were to be operating at an input current of 180 A at unity power-factor, what would then be the value of E_f? Under these conditions, calculate also the mechanical output and efficiency if mechanical, excitation and iron losses total 50 kW.

$$Z_s = \sqrt{0.66^2 + 6.6^2} = 6.63\,\Omega \text{ and } \alpha = \sin^{-1} 0.66/6.63 = 5^\circ.7$$

(a) *Type* (A). Substituting values in eqn (5.11) for electrical input:

$$\frac{2\,500\,000}{3} = -\frac{(6600/\sqrt{3} \times 4500)}{6.63}\sin(\delta + 5.7) + \frac{6600^2}{(\sqrt{3})^2} \times \frac{0.66}{6.63^2}$$

From which $\sin(\delta + 5.7) = -0.238$, so $(\delta + 5.7) = -13^\circ.76$; $\underline{\delta = -19^\circ.46}$.
Substituting in eqn (5.12) for mechanical output per phase:

$$P_m = -\frac{(6600/\sqrt{3} \times 4500)}{6.63}\sin(-19.46 - 5.7) - 4500^2 \times \frac{0.66}{6.63^2}$$

$$= 1099.6\,\text{kW} - 304\,\text{kW} = 795.6\,\text{kW}$$

Total air-gap power = $3P_m$ = 3 × 795.6 kW = 2387 kW.

Hence electromagnetic torque = $T_e = 3P_m/\omega_s = \dfrac{2387 \times 1000}{2\pi \times 50/4} = \underline{30\,390\text{ Nm}}$

$$\mathbf{I}_a = \frac{\mathbf{V} - \mathbf{E}_f}{\mathbf{Z}_s} = \frac{6600/\sqrt{3} - 4500(\cos - 19.46 + \text{j}\sin - 19.46)}{0.66 + \text{j}6.6} = \frac{-432.4 + \text{j}1499}{0.66 + \text{j}6.6}$$

$$= 218.7 + \text{j}87.3 = 235.5\ \underline{/21°.8\text{ A}}\ \underline{\cos\varphi = 0.93\text{ leading}}$$

Check input = $\sqrt{3} \times 6600 \times 218.7$ = 2500 kW

(b) *Type* (B)

$$\mathbf{E}_f = \mathbf{V} - \mathbf{Z}_s\mathbf{I}_a = \frac{6600}{\sqrt{3}} - (0.66 + \text{j}6.6)\ (180 + \text{j}0)$$

$$= 3691.7 - \text{j}1188 = 3878/\underline{-17°.8} = 6.717\text{ kV (line)}$$

Output = $3P_m$ − 'fixed' loss

$$= 3\left[\frac{-6600/\sqrt{3} \times 3878}{6.63}\sin(-17°.8 - 5°.7) - \frac{3878^2 \times 0.66}{6.63^2}\right] - 50\,000$$

$$= 2666.2 \qquad - 677.4 \qquad - 50\text{ kW}$$

$$= \underline{1938.7\text{ kW}}$$

$$\text{Efficiency} = \frac{1938.7}{\sqrt{3} \times 6.6 \times 180} = \underline{94.2\%}$$

As an exercise, the above figures, apart from efficiency, can be checked using the approximate circuit. It will be found that the load angles are within 1° accuracy, with similar small errors for the other quantities; see Tutorial Example T5.2.

Example 5.6

A 3300-V, 3-phase, 50-Hz, star-connected synchronous motor has a synchronous impedance of 2 + j15 Ω per phase. Operating with a line e.m.f. of 2500 V, it just falls out of step at full load. To what open-circuit e.m.f. will it have to be excited so that it will just remain in synchronism at 50% above rated torque. With this e.m.f., what will then be the input power, current and power factor at full load?

$$Z_s = \sqrt{2^2 + 15^2} = 15.1\ \Omega \quad \alpha = \sin^{-1} 2/15.1 = 7°.6$$

Type (C) *problem*, with pull-out torque equal to rated value; sin $(\delta - \alpha) = -1$.

Hence, rated air-gap power = $3P_m$ =

$$3\left[\frac{-3300 \times 2500}{\sqrt{3} \times \sqrt{3} \times 15.1} \times (-1) - \frac{2500^2}{(\sqrt{3})^2} \times \frac{2}{15.1^2}\right] \times 10^3\text{ kW}$$

$$= 546.4 \qquad - 54.8$$

$$= 491.6\text{ kW}$$

Since electromagnetic torque is proportional to this power, the new requirement is that pull-out should occur at 1.5 times this value; i.e. at 737.4 kW. Substituting again in eqn (5.12), but this time with E_f as the unknown:

$$737\,400 = \frac{3300 \times E_f\,(\text{line})}{15.1} - E_f^2 \times \frac{2}{15.1^2}$$

from which $E_f^2 - 24915E_f + 84\,067\,287 = 0$

$$\therefore E_f = \frac{24\,915 \pm \sqrt{24915^2 - 4 \times 84\,067\,287}}{2}$$

$= 20\,890\,\text{V}$ or $\underline{4024\,\text{V}}$; the lower value being feasible.

Substituting again in the output equation, this time set to full load:

$$491\,600 = \frac{-3300 \times 4024}{15.1}\sin(\delta - 7.6) - 4024^2 \times \frac{2}{15.1^2}$$

gives $\sin(\delta - 7.6) = -0.7205 = -46°.1$. Hence $\delta = -38°.5$

$$\mathbf{I}_a = \frac{\mathbf{V} - \mathbf{E}_f}{\mathbf{Z}_s} = \frac{3300/\surd 3 - 4024/\surd 3(\cos - 38.5 + \text{j}\sin - 38.5)}{2 + \text{j}15}$$

$$= \frac{(87 + \text{j}1446) \times (2 - \text{j}15)}{2^2 + 15^2} = 95.5 + \text{j}6.9\,\text{A} = \underline{95.7\,\angle{+3°.7}}$$

$\therefore$ input power $= \surd 3 \times 3300 \times 95.5 \times 10^{-3}$

$= \underline{545.8\,\text{kW}}$ at $\underline{\cos\varphi = 0.998 \text{ leading}}$

If the above figures are checked using the approximate circuit, they will again be well within 10%, apart from the output power, since I^2R losses are neglected in the approximation.

Example 5.7

A 3-phase, 4-pole, 400-V, 200-hp, star-connected synchronous motor has a synchronous reactance of 0.5 Ω per phase. Calculate the load angle in mechanical degrees and the input current and power factor when the machine is working at full load with the e.m.f. adjusted to 1 *per unit*. Neglect R_a but take the mechanical loss as 10 kW.

Type (A) *problem*. From eqn (5.10a) with e.m.f. the same as the terminal voltage:

$$\text{Power} = 200 \times 746 + 10\,000 = \frac{-3}{0.5} \times \left(\frac{400}{\surd 3}\right)^2 \sin\delta$$

from which $\sin\delta = -0.4975$, $\delta = -29°.8$ (elec.), and since $p = 2$,

$$\underline{\delta_{mech} = -14°.9}$$

Rated current $\mathbf{I}_{aR}$ $$= \frac{\mathbf{V} - \mathbf{E}_f}{\text{j}X_s} = \frac{1}{\surd 3}\left[\frac{400 - 400(\cos - 29.8 + \text{j}\sin - 29.8)}{\text{j}0.5}\right]$$

$= 229.8 - \text{j}61 = \underline{237.8\,\angle{-14°.9}} = 0.966$ p.f. lagging

Note: It is a pure coincidence that $\varphi = \delta_{mech}$.

Full-load power = $\omega_s T_e = 2\pi \times 50/3 \times 50 \times 10^3 = 5\,236\,000$ watts $= -3\,\dfrac{VE_f}{X_s}\sin\delta$

	(a) Unsaturated	(b) Saturated
X_m x_{al} X_s	10 Ω 2.5 Ω 12.5 Ω	6.67 Ω 2.5 Ω 9.17 Ω
E_f	$11\,000/\sqrt{3}$ V	$2/3 \times 11\,000/\sqrt{3}$ V
δ, from 5 236 000 =	$-3 \times \dfrac{11\,000^2}{(\sqrt{3})^2} \times \dfrac{\sin\delta}{12.5}$	$-3 \times \dfrac{11\,000^2}{(\sqrt{3})^2} \times \dfrac{2}{3} \times \dfrac{\sin\delta}{9.17}$
$\sin\delta$ = $\cos\delta$ = δ =	−0.5409 0.8411 −32°.7	−0.5952 0.8036 −36°.5
$\mathbf{I}_a$ from eqn (5.9a): $\dfrac{1100}{\sqrt{3}} \times \left[\dfrac{1-(\cos\delta + \mathrm{j}\sin\delta)}{\mathrm{j}X_s}\right]$ =	$\dfrac{11\,000}{\sqrt{3}} \times \left[\dfrac{1-(0.8411 - \mathrm{j}0.5409)}{\mathrm{j}12.5}\right]$	$\dfrac{11\,000}{\sqrt{3}} \times \left[\dfrac{1-(0.8036 - \mathrm{j}0.5952) \times 2/3}{\mathrm{j}9.17}\right]$
=	274.9 − j80.7	274.8 − j321.5
=	286.5 A	423 A
at $\cos\varphi$ =	0.96 lagging	0.65 lagging
Maximum power; at $\sin\delta = -1$	$\dfrac{11\,000^2}{12.5} = 9680$ kW	$\dfrac{11\,000^2}{9.17} \times \dfrac{2}{3} = 8797$ kW
For unity power factor $\mathbf{E}_f$ =	$11\,000 - \mathrm{j}12.5 \times 286.5 \times \sqrt{3}$ = 12.63 line kV	$11\,000 - \mathrm{j}9.17 \times 423 \times \sqrt{3}$ 12.89 line kV
giving an output $\sqrt{3}VI_a$ =	$\sqrt{3} \times 11 \times 286.5 = 5459$ kW	$\sqrt{3} \times 11 \times 423 = 8059$ kW
For z.p.f. lead, $\mathbf{E}_f = \mathbf{V} - \mathrm{j}X_s\mathrm{j}I_a$	$11\,000 + 12.5 \times 286.5 \times \sqrt{3}$ = 17.2 line kV	$11\,000 + 9.17 \times 423 \times \sqrt{3}$ = 17.72 line kV

Example 5.8

A 6-pole, 3-phase, star-connected synchronous motor has an unsaturated synchronous reactance of 12.5 Ω per phase, 20% of this being due to leakage flux. The motor is supplied from 11 kV at 50 Hz and drives a total mechanical torque of 50 × 10^3 Nm. The field current is so adjusted that the e.m.f. E_f read off the air-gap line is equal to the rated terminal voltage. Calculate the load angle, input current and power factor and also the maximum output power with this excitation, before pulling out of step. Neglect resistance throughout and assume that E_f is unchanged when the power increases to the maximum. The calculations are to be conducted (a) assuming saturation can be neglected and (b) assuming that all components which would be affected would be reduced by a factor of 1/3 due to saturation.

With the current calculated as above, to what value would the excitation have to be adjusted, in terms of E_f, so that the power factor would be unity? What would then be the output?

Finally, to what value would E_f have to be adjusted so that the machine could operate as a synchronous capacitor at the same armature current?

This again is related to *problem types* (A) *and* (B) but the effect of allowing for and of neglecting saturation is included. 10 Ω of the total synchronous reactance is due to mutual flux and therefore will be reduced if saturation is allowed for. Further, the e.m.f. will have to be reduced by the same amount.

The calculations are shown in the table on p. 146. It can be seen that the most pronounced effects of allowing for saturation follow from the change of power-factor-angle φ, which is much increased because E_f is no longer equal to but is less than the terminal voltage. If the field current had been adjusted to correct for this, then the discrepancies would have appeared in the other quantities, like load angle and maximum power.

Example 5.9

A 3-phase, 6-pole, 50-Hz, star-connected synchronous motor is rated at 500 kVA, 6600 V at unity power-factor. It has a synchronous impedance of j80 Ω per phase. Determine the mechanical torque for this rating neglecting all machine losses. If this torque can be assumed constant, what departure from rated armature current and excitation (in terms of $E_f \angle\delta$) are necessary for operation at (a) 0.9 p.f. lag and (b) 0.9 p.f. lead? How will the maximum torque be affected in both cases?

Problem type (D)

$$\text{Torque} = \frac{\text{Power}}{\text{Speed}} = \frac{500\,000}{2\pi \times 50/3} = \underline{4775\,\text{Nm}}$$

$$\text{Power component of current for this torque} = \frac{500\,000}{\sqrt{3} \times 6600} = 43.74\,\text{A}$$

$$\text{At 0.9 p.f. } (\sin\varphi = 0.436),\quad \mathbf{I}_a = \frac{43.74}{0.9}(0.9 \pm j0.436) = 43.74 \pm j21.2$$

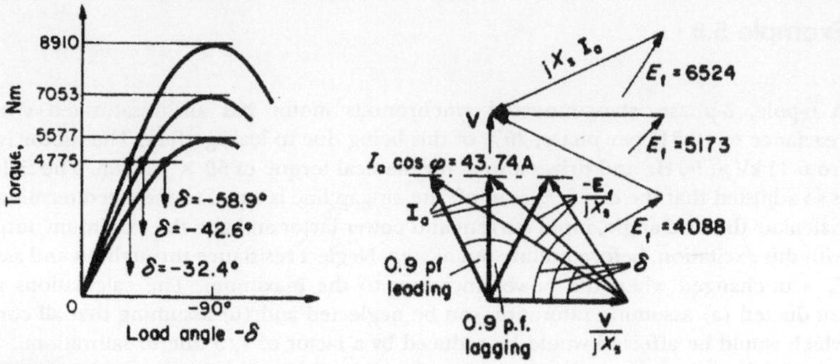

Figure E.5.9

Power factor	u.p.f.	0.9 lag	0.9 lead
jX_sI_a	$= j80 \times (43.74 + j0)$ $= j3499.2$	$j80 \times (43.74 - j21.2)$ $1696 + j3499.2$	$j80 \times (43.74 + j21.2)$ $-1696 + j3499.2$
$E_f = \dfrac{6600}{\sqrt{3}} - jX_sI_a$	$= 3810.5 - j3499.2$	$2114.5 - j3499.2$	$5506.5 - j3499.2$
	$= 5173 \angle -42°6$	$4088 \angle -58°9$	$6524 \angle 32°4$
$\sin -90°/\sin\delta$	1.477	1.168	1.866
Max $T_e = 4775$ $\times \dfrac{\sin - 90}{\sin\delta}$	= 7053 Nm	5577 Nm	8910 Nm

The last figures show that the higher excitations give higher overload capacities as well as the movement towards leading power factor. This is also indicated by the torque/load-angle and phasor diagrams above, which show the reduction of δ with increase of E_f.

Example 5.10

A 1000-kVA, 6.6-kV, 50-Hz, 3-phase, 6-pole, star-connected synchronous machine is connected to an infinite system. The synchronous impedance per-phase can be taken as constant at 0 + j50 Ω. The machine is operating as a motor at rated current and in such a manner, by excitation adjustment, that a 50% overload is possible before it pulls out of synchronism. What will be the necessary voltage E_f behind synchronous impedance to permit this overload margin? At what power and power factor will it be operating when the current is at the rated value? Neglect all machine power-losses.

If, with the same mechanical load, E_f was reduced by 30%, what would be the new status of the motor?

Type (C) Rated current $= \dfrac{1000}{\sqrt{3} \times 6.6} = 87.48$ A

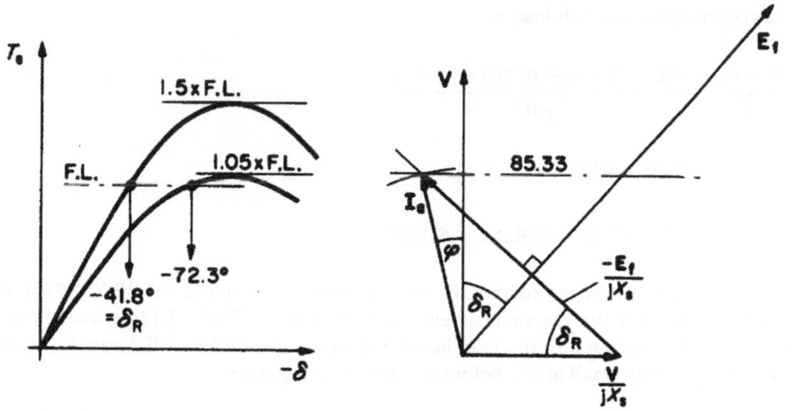

Figure E.5.10

In order to permit a load increase of 1.5 times, the value of sin δ at full load must be 2/3 so that a 50% increase will bring sin δ to unity.

The easiest way to solve this problem is to draw the phasor diagram as shown. Two sides of the current triangle are known, I_a and V/X_s, and the load angle δ_R is specified as $\sin^{-1} - 2/3 = -41°.8$. Either from the sine rule or resolving the two current components along E_f, the triangle can be solved. By the second method:

$$I_a \cos(\varphi + \delta_R) = \frac{V}{X_s} \sin \delta_R$$ – using positive δ_R for simplicity in this particular case,

$$\text{so: } \cos(\varphi + \delta_R) = \frac{6600/\sqrt{3}}{50} \times \frac{2/3}{87.48} = 0.5808$$

giving: $(\varphi + \delta_R) = 54°.49$ and hence $\varphi = 54.49 - 41.8 = 12°.69$

$\underline{\cos \varphi = 0.976}$

Power $= \sqrt{3} \times 6600 \times 87.48 \times 0.976 = 1000\,\text{kVA} \times 0.976 = \underline{976\,\text{kW}}$

E_f follows from eqn (5.10a):

$$976\,000 = -3\left(\frac{6600}{\sqrt{3}} \times \frac{E_f}{50}\right) \times \left(\frac{-2}{3}\right) \text{ from which } E_f = \underline{6403\,\text{V/phase}}$$

If E_f is reduced to 70% of 6403 V = 4482 V

$$976\,000 = -3\left(\frac{6600}{\sqrt{3}} \times \frac{4482}{50}\right) \sin \delta$$

from which $\sin \delta = 0.952$ and $\delta = \underline{72°.3}$. $\cos \delta = 0.306$

Maximum overload is now $\frac{\sin 90}{\sin 72.3} = \underline{1.05, \text{ i.e. } 5\% \text{ overload}}$

The armature current at full load is

$$\frac{\mathbf{V} - \mathbf{E}_f}{jX_s} = \frac{6600/\sqrt{3} - 4482(0.306 - j0.952)}{j50}$$

$$= 85.33 - j48.8$$

$$= 98.3\ \underline{/29°.8} \quad \underline{\cos\varphi = 0.868 \text{ lagging}}$$

The power component of the current, 85.33 A, is the same as before (87.48 × 0.976), since the load is unchanged. But the total current is greater at 98.3/87.48 = 1.12 *per unit*, the power factor is now 30° lagging and the load angle is greatly increased to 72°.3 which leaves the overload margin very small at 5% before pull-out; see diagrams.

Example 5.11

A 3-phase, 6600-V, 6-pole, 50-Hz, star-connected synchronous motor has a synchronous impedance of 0 + j30 Ω per phase. At rated load the armature current is 100 A at 0.9 p.f. leading. Neglecting losses:

(a) Determine the e.m.f. E_f, the load angle in mechanical degrees and the rated torque.
(b) What increase of excitation will be required to sustain a torque overload of 25 000 Nm without falling out of step?
(c) With this new excitation, what would be the values of current, power factor and load angle at rated torque?
(d) With this new excitation, what reduction of terminal voltage would be permissible so that the machine would just stay in synchronism at rated torque?
(e) With this new excitation, and rated terminal voltage, what would be the value of current if the load was removed altogether?

This is a *Type* (D) *problem* with further variations.

(a) $\mathbf{E}_f = \mathbf{V} - jX_s\mathbf{I}_a = \dfrac{6600}{\sqrt{3}} - j30 \times 100(0.9 + j0.436)$

$5118.5 - j2700 = 5787/\underline{27°.8}$

The load angle in mechanical degrees is $\delta/p = -27.8/3 = -9°.3$

$$\text{Rated torque} = \frac{\sqrt{3} \times 6600 \times 100 \times 0.9}{2\pi \times 50/3} = \underline{9827\,\text{Nm}}$$

(b) For maximum torque of 25 000 Nm $= -3 \times \dfrac{6600}{\sqrt{3}} + \dfrac{E_f}{30} \times \dfrac{1}{2\pi \times 50/3} \times \sin -90°$

$$E_f = 6870\,\text{V}$$

(c) For new load angle: $\underline{9827\,\text{Nm}} = -3 \times \dfrac{6600}{\sqrt{3}} \times \dfrac{6870}{30} \times \dfrac{\sin\delta}{2\pi \times 50/3}$

from which $\sin\delta = -0.393$; $\underline{\delta = -23°.1}$ elec.

For new current: $\mathbf{I}_a = \dfrac{6600/\sqrt{3} - 6870(0.92 - j0.393)}{j30}$

$= 90 + j83.7 = 122.9\ \underline{/42°.9}\quad \underline{\cos\varphi = 0.732\ \text{lead}}$

(d) For rated torque at reduced voltage:

$$9827\,\text{Nm} = 3 \times \frac{V}{\sqrt{3}} \times \frac{6870}{30} \times \frac{\sin 90°}{2\pi \times 50/3}$$

from which: $V = 2594\,\text{V} = 0.393$ *per unit*; $\underline{60.7\%\ \text{reduction}}$

(e) For zero load, $\delta = 0$, $\sin\delta = 0$ and $\cos\delta = 1$

$$\mathbf{I}_a = \frac{\mathbf{V} - \mathbf{E}_f(1 + j0)}{jX_s} = \frac{6600/\sqrt{3} - 6870}{j30} = \underline{j102}\ \text{A. Zero leading p.f.}$$

Example 5.12

If the excitation of a synchronous motor is so controlled that the power factor is always unity, show that the power is proportional to the tangent of the load angle. Neglect all machine losses.

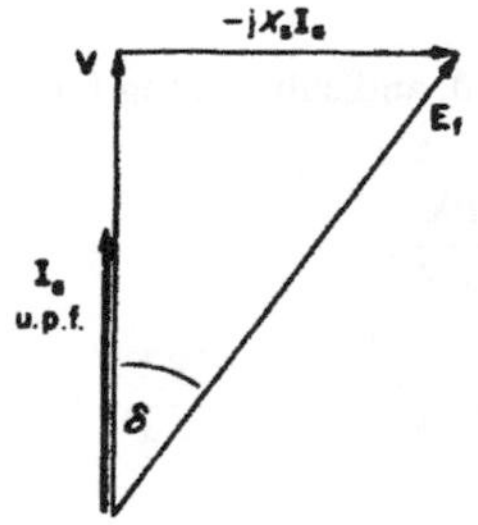

Figure E.5.12

Type (G) *problem.* Referring to the phasor diagram and taking **V** as reference, then

$\mathbf{I}_a$ will always be $I_a + j0$.

From $\mathbf{E}_f = \mathbf{V} - jX_s\mathbf{I}_a$ it can be seen that the load angle is:

$$\delta = \tan^{-1}\frac{-X_sI_a}{V}$$

But power is proportional to VI_a so $I_a = \dfrac{kP}{V}$

Hence $\tan\delta = \dfrac{-X_s kP}{V^2}$ and with V constant; <u>tan δ proportional to P</u>

Example 5.13

If for a synchronous motor the r.m.s. armature current is maintained constant by varying the terminal voltage V, develop the relationships comparable to those for constant voltage, which will yield solutions for three unknowns given three values out of V, I_a, φ, E_f, δ, and P (or T_e). Neglect machine losses.

Type (E) *problem*, again solved more easily by referring to the phasor diagram, this time with $\mathbf{I}_a$ as the reference phasor instead of **V**.

$$P_e = P_g = P_m = I_a V\cos\varphi = I_a E_f\cos(\varphi+\delta) \text{ per phase}$$

and:

$$\mathbf{V} = \mathbf{E}_f + jX_s\mathbf{I}_a$$
$$V(\cos\varphi + j\sin\varphi) = E_f[\cos(\varphi+\delta) + j\sin(\varphi+\delta)] + jX_s\mathbf{I}_a$$

Hence for example, given I_a, P and φ, $V = \dfrac{P}{I_a\cos\varphi}$

and $\mathbf{E}_f = V(\cos\varphi + j\sin\varphi) - jX_s\mathbf{I}_a = E_f\,\underline{/(\delta+\varphi)}$, hence δ

or, for a *Type* (F) *problem*, given I_a, P and δ find required voltage and excitation. Using the cosine rule on the phasor diagram:

$$(X_sI_a)^2 = V^2 + E_f^2 - 2VE_f\cos\delta, \text{ and substituting for } E_f \text{ from } P = \frac{VE_f}{X_s}\sin\delta:$$

$$= V^2 + \frac{P^2X_s^2}{V^2\sin^2\delta} - \frac{2PX_s}{\tan\delta}$$

$$0 = (V^2)^2 - \left[\frac{2PX_s}{\tan\delta} + X_s^2I_a^2\right]V^2 + \frac{P^2X_s^2}{\sin^2\delta}$$

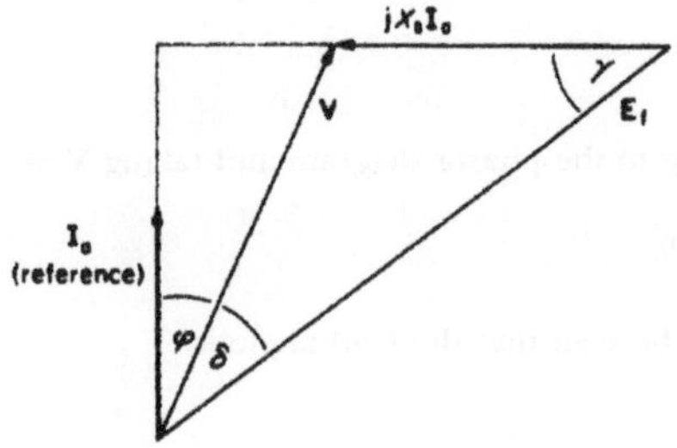

Figure E.5.13

This equation can be solved for V. E_f follows and using the sine rule yields γ and hence $\varphi = 90 - \delta - \gamma$.

Further discussion of the constant-current mode follows in Section 5.5.

Effect of supply voltage/frequency changes

The circuit, power and torque expressions derived may be applied at any frequency and voltage providing appropriate changes are made to the frequency-sensitive terms. For example, synchronous speed, reactance, induced e.m.f. (with constant excitation) are all proportional to frequency so must be changed from their normal values by the frequency ratio k, say. The approximate circuit becomes more inaccurate as frequency falls, due to the increasing significance of the RI_a voltage.

Example 5.14

If the supply frequency and voltage applied to a synchronous motor are both reduced, to fractions kf and kV, what will be the effect on the values of maximum torque and on maximum power? Take V, f, E_f, ω_s and X_s as the normal, rated parameter values.

If f becomes kf then the synchronous speed becomes $k\omega_s$ and therefore the induced e.m.f. E_f will become kE_f if the flux is maintained constant. The reactance too is proportional to frequency so X_s becomes kX_s. Substituting in the power equation (5.10a)

$$\text{Maximum power} = -3 \times \frac{kV \times kE_f}{kX_s} \times \sin(-90°) = \frac{3VE_f}{X_s} \sin 90 \times k$$

$$\text{Maximum } T_e = \frac{\text{power}}{\text{speed}} = \frac{3VE_f \sin 90 \times k}{k\omega_s X_s} = \frac{3VE_f}{\omega_s X_s}$$

The maximum torque is unchanged from the normal condition as might be expected since the flux and load angle are unchanged. Further, the voltages and the impedance are affected equally, by the terms of the question, so the current would be unchanged. The maximum power will be reduced by k from the normal value since although the torque is constant, the speed has been reduced by this k factor.

All the above reasoning neglects the effect of resistance. The equations are different if this is allowed for. Only the induced voltage terms are affected by the reduction in frequency (apart from the small resistance skin-effects), so the resistance assumes a more important role and at very low frequencies will dominate the current calculation. Expanding eqn (5.12) the electromagnetic torque is:

$$T_e = \frac{3P_g}{\omega_s} = \frac{3}{\omega_s}\left[\frac{-VE_f}{Z_s}(\sin\delta\cos\alpha - \cos\delta\sin\alpha) - \frac{E_f^2 R_a}{Z_s^2}\right]$$

Substituting $\sin\alpha = R_a/Z_s$ and $\cos\alpha = X_s/Z_s$, from the definition of α:

$$T_e = \frac{-3E_f}{\omega_s Z_s^2}(VX_s \sin\delta - VR_a \cos\delta + E_f R_a). \tag{5.13}$$

At any reduced frequency kf; E_f, ω_s and X_s are reduced in the same proportion and Z_s^2 becomes $R_a^2 + (kX_s)^2$. Under these conditions, the load angle for maximum torque can be deduced by differentiating the modified expression with respect to δ and leads to the statement that: $kX_s \cos\delta + R_a \sin\delta = 0$

from which:

$$\tan\delta = -k\,X_s/R_a.$$

This shows that at very low frequencies, the load angle for maximum torque approaches zero. Also, since $\tan\alpha = R/kX_s$, the value of $\sin(\delta - \alpha)$ in eqn (5.12) at maximum torque will be $\sin(-90°)$, as might be expected, the tangents of the two angles showing that they are complementary. With this value of $(\delta - \alpha)$ inserted in eqn (5.12) an expression for maximum torque at any frequency kf can be deduced:

$$\text{Max } T_e = \frac{-3}{k\omega_s}\left[\frac{VkE_f}{Z_s}(-1) + \frac{(kE_f)^2 R_a}{Z_s^2}\right]$$

$$= \frac{3E_f}{\omega_s}\left[\frac{V}{\sqrt{R_a^2 + (kX_s)^2}} - \frac{kE_f R_a}{R_a^2 + (kX_s)^2}\right] \tag{5.14}$$

which, with resistance neglected and with the applied voltage too reduced by the same fraction k, will give the same value of maximum torque as deduced from eqn (5.10b), as in Example 5.14, with $\sin\delta$ at its maximum. See also Example 5.15.

Eqn (5.14) becomes the general torque equation if the first term is multiplied by $-\sin(\delta - \alpha)$. As frequency falls till the reactance $kX_s \ll R_a$, the equation becomes:

$$\frac{-3E_f}{\omega_s}\left[\frac{V}{R_a}\sin(\delta - \alpha) - \frac{kE_f}{R_a}\right] = \frac{3E_f I_a}{\omega_s}$$

since δ is very small and α approaches 90°. The power factor approaches unity.

5.5 Constant-current operation

This mode was discussed in Section 4.3 for the induction motor. For the synchronous motor also, it is associated with variable frequency and special applications. Indeed, when the induction machine is in the dynamic-braking mode with a d.c. current-source for one winding, it is really operating as a synchronous generator, of variable frequency if the speed is changing. The power factor is always lagging, however, since it has a passive R. L. load.

Traditional approaches to electrical-machine theory have tended to assume nominally constant-voltage sources in deriving the various equations since this is the normal steady-state running condition. If, instead, currents are specified and controlled using a power-electronic converter and closed-loop current control, a different performance results. This is especially important when considering maximum-torque capability. The equivalent-circuit parameters are unchanged – apart from the greater likelihood of parameter changes due to saturation effects – but the viewpoint is different. For example, not until the performance is deduced from the specified currents does the primary impedance come into the calculation – for the required supply voltage. Variable frequency and changes of speed are the normal situation.

It is helpful as an introduction to return to the induction-motor phasor diagram which is shown on Figure 5.2a. The emphasis will be on the current time-phasors, which can also represent the m.m.f. space phasors to which they are proportional, being expressed in terms of the stator winding turns. $\mathbf{I}_1$ is for the stator (primary); $\mathbf{I}_2'$ is for the rotor (secondary) and $\mathbf{I}_m$, the magnetising current, represents the resultant m.m.f. The general expression for torque, eqn (4.4):

$$T_e = \frac{3}{\omega_s} \times \frac{I_2'^2 R_2'}{s} \text{ can be rearranged as:}$$

$$\frac{3}{\omega/\text{pole pairs}} \times X_m \times \frac{E_1}{X_m} \times \frac{E_1}{Z_2'} \times \frac{R_2'/s}{Z_2'},$$

$$\text{where } Z_2' = \frac{R_2'}{s} + jx_2'$$

$$= 3 \times \text{pole pairs} \times M' \times I_m \times I_2' \times \cos \varphi_2$$

and, since $\varphi_m = 90°$:

$$= 3 \times \text{pole pairs} \times M' \times I_m \times I_2' \times \sin(\varphi_m - \varphi_2) \qquad (5.15)$$

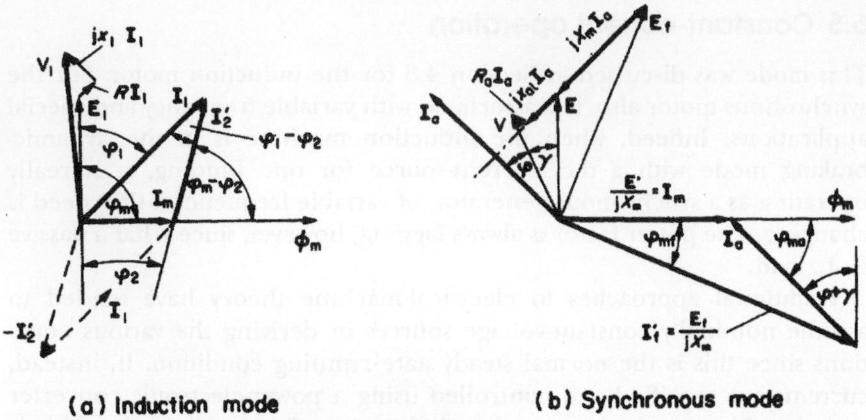

Figure 5.2 *Torque from phasor diagrams.*

where the angles have been measured from E_1 for convenience. Basically this equation has been obtained by dividing the air-gap power by the synchronous speed, $3P_g/\omega_s$, but it now expresses the torque as a product of the referred mutual-inductance, the r.m.s. currents referred to the same side of the air gap and the sine of the angle between the rotor and resultant m.m.f.s represented by $\mathbf{I}_2'$ and $\mathbf{I}_m$. For a.c. machines, this angle between the air-gap flux and the rotor m.m.f. is closest to the conventional load angle, which for synchronous machines is the angle δ between time-phasors **V** and $\mathbf{E}_f$.

From the geometry of the triangle, using the sine rule, it is readily shown that the same torque will be given by using the currents I_1 and I_2' or I_1 and I_m, with the appropriate correction for angular difference substituted; e.g.

$$T_e = 3 \times \text{pole pairs} \times M' I_1 I_2' \sin(\varphi_1 - \varphi_2) \tag{5.16}$$

as proved by an alternative approach in Reference 1. The angle here is the torque angle, between the stator and rotor m.m.f.s, as on the d.c. machine, between field and brush axes.

Turning now to the synchronous machine, the corresponding m.m.f. diagram (Figure 5.2b) is obtained on dividing the **E**, $\mathbf{E}_f$, $\mathrm{j}X_m\mathbf{I}_a$ voltage triangle by $\mathrm{j}X_m$ which gives currents referred to the armature turns. E/X_m, in phase with the mutual air-gap flux ϕ_m, is really the equivalent magnetising current I_m, like the induction motor and producing the air-gap e.m.f. E. The current E_f/X_m is really the equivalent field current I_f' in armature winding terms, producing E_f and is readily converted back to I_f if the coefficient relating E_f to I_f is known from the magnetising characteristic, i.e. k_f, for unsaturated conditions; see Section 5.1. The torque is given by:

$$T_e = \frac{\text{Air-gap power}}{\omega_s} = \frac{3EI_a\cos(\varphi+\gamma)}{\omega_s}$$

$$= 3 \times \frac{X_m}{\omega/\text{pole pairs}} \times \frac{E}{X_m} \times I_a \times \sin[90-(\varphi+\gamma)]$$

$$= 3 \times \text{pole pairs}, \times M' \times I_m \times I_a \times \sin\varphi_{ma} \qquad (5.17)$$

which really corresponds to the third equation which could be derived from Figure 5.2a, using $\mathbf{I}_1$ and $\mathbf{I}_m$. The angle φ_{ma} is that between $\mathbf{I}_m$ and $\mathbf{I}_a$, but any of the three angles, with the corresponding currents, could be used.

For constant-voltage constant-frequency drives, the air-gap flux is approximately constant since $V/f(\doteq E/f)$ is constant, therefore one value of M' is usually adequate for calculations. For current-fed drives, as already discovered with Examples 4.14–4.17, the flux may vary considerably and saturation changes which affect X_m have to be allowed for. Given the magnetising curve, which relates I_m (or I'_m) to E, at a particular frequency, this can be included since $X_m = E/I_m$. An iterative procedure may be necessary to match calculated values with the correct value of X_m and M'.

It is now appropriate and instructive to compare the induction machine and the synchronous machine in terms of their m.m.f. diagrams. Note that positive I'_2, the rotor m.m.f., is conventionally taken to be magnetising in the opposite sense to positive I_1, unlike the rotor m.m.f. I'_f (F_f), which is taken to be magnetising in the same sense as I_a (F_a) when both are positive. Reversing the phasor $\mathbf{I}'_2$ as shown brings Figure 5.2a into the same convention as Figure 5.2b. The important thing to realise is that the torque is a function of the m.m.f.s. and angles, for either synchronous or induction machines and regardless of the speed (though the condition must be steady state, with the m.m.f.s in synchronism). Synchronous and induction machines can thus have a common basis when considering the torque produced and its control. If the three current magnitudes are predetermined, then the geometry of the triangle constrains the angles to certain values in accordance with the cosine rule. Alternatively, if two of the current magnitudes are specified and the angle of one current with respect to another, then again the geometry determines the remaining current and angles.

This last statement is important when considering the maximum-torque conditions. Clearly, from the equations derived, for a fixed M and for any two specified currents, maximum torque occurs when the angle between them is 90°. For the induction motor, Figure 5.2a, this cannot occur with an uncontrolled (cage) rotor, since $\mathbf{I}'_2$ must always lag behind $\mathbf{E}_1$. The torque can only be obtained with an angle ($\varphi_m - \varphi_2$) less than 90°; or considering the angle between $\mathbf{I}_m$ and $-\mathbf{I}'_2$, an angle greater than 90° which in fact has the same sine. This is a consequence of the induction mode, with the rotor current dependent on induction from the stator currents. For the d.c. machine, a torque angle of 90° is the normal condition, with field and armature m.m.f.s maintained in quadrature by the angular-position switching action of brushes and commutator.

A recent development, 'vector' control, discussed in Section 7.5, attempts to match this unique commutator-switching feature, during transient operation of cage induction motors.

For the synchronous machine, the angles are functions of load, voltage, field current and frequency. But supposing a shaft-position detector was incorporated, so that specified stator and rotor m.m.f.s could be held at a fixed relative angle, this could be made 90° and again the maximum torque could be obtained (Figure 5.3a). In practice, however, this would mean that the flux would have to adapt as the dependent variable, and saturation effects on the value of M' would have to be allowed for. Maximum torque then occurs at a torque angle greater than 90° as will be seen. The normal, voltage-fed machine, with flux per pole approximately

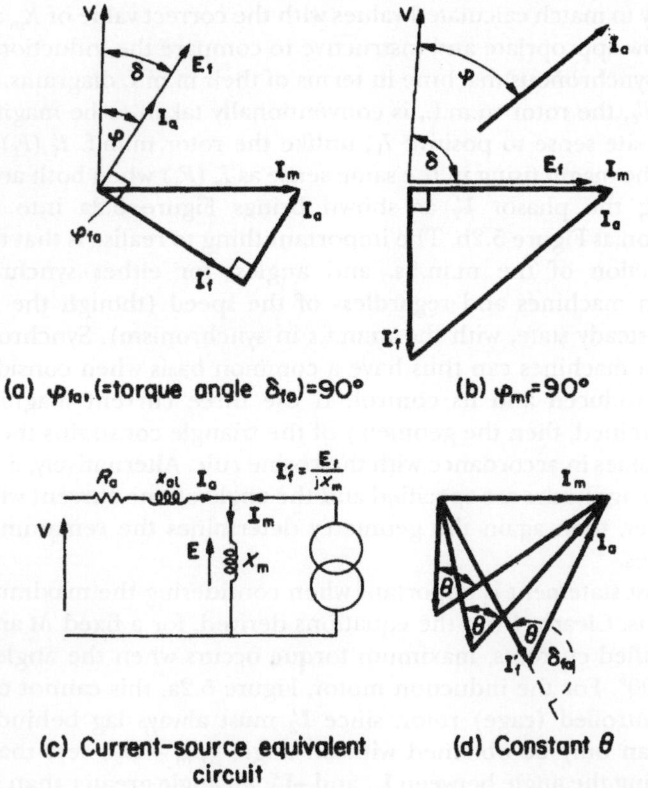

Figure 5.3 *Maximum-torque conditions for synchronous machine.*

constant ($\phi \propto E/f \simeq V/f$), and with a particular field m.m.f., will have maximum torque when the load angle δ between $\mathbf{V}$ and $\mathbf{E}_f$ is virtually 90°, δ being the same as the angle between $\mathbf{I}_m$ and $\mathbf{I}'_f$ neglecting leakage impedance; see Figure 5.3b. With this as the constraint, I_a is the dependent variable and will be higher in terms of its m.m.f. than either I_m or I'_f. The third possibility is to specify the flux level (I_m) and the armature current I_a, with an angle of 90° between them for maximum torque. $\mathbf{I}_a$ would then be vertical on Figure 5.3b and I'_f would have to be the highest in terms of its m.m.f. This might be the most appropriate option since the field power is very much less than the armature power and would therefore be cheaper to control.

The current-source equivalent circuit for the synchronous machine is shown on Figure 5.3c. It follows from Figure 5.2b, though the current source due to the field could also be obtained from standard circuit techniques which would convert a voltage source E_f behind an impedance jX_m to a current source $\mathbf{I}'_f$, ($\mathbf{E}_f/jX_m$), at an angle which is known only with respect to $\mathbf{V}$ as (δ – 90°). But V itself is not specified if I_a is to be predetermined and controlled. V awaits the solution of performance from the current diagram. The angles of the other currents are also unknown, though it is usually convenient to take $\mathbf{I}_a$ as the reference phasor, Example 5.13, and $\mathbf{I}_m$ is known to lag $\mathbf{E}$, ($X_m I_m$), by 90°.

If current-controlled sources are to be used, it has already been found when discussing the induction motor, Section 4.3, that saturation effects become very important. Consider Figure 5.3d. This shows a particular $\mathbf{I}_m$ and flux, which may or may not result in pronounced saturation effects, and the two currents $\mathbf{I}'_f$ and $\mathbf{I}_a$ with an angle θ which is actually equal to (180 – the torque angle δ_{fa}), Figure 5.2b. A few simple drawings will quickly demonstrate that maximum torque occurs when I'_f and I_a are equal.[7] Actually this follows from eqn (5.17) which has the same form as the area-of-a-triangle formula, $\frac{1}{2}a \times b \sin C$, and maximum area occurs for the condition stated. A comparison of control strategies could as a first approximation consider the criterion of torque per ampere of *total* current, which based on eqn (5.17) re-formed in terms of I'_f and I_a would be:

$$T_e/A = \frac{3 \times \text{pole pairs} \times M' I_a I'_f \sin\theta}{I_a + I'_f} \propto \frac{M' I^2 \sin\alpha}{2I} \propto \frac{M' I \sin\theta}{2} \qquad (5.18)$$

if $I_a = I'_f = I$. For optimum torque angle δ_{fa} (= 180 – θ) = 90°, then θ = 90° and the maximum torque per ampere would be directly proportional to I. However, this assumes that M' is unaffected by the value of I, which is only a reasonable assumption for a low-flux, unsaturated state. But if I was the rated value and θ = 90°, I_m would be $\sqrt{2}I$ and M' would therefore be appreciably lower than its unsaturated value. Suppose now that θ is reduced from 90°; I_m will be less and so M' will increase. If this increase is greater than the decrease of sin θ, the torque will actually rise. What this means is that in the saturated state the maximum torque occurs at a torque angle greater than 90°,[7] though the effect is not very marked until the magnetising current is approaching 1.5 *per unit* or more, and costs of excessive currents must be taken into account.

In the next example, a simplification will be made by expressing all the relevant equations in *per-unit*, with rated **kVA**, rated voltage and the synchronous speed at rated frequency, as base values. The equations are unchanged apart from the omission of constants like '3', $\omega_{s(\text{base})}$ and pole pairs. The conversion is simple; for example, base torque is: (rated kVA × $10^3)/\omega_{s(\text{base})}$ and on dividing the torque expressions by this, the voltages V and E_f by V_{base} and remembering that $Z_{\text{base}} = (V_{\text{base}})^2/\text{Power}_{\text{base}}$ (see Example 4.2d, p. 86), the following equations result, E_f and X_s being the values at base frequency:

For the voltage equation at any frequency k × f_{base}:

$$\mathbf{V} = \mathrm{k}\mathbf{E}_f + R_a\mathbf{I}_a + \mathrm{j}\mathrm{k}X_s\mathbf{I}_a \tag{5.19}$$

For eqn (5.12, p. 142):

multiply the first term of eqn (5.20) below by $-\sin(\delta - \alpha)$.

For eqn (5.14, p. 154):

$$\text{Max. } T_e = E_f\left[\frac{V}{\sqrt{R^2{}_a + (\mathrm{k}X_s)^2}} - \frac{\mathrm{k}E_fR_a}{R^2{}_a + (\mathrm{k}X_s)^2}\right] \tag{5.20}$$

From eqn (5.17):

$$T_e = X_m I'_f I_a \sin\delta_{fa} \tag{5.21}$$

now re-formed in terms of I'_f, I_a and the torque angle δ_{fa}. Note that since X_m is in *per unit* = $\omega_{\text{base}} M'/(V_{\text{base}}/I_{\text{base}})$, it is proportional to the mutual inductance and k does not appear in the equation, nor in eqn (5.17).

Example 5.15

A polyphase, cylindrical-rotor synchronous machine has a *per-unit* synchronous impedance Z_s = 1. The resistance when allowed for can be taken as 0.05. The leakage impedance can be taken as zero ($X_s = X_m$). Determine:

(a) the *per-unit* E_f for operation as a motor at unity power-factor and rated armature current, voltage and frequency,

(b) the *per-unit* torque at this rated condition calculated from (i) $VI_a \cos\varphi - I^2{}_a R_a$, (ii) eqn (5.12) and (iii) eqn (5.21);
(c) the maximum torque with rated voltage and the same value of E_f;
(d) the armature current at condition (c);
(e) the air-gap e.m.f. E at condition (c);
(f) I_m and I'_f at condition (c);
(g) the phasor diagram for condition (c), checking all the angles of the m.m.f. triangle by the cosine rule on the current magnitudes;
(h) the voltage and frequency required to sustain condition (c) at a speed of 0.1 *per unit*;
(i) the answers to (a)–(h) if resistance is neglected;
(j) the values of torque and torque per ampere in *per unit*, for the following conditions when supplied from current sources giving $I_a = I'_f = 2$. (i) $I_m = 1$; (ii) $I_m = 2$; (iii) $I_m = 3$. Neglect the armature resistance and use the approximation to the magnetisation curve employed in Example (4.16) for determining the reduction in *per-unit* mutual inductance – $X_m = 1$ at $I_m = 1$. (M' per unit $= \omega_{base} M' I_{base}/V_{base} = X_m/Z_{base}$.)

If $R_a = 0.05$, then $X_s = \sqrt{1^2 - 0.05^2} = 0.9987$ and $\alpha = \tan^{-1} R_a/X_s = 2°.87$

(a) $\mathbf{E}_f = \mathbf{V} - \mathbf{I}_a(R_a + jX_s) = 1 - (1 + j0)(0.05 + j0.9987) = \underline{1.3784 \angle{-46°.43}}$

(b) (i) $T_e = 1 \times 1 \times 1 - 1^2 \times 0.05 \qquad = 0.95$

(ii) $$T_e = \frac{-1 \times 1.3784}{1} \sin(-46.43 - 2.87) - \frac{1.3784^2 \times 0.05}{1^2} = \underline{0.95}$$

(iii) $$\mathbf{I}'_f = \frac{\mathbf{E}_f}{jX_m} = \frac{1.3784 \angle{-46.43}}{0.9987 \angle{90}} = \underline{1.38 \angle{-136.43}}$$

so $T_e = X_m I'_f I_a \sin\delta_{fa} = 0.9987 \times 1.38 \times 1 \times \sin 136°.43 = \underline{0.95}$

(c) $T_{e(max)}$, from eqn (5.12) with $(\delta - \alpha) = -90°$

$$= \frac{-1 \times 1.3784}{1} \times (-1) - \frac{1.3784^2 \times 0.05}{1^2} = \underline{1.283}$$

(c.) $$\mathbf{I}_a = \frac{\mathbf{V} - E_f \angle{-(90-\alpha)}}{Z_s \angle{90 - \alpha}} = \frac{1 - 1.3784(0.05 - j0.9987)}{0.05 - j0.9987} = \underline{1.662 \angle{-31°.2}}$$

(e) $\mathbf{E} = \mathbf{V} - \mathbf{I}_a R_a = 1 - 1.662\angle{-31.2} \times 0.05 \qquad = \underline{0.93\angle{2°.65}}$

(f) $$\mathbf{I}'_f = \frac{E_f \angle{-90 + 2.87}}{0.9987 \angle{90}} = \underline{1.38\angle{-177°.13}}$$

$$\mathbf{I}_m = \frac{\mathbf{E}}{jX_m} = \frac{0.93 \angle{2.65}}{0.9987 \angle{90°}} = \underline{0.931 \angle{-87°.35}}$$

The angle between $\mathbf{I}'_f$ and $\mathbf{I}_m$ (φ_{mf}) is thus $-177.13 - (-87.35) = 89°.78$ which is very nearly at the optimum value of 90°, even though it is the result of a maximum-torque condition deduced from the voltage-source equations. These equations allow for impedance drop and maximise the air-gap power rather than considering the optimum angle as for the current-source approach. In fact, if the leakage reactance had been allowed for, this angle would have progressively reduced from 90° as a larger share of X_s was ascribed to leakage reactance. The

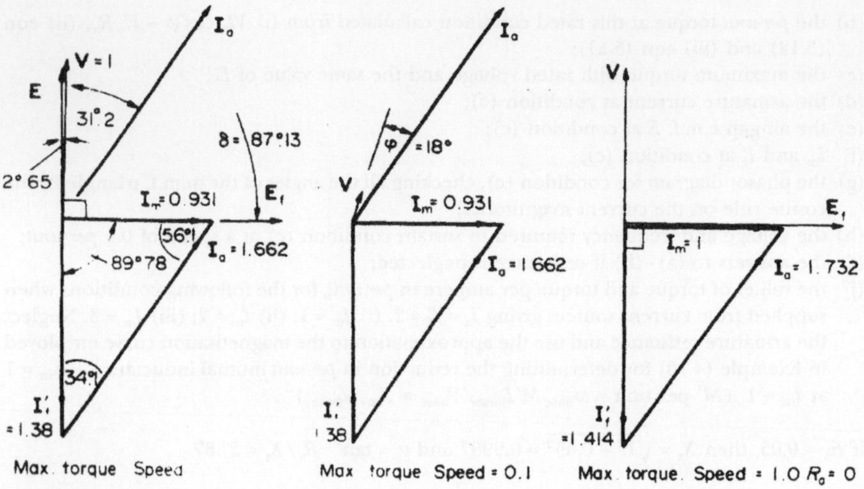

Figure E.5.15

maximum torque would have been unchanged since the value depends on the total synchronous impedance. This will be verified if the calculations are repeated with x_{al} typically 0.1 *per unit*, leaving X_m as 0.8987 *per unit*.

(g) The phasor diagram has been constructed from the above information and is shown on the first figure. The check on the angles from the calculated current magnitudes is left as an exercise to familiarise the calculation procedure when the currents are specified, as indeed they are in the next question and part (j).

(h) The condition here is a speed and frequency of 0.1 per unit which means that k in the variable-frequency expressions of eqns (5.19) and (5.20) is 0.1. The m.m.f. diagram is to be the same, to give the same torque – which is not quite the maximum for this particular flux (I_m) and field current, since the angle between them is slightly less than 90°. It is convenient to take $\mathbf{I}_a$ as a reference phasor; it will be 1.662 + j0. Hence, from eqn (5.19): and the phasor diagram from which the angles can be deduced, e.g.

$$\delta_{fa} = 89.78 + (90 - 31.2 - 2.65) = 145°.93:$$

$$\mathbf{V} = \mathrm{k}\mathbf{E}_f + R_a\mathbf{I}_a + \mathrm{jk}X_s\mathbf{I}_a = \mathrm{jk}\mathbf{I}_f'X_m + R_a\mathbf{I}_a + \mathrm{jk}X_s\mathbf{I}_a$$

$$= \mathrm{j}0.1 \times 1.38\underline{/-145°.93} \times 0.9987 + 0.05 \times 1.662 + \mathrm{j}0.1 \times 0.9987 \times 1.662$$

$$= 0.1378\underline{/-55°.93} \qquad + 0.0831 \qquad + \mathrm{j}0.166$$

$$= 0.16 + \mathrm{j}0.052 = \underline{0.168\underline{/+18°}}$$

Note that in the above calculations, Cartesian and polar forms have been used as convenient but care must be taken in combining them finally. The second figure shows the phasor diagram, the current $\mathbf{I}_a$ being at a slightly lagging power-factor of 0.95. The load angle δ between **V** and $\mathbf{E}_f$ is $-[18 - (-55.93)] = -73°.93$ which is $\ll 90°$ even though torque is maximum. δ falls further if k is reduced further.

The calculations neglecting resistance are much simpler and will now be outlined in the same sequence as previously.

(a) $\mathbf{E}_f = 1 - 1 \times j1 = 1 - j1 = \underline{1.414\angle -45°}$

(b) (i) $T_e = 1 \times 1 \times 1 \qquad = \underline{1}$

(ii) $$T_e = \frac{-1 \times 1.414 \times \sin(-45°)}{1} = \underline{1}$$

(iii) $$T_e = 1 \times \frac{1.414}{1} \times 1 \times \sin 45° = \underline{1}$$

(c) $$T_{e(max)} = \frac{-1 \times 1.414}{1}(-1) = \underline{1.414}$$

(d) $$\mathbf{I}_a = \frac{1 - 1.414\angle -90°}{j1} = 1.414 - j1 = \underline{1.732\angle -35°.3}$$

(e) $E = V = \underline{1}$

(f) $$\mathbf{I}_f' = \frac{1.414\angle -90°}{j1} = \underline{-1.414} \quad \mathbf{I}_m = \frac{1}{j1} = \underline{-j1}$$

(g) The phasor diagram is shown on the 3rd figure and should be compared with the one including resistance. The differences are about 5% but note that the angle is now exactly 90.

(h) With $\mathbf{I}_a$ as the reference phasor, $\mathbf{I}_f'$ is $1.414\angle -180° - (-35°.3) = 1.414\angle -144°.7$

So $\mathbf{V} = jk\mathbf{I}_f'X_m + jkX_s\mathbf{I}_a$

$$= j0.1 \times 1.414\angle -144°.7 \times 1 + j0.1 \times 1 \times 1.732$$

$$= 0.1414(0.578 - j0.816) \quad + j0.1732$$

$$= 0.082 + j0.058 = \underline{0.1\angle +35°.3}$$

This result, which shows that V is proportional to k when resistance is neglected, has already been illustrated in Example 5.14, but is now confirmed. With R_a included, a much higher voltage is required to provide for the increasingly dominating effect of resistance at low frequencies, as shown in the first part of this question.

(j) Allowance for saturation of mutual inductance is now to be made and current sources $I_a = I_f' = 2$ are to be supplied at different phase angles to permit I_m to vary from rated value of 1, to 3 *per unit*. The empirical expression relating flux and magnetising current is, from Example 4.16:

$$I_m = \frac{0.6\phi_m}{1 - 0.4\phi_m} \quad \text{from which} \quad \phi_m = \frac{I_m}{0.6 + 0.4I_m}$$

The average slope of the flux/I_m curve is proportional to the mutual inductance. When $I_m = 1$, $\phi_m = 1$ and the slope is proportional to 1. For $I_m = 2$, $\phi_m = 2/1.4$ and for $I_m = 3$, $\phi_m = 3/1.8$.

The average slopes for the last two are 1/1.4 and 1/1.8 respectively. These are the coefficients by which M' must be reduced. The angle θ between I_a and I'_f used on Figure 5.3d is obtained from the cosine rule as:

$$\cos^{-1}\theta = \frac{I_f'^2 + I_a^2 - I_m^2}{2I_f' I_a}$$

and the torque expression of eqn (5.21) will complete the calculation. The results are tabulated below:

	(i)	(ii)	(iii)
M'	1	1/1.4	1/1.8
θ	29°	60°	97°
$T_e = 2 \times 2 \times M' \sin\theta$	1.93	2.47	2.2
Torque/amp $T_e/4$	0.48	0.62	0.55

A few more calculations will show that the maximum torque occurs at an angle θ rather less than 90° which means that the torque angle δ_{fa}, between stator and rotor m.m.f.s is rather greater than the unsaturated value of 90°. The load angle between $\mathbf{I}_m$ and $\mathbf{I}'_f$ is still less than 90°. The maximum torque per ampere is somewhat less than the unsaturated value given by eqn (5.18) which is $I/2$, equal to 1 in this case.

5.6 Operating charts

Under certain circumstances the phasor diagrams for a.c. circuits under variable conditions give rise to phasor loci which are circular. Such circle diagrams are useful visual aids and assist in rapid calculations. They are less important nowadays with improved calculating facilities which can operate directly on the equations, which are really the basis on which the diagrams are constructed. No examples of induction machine circle diagrams have been given in this book though they are dealt with in Reference 1. However, for synchronous machines they are still of general interest because they delineate the various regions of operation in a manner especially useful for power-systems studies. One example will therefore be given here, as a simple extension of the phasor loci shown on Figure 5.1.

Example 5.16

A 3-phase synchronous machine has a synchronous reactance X_s = 1.25 *per unit* and on full load as a generator it operates at 0.9 power-factor lagging. The machine losses may be neglected.

(a) Determine the rated excitation in *per unit* to sustain this condition.
(b) What excitation would be required to operate at full-load current but 0.707 p.f. lagging?
(c) With this maximum excitation, what would be the maximum motoring torque in *per unit* – expressed in terms of full-load torque?
(d) What *per-unit* current would be drawn for condition (c)?

(e) What power can be developed as a motor when running at 0.8 power-factor leading, without exceeding rated excitation?

(f) With the maximum excitation of condition (c) what will be the kVAr rating as a synchronous capacitor?

(g) What will be the maximum kVAr rating as a synchronous reactor with the excitation reduced to a minimum of 1/3 of the rated value?

The problem can be solved using the approximate circuit as in the previous examples. With the circle diagrams, the solution is much quicker though less accurate.

(a) The two current phasors; $\mathbf{V}/jX_s = 1/j1.25$, lagging 90° behind **V**, and $\mathbf{I}_a = 1$ *per unit* at cos $\varphi = 0.9$ lagging as a generator, are first drawn. With a motor convention, the $\mathbf{I}_a$ phasor is reversed at angle 25°.8 lagging – **V**. The closing phasor of the triangle is the rated excitation divided by jX_s, which by measurement is 1.52 *per unit.* Hence rated $E_f = 1.52 \times 1.25 = \underline{1.9 \textit{ per unit.}}$

(b) More excitation will be required to operate at rated current and a lower power-factor. Drawing an arc from the origin at $I_a = 1$ *per unit* to angle 45° for 0.707 power-factor identifies the end of the $\mathbf{E}_f/jX_s$ phasor which by measurement is 1.64 *per unit.* Hence $E_f = 1.64 \times 1.25 = \underline{2.05 \textit{ per unit.}}$

(c) An arc drawn at this new $\mathbf{E}_f/jX_s$ radius into the motoring region until $\mathbf{E}_f/jX_s$ makes an angle $\delta = 90°$ with the $\mathbf{V}/jX_s$ phasor, defines the point of maximum torque. The rated torque corresponds to rated $I_a \cos\varphi = 0.9$ *per unit* and so the maximum torque is $1.64/0.9 = \underline{1.82 \textit{ per unit.}}$

(d) A current phasor drawn from the origin to the point of maximum torque has a value of $\underline{1.85 \textit{ per unit}}$, at angle $\varphi = 25°.8$ lagging; $\cos\varphi = 0.9$.

(e) A motoring-current phasor drawn at $\varphi = \cos^{-1} 0.8$ leading = 36°.9 intersects the rated excitation circle at $I_a = 0.9$ *per unit,* and $I_a \cos\varphi = \underline{0.72 \textit{ per unit}}$ is the power and torque for this condition.

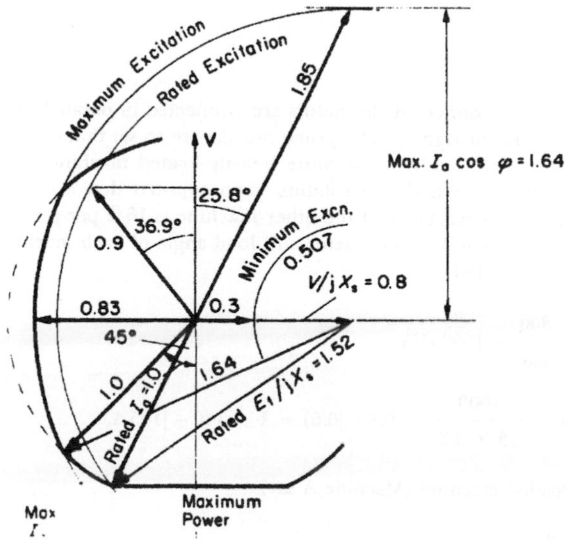

Figure E.5.16

(f) For a 'motor' at zero p.f. leading, the maximum excitation circle intersects the zero-power axis at $\mathbf{I}_a = j0.83$ *per unit*, so kVAr as a synchronous capacitor is 0.83 *per unit*.
(g) For zero power-factor lagging, the excitation circle drawn at 1/3 of the rated $\mathbf{E}_f/jX_s = 1/3 \times 1.52 = 0.507$ intersects the zero-power axis at $-j0.3$, so kVAr as a synchronous reactor would be 0.3 *per unit*. This rating is always less than as a capacitor because of stability considerations at low excitation.

Boundaries are shown on Figure E.5.16 indicating the operating limits for the various modes. *Per-unit* notation is especially useful for the portrayal of so many operating modes and conditions. Resistance can be incorporated quite simply by drawing the excitation circles from a $\mathbf{V}/\mathbf{Z}_s$ phasor, which will be at an angle less than 90° behind **V**, and the radii will then be $|\mathbf{E}_f/\mathbf{Z}_s|$.

5.7 Multi-machine problems

This area of study for synchronous machines is very important for power systems. Only a few simple examples will be given here to illustrate parallel operation of generators and the use of synchronous motors to improve an industrial-system power factor. Examples up to now have assumed that the supply-system has zero impedance, giving constant terminal voltage V; and infinite inertia, giving constant frequency. These are the characteristics of the so-called 'infinite' busbar system. For two paralleled generators, the interaction of the individual machine impedances and e.m.f.s on the sharing of kW and kVAr is instructive as an introduction to the power-flow problem.

Example 5.17

Two 3-phase, 3.3-kV, star-connected alternators are connected in parallel to supply a load of 800 kW at 0.8 power factor lagging. The prime movers are so set that one machine delivers twice as much power as the other. The more heavily loaded machine has a synchronous reactance of 10 Ω per phase and its excitation is so adjusted that it operates at 0.75 p.f. lagging. The synchronous reactance of the other machine is 16 Ω per phase.

Calculate the current, e.m.f., power factor and load angle of each machine. The internal resistances may be neglected.

$$\text{Total kVA} = \frac{800}{0.8} = 1000\,\text{kVA}.$$

$$\text{Total current} = \frac{1000}{\sqrt{3} \times 3.3} \times (0.8 - j0.6) = \mathbf{I} = 140 - j105\,\text{A}.$$

For the heavier loaded machine (Machine A say)

$$I_A \cos \varphi_A = \frac{2}{3} \times 140 = 93.3\,\text{A}$$

and

$$I_A \sin \varphi_A = I_A \cos \varphi_A \times \tan \varphi_A = 93.3 \times 0.882$$

$$= 82.3\,\text{A}.$$

$$\therefore \mathbf{I}_A = \underline{93.3 - j82.3.}$$

$$\therefore \mathbf{I}_B = \mathbf{I} - \mathbf{I}_A = \underline{46.7 - j22.7.}$$

$$\therefore \mathbf{I}_A = \underline{124.4\,\text{A}} \text{ at } \cos \varphi_A = \underline{0.75.}$$

$$\mathbf{I}_B = 51.9\,\text{A at } \cos \varphi_B = 0.9.$$

$$\mathbf{E}_A = \mathbf{V} + jX_A\mathbf{I}_A = \frac{3300}{\sqrt{3}} + j10(93.3 - j82.3) = 2728 + j933 = \underline{2888/18°.0.} \text{ per phase}$$

$$\mathbf{E}_B = \mathbf{V} + jX_B\mathbf{I}_B = \frac{3300}{\sqrt{3}} + j16(46.7 - j22.7) = 2268 + j747 = \underline{2388/18°.2.} \text{ per phase}$$

Example 5.18

Two star-connected, non-salient-pole, synchronous generators of identical rating operate in parallel to deliver 25 000 kW, 0.9 power-factor lagging-current at 11 kV. The line induced e.m.f. of Machine A is 15 kV and the machine delivers 10 MW, the remaining power being supplied by Machine B. Determine for each machine:

(a) the load angle in electrical degrees
(b) the current and power factor
(c) the kVA

Find also the induced e.m.f. of Machine B. Take the synchronous reactance for each machine as 4.8 Ω per phase and neglect all machine losses.

$$\text{Total kVA} = \frac{25\,000}{0.9} = 27\,777\,\text{kVA}.$$

$$\text{Total current} = \frac{27\,777}{\sqrt{3} \times 11}(0.9 - j0.436) = 1312 - j635.7.$$

$$\text{Load angle } \delta_A \text{ from } 10\,\text{MW} = \frac{3 \times 11\,\text{kV} \times 15\,\text{kV}}{4.8 \times \sqrt{3} \times \sqrt{3}} \sin \delta_A$$

from which: $\sin \delta_A = 0.291$; $\delta_A = \underline{16°.9}$; $\cos \delta_A = 0.957$.

$$\text{Hence } \mathbf{I}_A = \frac{\mathbf{E}_f - \mathbf{V}}{jX_s} = \frac{1000}{\sqrt{3}} \times \frac{15(0.957 + j0.291) - 11}{j4.8} = \underline{525 - j403.5.}$$

$$\text{Hence } \mathbf{I}_B = \mathbf{I} - \mathbf{I}_A = \underline{787 - j232.2.}$$

$$\mathbf{E}_B = \mathbf{V} + jX_s\mathbf{I}_B = 11\,000 + \sqrt{3} \times j4.8(787 - j232.2) = 12.93 + j6.54 \text{ Line kV}$$

$$= 14.49\underline{/26°.8}.$$

From above:

(a) $\delta_A = \underline{16°.9}$. $\delta_B = \underline{26°.8}$.

(b) $I_A = \underline{662\,A}$ at $\cos\varphi_A = \underline{0.793}$. $I_B = \underline{820.5\,A}$ at $\cos\varphi_B = \underline{0.959}$.

(c) $kVA_A = \sqrt{3} \times 11 \times 662 = \underline{12\,600}$. $kVA_B = \sqrt{3} \times 11 \times 820.5 = \underline{15\,630}$.

Induced e.m.f. for Machine B = $\underline{14.49 \text{ Line kV}}$.

Example 5.19

A 6.6-kV industrial plant has the following two induction motor drives and a main transformer for the other plant

	Ind. Mtr A	Int. Mtr B	Transformer
Rated output (kW)	100	200	300
Full-load efficiency	94%	95%	99%
Full-load power factor	0.91	0.93	0.98 lagging

A star-connected synchronous machine rated at 250 kVA is to be installed and so controlled that when all the equipment is working at full rating, the overall power-factor will be unity. At the same time, the synchronous machine is to draw rated current and deliver as much mechanical power as possible within its current rating. If its efficiency can be taken as 96% and its synchronous impedance 0 + j100 Ω, calculate the required e.m.f. behind synchronous impedance to sustain this condition.

As an alternative strategy, consider what rating of synchronous machine, operating the same motoring load, would be required to bring the overall works power-factor to 0.95 leading.

First, the total works load can be calculated by summing

$$\mathbf{I} = \sum \frac{\text{Power}}{\eta \times \sqrt{3} \times V \times \text{p.f.}} (\cos\varphi + \mathrm{j}\sin\varphi)$$

$$\text{Ind. Mtr A. } \mathbf{I}_A = \frac{100}{0.94 \times \sqrt{3} \times 6.6 \times 0.91}(0.91 - \mathrm{j}0.4146) = 9.306 - \mathrm{j}4.24$$

$$\text{Ind. Mtr B. } \mathbf{I}_B = \frac{200}{0.95 \times \sqrt{3} \times 6.6 \times 0.93}(0.93 - \mathrm{j}0.3676) = 18.42 - \mathrm{j}7.28$$

$$\text{Transformer. } \mathbf{I}_T = \frac{300}{0.99 \times \sqrt{3} \times 6.6 \times 0.98}(0.98 - \mathrm{j}0.199) = \underline{26.51 - \mathrm{j}5.38}$$

$$\text{Total current} = I_P - \mathrm{j}I_Q = \underline{54.24 - \mathrm{j}16.9}$$

$$\text{Rated synchronous machine current} = \frac{250}{\sqrt{3} \times 6.6} = 21.87\text{ A}$$

Reactive component of current must be j16.9 to bring p.f. to unity.

$$\therefore \text{ active component of current} = \sqrt{21.87^2 - 16.9^2} = 13.88\text{ A.}$$

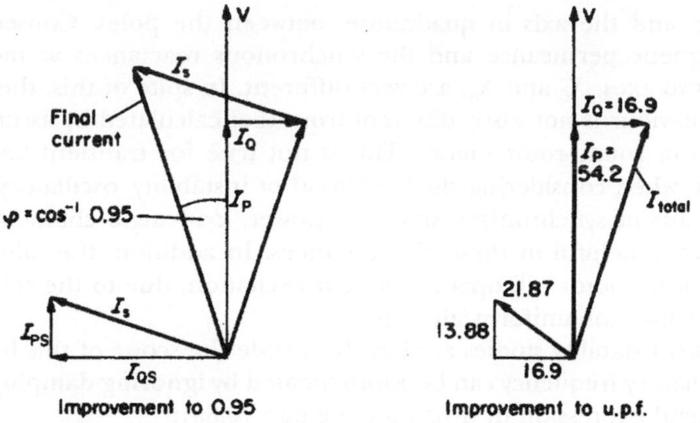

Figure E.5.19

So synchronous motor output power = $\sqrt{3} \times 6.6 \times 13.88 \times 0.96$ = 152.3 kW = 204 hp.

$$\mathbf{E}_f = \mathbf{V} - jX_s\mathbf{I}_a = \frac{6600}{\sqrt{3}} - j100(13.88 + j16.9) = 5500.5 - j1388 = \underline{5673\text{V}} = \underline{1.49\,\text{p.u.}}$$

For a power factor of 0.95 leading and with the same power component of synchronous-motor current, the phasor diagram shows that the following relationship holds:

$$\tan\varphi = \frac{I_{QS} - I_Q}{I_{PS} + I_P} = \frac{I_{QS} - 16.9}{13.88 + 54.24},$$

and since $\tan\varphi = \tan(\cos^{-1} 0.95) = 0.3286 = (I_{QS} - 16.9)/68.12$,

$I_{QS} = 39.29$ A.

Hence required synchronous machine rating = $\sqrt{3} \times 6.6\sqrt{(39.29^2 + 13.88^2)}$ = 476 kVA, requiring a machine of nearly twice the previous size.

5.8 Salient-pole and reluctance-type machines, synchronising power

All the examples so far posed have been solved using round-rotor or cylindrical-rotor theory; namely assuming that synchronous machines are built like induction machines with uniform air gap such that a sinusoidal m.m.f. distribution will produce a sinusoidal flux distribution. The majority of synchronous machines do not have such a symmetrical magnetic geometry, though the high-power turbo-generators are close to this. Most synchronous-motor drives have salient-pole rotors, giving two axes of symmetry per pole pair; the direct axis, along the main field winding, and the axis in quadrature, between the poles. Consequently, the magnetic permeance and the synchronous reactances as measured on the two axes, X_d and X_q, are very different. In spite of this, the steady-state behaviour is not vastly different from that calculated by assuming $X_s = X_d$, as in round-rotor theory. This is not true for transient behaviour however, when considering the likelihood of instability, oscillatory behaviour or loss of synchronism, since the power/load-angle characteristic is much more helpful in these circumstances. In addition, the salient-pole synchronous motor will operate without excitation, due to the reluctance effect of the non-uniform air gap.

Transient stability studies are largely outside the scope of this book but the oscillatory frequency can be approximated by ignoring damping. From the general expression in dynamics, see eqn (6.8a):

$$\text{Natural, undamped frequency} = \sqrt{(\text{stiffness/inertia})}\ \text{rad/s}.$$

The effect of damping terms is discussed in Section 6.1, to which reference might usefully be made. For a synchronous machine, the stiffness is the rate of change of torque per mechanical radian, obtained by differentiating the torque/load-angle curve at the operating point. Sometimes the term 'synchronising power' is used to define the power ΔP, brought into play on a change of angle $\Delta\delta$, and this in turn is obtained by differentiating the power/load-angle curve to get $dP/d\delta$.

Example 5.20

A 3-phase, 5000-kVA, 11-kV, 50-Hz, 1000-rev/min, star-connected synchronous motor operates at full load, 0.8 p.f. leading. The synchronous reactance is 0.6 *per unit* and the resistance may be neglected. Calculate for these conditions, the synchronising torque per mechanical degree of angular displacement. The air gap may be assumed to be uniform around the periphery.

Working in *per unit*; $\mathbf{E}_f = 1 - jX_s\mathbf{I}_a = 1 - j0.6 \times 1(0.8 + j0.6)$

$$= 1.36 - j0.48 = 1.44\underline{/19°.44}.$$

$$\text{Power/angle equation } P = \frac{VE}{X_s}\sin\delta = \frac{1 \times 1.44}{0.6}\sin\delta = 2.4\sin\delta.$$

Hence, synchronising power from $dP/d\delta = 2.4\cos\delta = 2.4\cos 19.44 = 2.263$ *p.u.*/radian

$$1\ \textit{per-unit}\ \text{torque} = \frac{5\,000\,000}{2\pi \times 1000/60} = 47\,746\ \text{Nm}.$$

$$\therefore 2.263\ \textit{per unit} = 2.263 \times 47\,746 \times \frac{2\pi}{360} \times 3 = \underline{5657\ \text{Nm/mechanical degree}}$$

since there must be 6 poles, if the synchronous speed is 1000 rev/min at 50 Hz.

Example 5.21

A salient-pole synchronous motor has $X_d = 0.9$ and $X_q = 0.6$ *per unit* and is supplied from rated voltage and frequency. Calculate the current, power factor and power for a load angle of −30° (motoring) and for excitation e.m.f.s (E_f) of 1.5, 1.0, 0.5 and 0 *per unit*, the latter case being the reluctance motor (zero excitation). What would be the new values if, as a reluctance motor, the rotor was redesigned to give $X_d = 0.75$ and $X_q = 0.25$? What would then be the maximum torque? Armature resistance may be neglected throughout.

The phasor diagram of Figure 5.4a is shown for the overexcited condition as a motor. The equation is similar to that for the round-rotor machine but the resolution of the armature m.m.f. $\mathbf{F_a}$ into direct-axis ($\mathbf{F_{ad}}$)

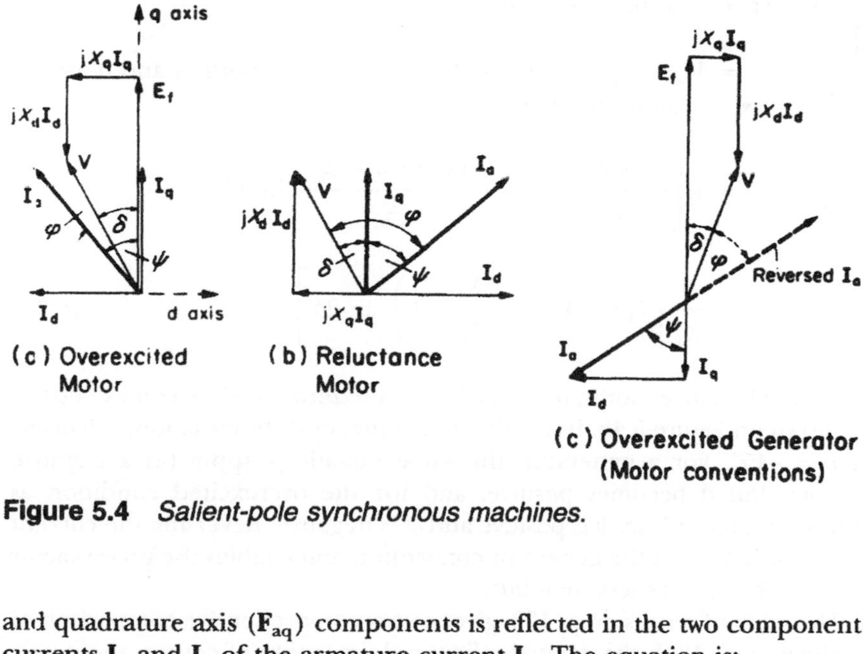

Figure 5.4 *Salient-pole synchronous machines.*

and quadrature axis ($\mathbf{F_{aq}}$) components is reflected in the two component currents $\mathbf{I_d}$ and $\mathbf{I_q}$ of the armature current $\mathbf{I_a}$. The equation is:

$$\mathbf{V} = \mathbf{E_f} + jX_q\mathbf{I_q} + jX_d\mathbf{I_d} \text{ (see Reference 1).}$$

Remembering that δ is negative for a motor, the angle by which $\mathbf{I_q}$ (which is in phase with $\mathbf{E_f}$) leads $\mathbf{I_a}$ must be defined as:

$$\psi = \delta - \varphi \text{ and is negative on Figure 5.4a.}$$

From the geometry of the diagram and deliberately arranging the equations so that positive I_d will be magnetising along the d axis (as for F_f):

$$X_q I_q = -V \sin \delta \qquad \therefore I_q = \frac{-V \sin \delta}{X_q} \tag{5.22}$$

$$X_d I_d = V \cos \delta - E_f \qquad \therefore I_d = \frac{V \cos \delta - E_f}{X_d} \tag{5.23}$$

I_d is negative on Figure 5.4a corresponding to the demagnetising action for this condition. $I_d = I_a \sin \psi$ and $I_q = I_a \cos \psi$ but for generating operation this must be made negative if using motoring conventions.

$$\tan \psi = I_d / I_q \text{ and the total armature current } I_a = \sqrt{I_d^2 + I_q^2}.$$

Power = (torque in *per unit*)

= $VI_a \cos \varphi = V(I_q \cos \delta + I_d \sin \delta)$ and both I_d and δ are negative in Figure 5.4a.

$$= V\left[\frac{-V \sin \delta}{X_q} \cos \delta + \frac{(V \cos \delta - E_f)}{X_d} \sin \delta\right]$$

$$= \frac{-V}{X_d}\left[E_f \sin \delta + \frac{V}{2}\left(\frac{X_d}{X_q} - 1\right) \sin 2\delta\right] \tag{5.24}$$

For the reluctance motor, terms in E_f become zero and I_d becomes positive as shown on Figure 5.4b. Its maximum torque, by differentiation, will occur at $\delta = -45°$. For a generator, the same equations apply (as a negative motor), but δ becomes positive, and for the overexcited condition as shown on Figure 5.4c, ψ is positive and φ is negative. Reversing the current phasor brings back the generator convention and enables the power-factor angle to be seen as less than 90°.

The table of results on p. 173, gives a very comprehensive comparison of performance. With the excited salient-pole machine of course, the power factor can be controlled from lagging to leading. As excitation is reduced step by step to zero, the power factor falls drastically as does the power and torque. The output as a reluctance motor is very small. However, with a higher X_d/X_q ratio, the performance is much improved, at the expense of higher currents. This is due to the fact that the magnetising current must now come from the a.c. supply as for an induction machine. The table also shows the value and simplicity of the *per-unit* system in making these kinds of comparisons. As an exercise, the maximum torque for the excited machine could be worked out for further comparisons. The next example provides the maximum torque expression.

In the problem, $\delta = -30°$ so $\sin\delta = -0.5$, $\cos\delta = 0.866$ and $\sin 2\delta = -0.866$

D.C. excited salient-pole motor, $X_d = 0.9$ $X_q = 0.6$					Reluctance motor $X_d = 0.75$, $X_q = 0.25$	
E_f	1.5	1.0	0.5	0	$\delta = -30°$	$\delta = -45°$
$I_q = \dfrac{-V\sin\delta}{X_q}$	$\dfrac{0.5}{0.6} = 0.833$	0.833	0.833	0.833	2	2.83
$I_d = \dfrac{V\cos\delta - E_f}{X_d}$	$\dfrac{0.866 - 1.5}{0.9} = -0.704$	−0.149	+0.406	+0.96	+1.155	+0.943
$I_a = \sqrt{I_d^2 + I_q^2}$	1.091	0.846	0.927	1.271	2.309	2.98
$\psi = \tan^{-1} I_d/I_q$	−40°.2	−10°.1	+25°.9	+49°	+30°	+18°.4
$\varphi = \delta - \psi$	10°.2	−19°.9	−55°.9	−79°	−60°	−63°.4
$\cos\varphi$	0.98 lead	0.94 lag	0.56 lag	0.19 lag	0.5 lag	0.45 lag
$VI_a\cos\varphi$	1.073	0.795	0.52	0.242	1.155	1.334
Excn. power $\dfrac{V}{X_d} E_f \sin\delta$ +	0.833	0.555	0.277	0	0	0
Reluctance power $\dfrac{V^2}{2X_d}\left(\dfrac{X_d}{X_q} - 1\right)\sin 2\delta$	0.241	0.241	0.241	0.241	1.155	1.333
= Total power (check)	1.07	0.796	0.518	0.241	1.155	1.333

Example 5.22

A 6.6-kV, 5-MVA, 6-pole, 50-Hz, star-connected synchronous generator has $X_d = 8.7\,\Omega$ per phase and $X_q = 4.35\,\Omega$ per phase. Resistance may be neglected. If the excitation is so adjusted that $E_f = 11$ kV (line), and the load angle is 30° (elec.), determine:

(a) the power factor, output current and power in *per unit*;
(b) the load angle at maximum torque;
(c) the ratio between maximum torque and that occurring with $\delta = 30°$;
(d) the stiffness in newton metres per mechanical radian for a load angle of 30°;
(e) the frequency of small undamped oscillations if the total coupled inertia is 8200 kg m^2, the mean load angle being 30°;
(f) the stored-energy constant.

Rated armature current $I_a = \dfrac{5000}{\sqrt{3} \times 6.6} = 437.4\,\text{A} = 1$ *per-unit* current.

$$\text{(a)}\ I_q = \frac{-V \sin\delta}{X_q} = \frac{-6600/\sqrt{3} \times \sin 30°}{4.35} = -438\,\text{A}.$$

$$I_d = \frac{V\cos\delta - E_f}{X_d} = \frac{6600 \cos 30° - 11\,000}{\sqrt{3} \times 8.7} = -350.7\,\text{A}.$$

$$I_a = \sqrt{I_d^2 + I_q^2} = \sqrt{438^2 + 350.7^2} = \underline{561.1\,\text{A}} = \underline{1.283\ \textit{per unit}}.$$

$$\tan\psi = \frac{-350.7}{-438}\ \text{so}\ \psi = 38°.7 = \delta - \varphi = 30 - \varphi\ \text{so}\ \varphi = -8°.7,\ \cos\varphi = \underline{0.988\ \text{lagging}}$$

$$\text{Power} = VI_a \cos\varphi = 1 \times 1.283 \times 0.988 = \underline{1.268\ \textit{per unit}}$$

$$\text{(b) Torque} = \frac{-3}{\omega_s} \times \frac{V}{X_d} \times \left[E_f \sin\delta + \frac{V}{2}\left(\frac{X_d}{X_q} - 1\right) \sin 2\delta \right],$$

$$\frac{dT}{d\delta} = 0\ \text{when}\ E_f \cos\delta = -V\left(\frac{X_d}{X_q} - 1\right)\cos 2\delta = -V\left(\frac{X_d}{X_q} - 1\right)(2\cos^2\delta - 1).$$

$$\text{Forming the quadratic: } 2\cos^2\delta + \frac{E_f}{V}\left(\frac{X_q}{X_d - X_q}\right)\cos\delta - 1 = 0.$$

$$\text{From which: } \cos\delta = \frac{-E_f}{4V}\left(\frac{X_q}{X_d - X_q}\right) \pm \sqrt{\left(\frac{E_f}{4V}\right)^2\left(\frac{X_q}{X_d - X_q}\right)^2 + \frac{1}{2}}.$$

$$\text{Substituting: } \cos\delta = \frac{-11}{4 \times 6.6}\left(\frac{4.35}{4.35}\right) \pm \sqrt{\frac{121}{26.4^2} + 0.5}$$

$$= -0.4135 \qquad \pm 0.8207$$

$$= 0.4072 = \underline{\cos 66°};\ \sin\delta = 0.914;\ \sin 2\delta = 0.743.$$

(c) The ratio of torques is obtained by substituting the appropriate values of $\sin\delta$ and $\sin 2\delta$ for $\delta = 30°$ *and* $\delta = 66°$ in the torque expression. Some constants cancel.

$$\frac{T_{max}}{T_{rated}} = \frac{11 \times 0.914 + \frac{6.6}{2}(2-1) \times 0.743}{11 \times 0.5 + \frac{6.6}{2}(2-1) \times 0.866} = \underline{1.496}.$$

(d) $$\text{Torque at angle } \delta = \frac{-3}{2\pi \times 50/3} \times \frac{6600}{\sqrt{3} \times 8.7} \times \left[\frac{11\,000}{\sqrt{3}}\sin\delta + \frac{6600}{2 \times \sqrt{3}} \times (2-1)\sin 2\delta\right]$$

$$\frac{dT_e}{d\delta} = \frac{3}{100\pi} \times \frac{6600}{8.7} \times (11\,000\cos 30° + 6600 \times 1 \times \cos 60°)$$

$$= 92\,917\,\text{Nm per electrical radian}$$

$$= 92\,917 \times 6/2 = \underline{2.79 \times 10^5\,\text{Nm/mechanical radian}}.$$

(e) $$\text{Undamped natural frequency} = \sqrt{\frac{\text{stiffness}}{\text{inertia}}} = \sqrt{\frac{2.78 \times 10^5}{8200}} = \underline{5.83\,\text{rad/s}}$$

$$= \frac{1}{2\pi} \times 5.83 = \underline{0.928\,\text{Hz}}.$$

(f) The stored-energy constant is defined as: $$\frac{\text{Stored energy at rated speed}}{\text{Rated voltamperes}} = \frac{\frac{1}{2}J\omega_s^2}{\text{MVA} \times 10^6}$$

$$= \frac{\frac{1}{2} \times 8200 \times (100\pi/3)^2}{5\,000\,000} = \underline{9\,\text{seconds}}.$$

Stepper and switched reluctance (SR) motors[8]

These motors do not offer much possibility for numerical examples within the scope of this text, since they would involve the complexities of electromagnetic design, stability analysis or the special disciplines of power-

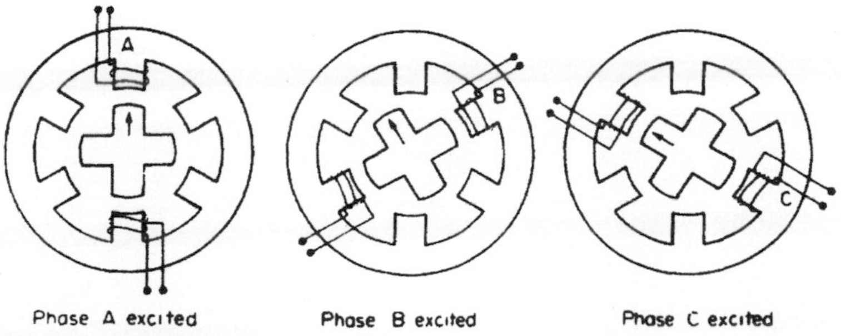

Figure 5.5 *Stepper motor.*

electronic control. However, they are important in certain Electrical Drive applications and a brief enunciation of the operating principle can be made with reference to Figure 5.5. Sequential switching of the 'phases', not necessarily three, brings the rotor in steps to the various stable positions of minimum reluctance – only 12 per revolution in the simplified case shown. Clearly, the convenience of digital control and the precision of angle offer advantages for certain drives, e.g. machine tools and computer printers. The addition of a rotor-position detector to govern the frequency of switching produces, in the SR motor, a rugged, variable-speed drive with controllable characteristics matching d.c. or other rotor-switched machines.

6 Transient behaviour; closed-loop control

So far, nearly all the examples have been concerned with steady-state behaviour. It is important however to introduce the ideas underlying the equally significant behaviour which occurs during the approach to, or the retreat from, a particular steady state. The transient state is a very interesting field for mathematical and computer experts, but it is still possible to study many practical aspects of machine transients without going much beyond the theory of the first-order differential equation. Thermal, electrical and mechanical transients are all partly covered by such simple equations, though clearly these will have to become more complex as more elements and control circuits are included in the system of which the machine may be only the main power unit. Usually the machine has a much larger time-constant than the control and power-electronic time-constants, so has a dominant effect and will be the area of study in this chapter as a simple introduction to the topic. Even for the machine itself, the mechanical time-constant is usually much greater than the electrical time-constants and so the electrical and mechanical responses can often be studied separately. The meaning of this is that the electrical-system changes take place at virtually constant speed and the mechanical changes take place after the electrical system has virtually reached its steady state. This particular problem will be discussed in illustrative examples but of course in the space available, the coverage of the transient state can only be limited and selective.

The manner of response in which a system gets from one state to another is often highly important in drive design, the principal features being speed of response, stability and accuracy in meeting an input demand. The stress placed on components is also important. The speed of response may be represented by time constant, settling time or rise time; stability may be represented by the damping ratio, which is a measure of how many oscillations, if present, exist before these die away; accuracy reflects whether the steady state reached is exactly the required value or if a steady-state error exists. The response is made up of two parts: a steady-state response which remains until conditions change and a transient response which should decay to zero.

6.1 Transient equations

First-order equation

Consider a system which has a variable θ and an equation which can be expressed in the form:

$$\tau \frac{d\theta}{dt} + \theta = \theta_f \qquad (6.1)$$

where τ is a constant known as the 'time-constant' and θ_f is the 'final value' of θ, i.e. the value once steady state ($d\theta/dt = 0$) has been reached. The solution of eqn (6.1) for a suddenly applied step input θ_f at $t = 0$ may be found from differential equation theory or using the Laplace transform as follows.

The Laplace transform of θ is $\theta(p)$, where 'p' is the Laplace operator; in control texts, 's' is normally used but 's' has already been used to represent slip and to denote stator quantities. From Appendix A, the Laplace transform of the constant term θ_f is θ_f/p. The Laplace transform of $d\theta/dt$ is $(p\theta(p) - \theta_0)$, where θ_0 is the initial value of θ when the step change in input is applied. Note that, as this is a linear equation, the change in output due to a particular change in input will always be the same, whatever the initial value.

Substituting the Laplace transformed values,

$$\tau p\theta(p) - \tau\theta_0 + \theta(p) = \theta_f/p$$

$$\therefore \quad \theta(p)\{1 + \tau p\} = \theta_f/p + \tau\theta_0$$

$$\therefore \quad \theta(p) = \frac{\theta_f}{p(1 + \tau p)} + \frac{\tau\theta_0}{1 + \tau p}$$

which can be written as:

$$\theta(p) = \frac{\theta_f(1/\tau)}{p(p + 1/\tau)} + \frac{\theta_0}{p + 1/\tau}$$

which corresponds to:

$$\theta(p) = \frac{\theta_f a}{p(p + a)} + \frac{\theta_0}{p + a}$$

in the table of Laplace transforms in Appendix A. The inverse transform is:

$$\theta = \theta_f(1 - e^{-t/\tau}) + \theta_0 e^{-t/\tau} \tag{6.2a}$$

This can be seen to have two components, a final value θ_f and a transient component

$$\{\theta_0 - \theta_f\}\ e^{-t/\tau}$$

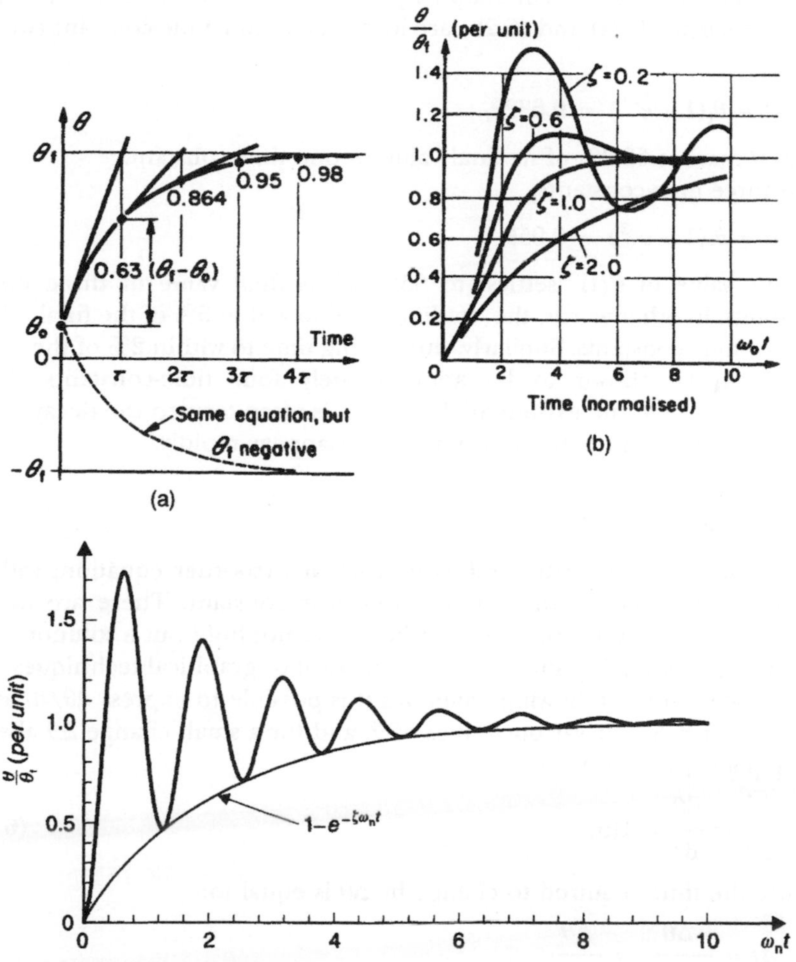

Figure 6.1 *Transient response. (a) 1st-order equation; (b) 2nd-order equation; (c) Exponential decay of sine wave.*

which decays as t increases. The response is exponential, as shown in Figure 6.1a, which also shows the response of the system to a negative step $(-\theta_f)$.

An alternative arrangement of eqn (6.2a) is sometimes useful when θ_0 is not zero. It considers the response as an exponential rise (or fall) of the *change* $(\theta_f - \theta_0)$, superimposed on the initial value θ_0, namely:

$$\theta = \theta_0 + (\theta_f - \theta_0)(1 - e^{-t/\tau}) \tag{6.2b}$$

Either form of solution may be quoted, once the system equation has been organised into the standard form of eqn (6.1) by reducing the coefficient of the variable to unity. For simplicity, consider the case where θ_0 is zero, i.e. where eqns (6.2a) and (6.2b) are identical. In one time-constant (when $t = \tau$),

$$\theta = \theta_f(1 - e^{-1}) = 0.632\theta_f$$

Thus θ reaches 63.2% of its final value in one time-constant.

In three time-constants,

$$\theta = \theta_f(1 - e^{-3}) = 0.95\theta_f$$

i.e. the value of $\theta(t)$ 'settles' to 95% of its final value in three time-constants. In other words, the 'settling time' to within 5% of the final value is three time-constants. Similarly, the settling time to within 2% of the final value can be shown to be approximately four time-constants. The coefficient of t in the exponential, i.e. $1/\tau$, is often termed the decay rate. For a high decay rate, the transient will disappear rapidly.

Non-linear response

The solution just described refers to a linear first-order equation, with τ constant and also the driving function θ_f is constant. There are many practical situations where these conditions do not hold but a solution can still be organised fairly simply using numerical or graphical techniques. As will be seen in the following examples it is possible to express $d\theta/dt$ as a function of θ. So for various values of θ, and for a small change $\Delta\theta$ about these points, we can write:

$$\frac{\Delta\theta}{\Delta t} \simeq \frac{d\theta}{dt} = f(\theta). \tag{6.3}$$

Hence the time required to change by $\Delta\theta$ is equal to:

$$\Delta t = \frac{\Delta\theta}{d\theta/dt} = \frac{\Delta\theta}{f(\theta)}.$$

The method could be applied to the linear equation but of course there is no point in doing this since an analytical solution is available here.

Second-order equation

The next stage of complexity to be described is the situation where there are two energy stores in the system. For the first-order equation there is only one, like the electromagnetic energy in an inductor or the electrostatic energy in a capacitor. If both these elements are in circuit, or any other two such stores, then the second differential coefficient comes into the system equation. Depending on the system parameters, the response can have rather similar characteristics (if heavily damped), or can be quite different in that the response can be oscillatory. The general form of the second-order system differential equation is expressed in terms of the undamped natural frequency of oscillation ω_n and the damping ratio ζ:

$$\frac{1}{\omega E_n^2}\frac{d^2\theta}{dt^2} + \frac{2\zeta}{\omega_n}\frac{d\theta}{dt} + \theta = \theta_f \tag{6.4}$$

The significance of ω_n and ζ will be made clearer later. Taking the Laplace transform,

$$\frac{p^2\theta(p)}{\omega_n^2} + \frac{2\zeta p\theta(p)}{\omega_n} + \theta(p) = \theta_f(p)$$

$$\therefore \frac{\theta(p)}{\theta_f(p)} = \frac{\omega_n^2}{p^2 + 2\zeta\omega_n p + \omega_n^2} \tag{6.5}$$

If $\zeta > 1$, the denominator factorises, e.g.

$$\frac{\theta(p)}{\theta_f(p)} = \frac{\omega_n^2}{(p + a)(p + b)}$$

where $a = \zeta\omega_n + \omega_n\sqrt{(\zeta^2 - 1)}$ and $b = \zeta\omega_n - \omega_n\sqrt{(\zeta^2 - 1)}$. Also, $\omega_n^2 = ab$.

If θ_f is a step input, then:

$$\theta_f(p) = \theta_f/p$$

and $$\theta(p) = \frac{\theta_f ab}{p(p + a)(p + b)}$$

The inverse transform, as obtained from the table of transforms in Appendix A, is:

$$\theta = \theta_f\left\{1 - \frac{be^{-at}}{b - a} + \frac{ae^{-bt}}{b - a}\right\}$$

This solution contains a steady-state portion of value θ_f and a transient portion consisting of exponential terms.

If $\zeta = 1$, the denominator of the right-hand side of eqn (6.5) can be factorised to give:

$$\frac{\theta(p)}{\theta_f(p)} = \frac{\omega_n^2}{p(p+\omega_n)^2}$$

If, again, θ_f is a step input $\theta_f(p) = \theta_f/p$,

$$\theta(p) = \frac{\theta_f\omega_n^2}{p(p+\omega_n)^2}$$

This corresponds to the transform $\dfrac{a^2}{p(p+a)^2}$

(with $a = \omega_n$) in the table of Laplace transforms in Appendix A. The inverse transform is:

$$\theta = \theta_f(1 - e^{-\omega_n t} - \omega_n t e^{-\omega_n t})$$

The transient portion again consists of exponential terms.

If $\zeta < 1$, eqn (6.5) is:

$$\frac{\theta(p)}{\theta_f(p)} = \frac{\omega_n^2}{p^2 + 2\zeta\omega_n p + \omega_n^2}$$

If, again, θ_f is a step input $\theta_f(p) = \theta_f/p$,

$$\theta(p) = \frac{\theta_f\omega_n^2}{p(p^2 + 2\zeta\omega_n p + \omega_n^2)}$$

but the denominator will not factorise into real factors. From the last line of the table of Laplace transforms in Appendix A:

$$\theta = \theta_f\left(1 - \frac{e^{-\zeta\omega_n t}}{\sqrt{1-\zeta^2}}\sin(\omega_n\sqrt{1-\zeta^2}\,t + \cos^{-1}\zeta)\right)$$

The transient portion of this solution consists of a sinusoidal oscillation of frequency $\omega_n\sqrt{(1-\zeta^2)}$, decaying exponentially at a decay rate $\zeta\omega_n$. This means that the sine wave decays within an exponential 'envelope' of time-constant $1/\zeta\omega_n$, as shown in Figure 6.1c. The decaying oscillation will then settle to within approximately 5% of the steady-state value in a time $3/\zeta\omega_n$ seconds.

If the 'damping ratio' ζ is zero, the transient term will be

$$\theta_f \sin(\omega_n t + \pi/2)$$

which is a continuous oscillation at the 'natural frequency' ω_n.

The damping ratio ζ is a measure of the amount of oscillation in the response of the system to a step change in input and its value is adjusted to suit the particular application. For control of shaft speed, current, etc., ζ of 0.4 to 0.6 is usually appropriate. For position control, where 'overshoot' of the final value is normally undesirable, ζ of approximately 1.0 is preferable.

Although systems are in general higher than second order, short time-constants can often be neglected and it is usual to approximate systems to second order if possible, to make use of simple measures of stability (e.g. damping ratio) and speed of response (e.g. $\zeta\omega_n$, which is the rate of decay of the transient).

If a system has two time-constants, one long, one short, the longer time-constant is dominant, as its transient takes longer to decay. In Example 6.12, the shorter time-constant is neglected to simplify the transfer function to second order and allow ζ and ω_n to be estimated.

6.2 Transfer functions

In a linear control system, the 'transfer function' is a convenient way to relate input and output. Although other representations, e.g. state space, are common in control theory, the transfer function remains popular in industry. The transfer function is defined as:

output/input, i.e. output = input × transfer function

where the input, output and transfer function are all expressed in terms of the Laplace operator 'p' instead of time. In the first-order example on p.178, taking $\theta_0 = 0$ for simplicity, and allowing θ_f to be any function of time which has a Laplace transform $\theta_f(p)$,

$$\tau p\theta(p) + \theta(p) = \theta_f(p)$$

$$\therefore \theta(p)\{1 + \tau p\} = \theta_f(p)$$

$$\therefore \frac{\theta(p)}{\theta_f(p)} = \frac{1}{(1 + \tau p)} = \text{transfer function}$$

$$= G(p), \text{ say.}$$

In many systems, the output quantity has to be regulated by modifying the input as the output changes. To achieve this, a measure of the controlled variable $Y(\mathrm{p})$, say, is fed back and subtracted from the input $X(\mathrm{p})$, in order that an excessive output leads to reduction of the input. The control signals X and Y are usually low voltages in analogue systems or numbers in digital systems. Figure 6.2a shows the simple example of closed-loop control of current. The current $I(\mathrm{p})$ is measured by a sensor which produces a voltage output $Y(\mathrm{p})$ proportional to $I(\mathrm{p})$. The difference between a reference input voltage $X(\mathrm{p})$ and $Y(\mathrm{p})$ is an 'error' voltage which indicates an error in current value. The 'error' voltage is fed into an amplifier which produces the voltage $V(\mathrm{p})$. The load has a transfer function $1/(R + \mathrm{p}L)$ which represents a resistance and inductance in series, as in Example 6.1. The 'forward loop' elements (transfer functions of amplifier, load and sensor) are multiplied together to form $G'(\mathrm{p})$ in Figure 6.2b. Note that it is usually more convenient to define the measure of the controlled variable as the output. In Figure 6.2a, current is the controlled variable but its measured value, Y, is defined as the output. The transfer function of a current sensor will normally consist of a numerical constant and perhaps a simple filter.

From Figure 6.2b,

$$Y(\mathrm{p}) = G'(\mathrm{p})\{X(\mathrm{p}) - Y(\mathrm{p})\}$$

$$\therefore \frac{Y(\mathrm{p})}{X(\mathrm{p})} = \frac{G'(\mathrm{p})}{1 + G'(\mathrm{p})}$$

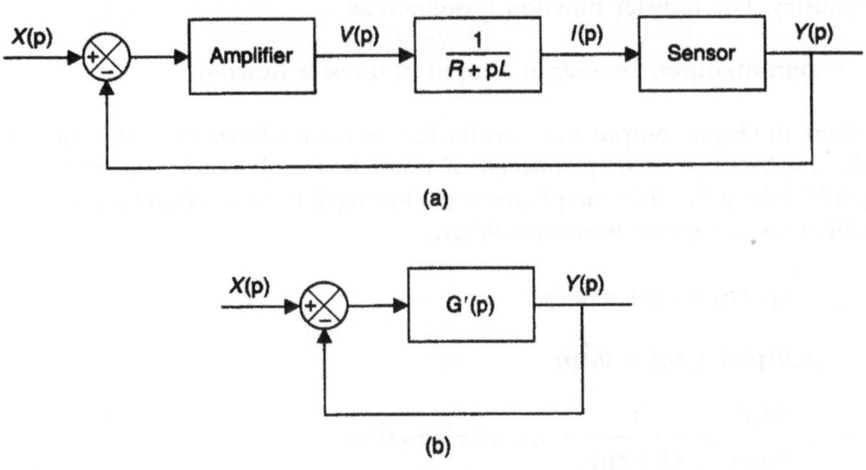

Figure 6.2 *Block diagrams: (a) closed-loop current control; (b) unity-feedback system.*

This is a *closed-loop* transfer function. G′(p) is known as the open-loop transfer function of this closed-loop system. Note that there is no block in the feedback loop of Figure 6.2b. This is known as a 'unity-feedback' system. It is often convenient to redraw block diagrams into this form to obtain the simplest relationship between open- and closed-loop transfer functions.

Let G′(p) be $\dfrac{K}{1+\tau p}$, for example.

$$\therefore \frac{Y(p)}{X(p)} = \frac{G'(p)}{1+G'(p)} = \frac{K/(1+\tau p)}{1+K/(1+\tau p)}$$

$$= \frac{K}{1+\tau p+K} = \frac{K/(K+1)}{1+p\tau/(K+1)}$$

The time-constant has been reduced from τ to $\tau/(K+1)$. Using a closed-loop system will generally reduce the time-constant from its open-loop value. If $X(p)$ is known, e.g. a step change in input, $Y(p)$ can be determined and $Y(t)$ found from a table of Laplace transforms.

For a second-order system, a transfer function can similarly be developed. Eqn (6.5) is the transfer function relating output $\theta(p)$ to an input $\theta_f(p)$. This is the 'standard form' for a second-order transfer function for a closed-loop control system. By comparing the denominator of a transfer function with that in eqn (6.5), ζ and ω_n may be found.

Note that a control system may be used to regulate any physical quantity. In electrical drives, this may be voltage, current, torque, speed, flux, shaft position or winding temperature, for example, or the output of a process of which the drive forms part, e.g. pressure, flow, density, etc.

6.3 Mechanical system

For the electrical engineer, it is often useful to visualise mechanical quantities by their equivalence to electrical variables. For example, current is analogous to force or torque; both are measured by apparatus placed in series with the quantity being measured. Voltage is analogous to speed, as both quantities are measured in relation to a fixed reference. Table 6.1 shows a range of analogies.

Referring to Table 6.1, capacitive energy storage, which is measured in terms of voltage, obeys the linear equation $i = C\,dV/dt$. The translational (moving in a straight line) mechanical equivalent is $F = M\,dv/dt$, i.e. force = mass × acceleration. For a rotating component, mass M (kg) is replaced by inertia J (kgm^2), force F (N) by torque T (Nm) and acceleration dv/dt

Table 6.1 Analogies between physical quantities

	Electrical system	Mechanical-translational system	Mechanical-rotational system
ACROSS (effort) variable	Potential difference v	Linear velocity v	Angular velocity ω
THROUGH (flow) variable	Current i	Force F	Torque T
CAPACITANCE (across-type storage)	Capacitance C $i = C\dfrac{dv}{dt}$	Mass M $F = M\dfrac{dv}{dt}$	Inertia J $T = J\dfrac{d\omega}{dt}$
INDUCTANCE (through-type storage)	Inductance L $v = L\dfrac{di}{dt}$	Translational spring K $v = \dfrac{1}{K}\dfrac{dF}{dt}$	Rotational spring K $\omega = \dfrac{1}{K}\dfrac{dT}{dt}$
RESISTANCE (dissipation)	Resistance R $v = iR$	Frictional damper C $v = \dfrac{1}{C}F$	Torsional damper C $\omega = \dfrac{1}{C}T$

(m/s^2) by angular acceleration $d\omega_m/dt$ (rad/s^2). For energy stored by current in an inductance, the translational mechanical analogue of $V = L\, di/dt$ is $v = (1/K)\, dF/dt$, i.e.

$$dF/dt = Kv$$

Integrating this, where $v = dx/dt$, gives $F = Kx$, which is the equation of a spring, K being the stiffness of the spring in N/m. This form of mechanical energy storage therefore corresponds to potential energy. For a rotational spring, e.g. in a mechanical meter movement,

$$\omega = (1/K)\, dT/dt$$

and integrating, where $\omega = d\theta/dt$ gives $T = K\theta$.

In this case, K is in Nm/rad. Energy dissipation in a linear electrical circuit obeys Ohm's law $V = iR$. In a linear mechanical system, there is a lost force or torque proportional to speed, $F = Cv$ in a damper or $T = C\omega$ in a torsional damper, where C is a constant, its units being N per m/s or Nm per rad/s, respectively.

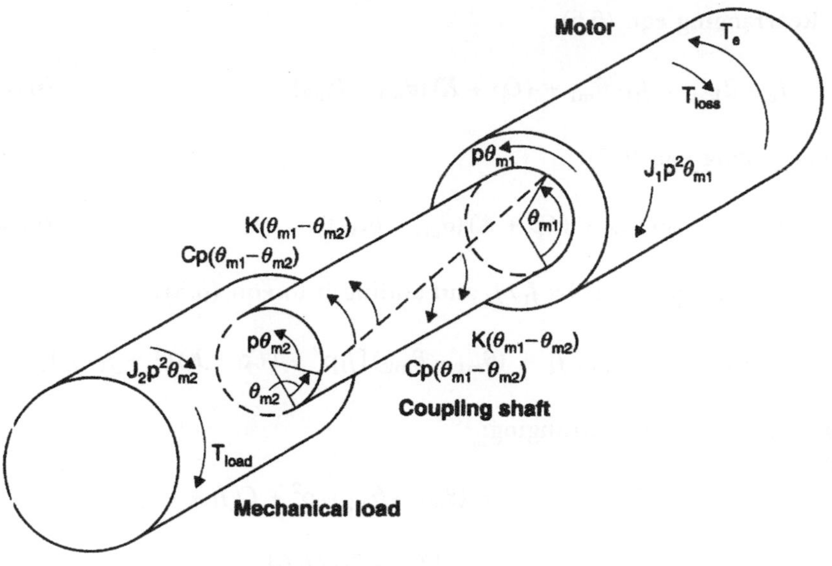

Figure 6.3 *Dynamics of rotating shaft in torsion.*

Figure 6.3 shows a schematic diagram of a motor driving a mechanical load through a coupling shaft. The torque is transmitted because the shaft tends to twist and carry the load round in the same direction. The angle of twist at the motor end θ_{m1} on steady state will be greater than the angle of twist θ_{m2} at the load end, because of the shaft flexibility, or resilience. Within the elastic limit, the torque transmitted is proportional to the difference $(\theta_{m1} - \theta_{m2})$, the resulting angle of shaft twist. In the transient state, there is another (viscous) damping force exerted by the shaft due to rate of change of shaft twist-angle, $p(\theta_{m1} - \theta_{m2})$. The inertia of the motor, J_1 kgm^2 opposes the acceleration $d\omega_m/dt = p^2\theta_{m1}$. The electromagnetic torque T_e is therefore opposed by the loss torque, the inertia torque and the shaft torques giving the following equation, where K is the shaft stiffness (Nm/radian), and C is in Nm per rad/s:

$$T_e = T_{loss} + J_1p^2\theta_{m1} + Cp(\theta_{m1} - \theta_{m2}) + K(\theta_{m1} - \theta_{m2}). \qquad (6.6)$$

At the load end, the shaft torques tending to turn the load in the same direction as the motor are opposed by the load inertia-torque and the load torque itself. Hence:

$$Cp(\theta_{m1} - \theta_{m2}) + K(\theta_{m1} - \theta_{m2}) = J_2p^2\theta_{m2} + T_{load}. \qquad (6.7)$$

Rearranging eqn (6.6):

$$T_e - T_{loss} = J_1 p^2\theta_{m1} + (Cp + K)(\theta_{m1} - \theta_{m2}) \tag{6.6a}$$

Rearranging eqn (6.7):

$$T_{load} = -J_2 p^2\theta_{m2} + (Cp + K)(\theta_{m1} - \theta_{m2}) \tag{6.7a}$$

Multiplying eqn (6.7a) by J_1/J_2 and adding it to eqn (6.6a) gives:

$$T_e - T_{loss} + J_1 T_{load}/J_2 = (\theta_{m1} - \theta_{m2})[J_1 p^2 + (Cp + K)(1 + J_1/J_2)]$$

Dividing by J_1 and rearranging:

$$(T_e - T_{loss})/J_1 + T_{load}/J_2 = (\theta_{m1} - \theta_{m2})[p^2 + C(J_1 + J_2)p/J_1J_2$$

$$+ K(J_1 + J_2)/J_1J_2]$$

Expressing this as a transfer function relating the angle of shaft twist ($\theta_{m1} - \theta_{m2}$) to the torque terms:

$$\frac{(\theta_{m1} - \theta_{m2})}{(T_e - T_{loss})/J_1 + T_{load}/J_2} = \frac{1}{p^2 + C(J_1 + J_2)p/J_1J_2 + K(J_1 + J_2)/J_1J_2}$$

This is in the standard form for a second order transfer function. Natural frequency ω_n and damping ratio ζ may be found by comparing coefficients with the denominator of eqn (6.5). ω_n can be deduced from:

$$\omega_n^2 = \frac{K(J_1 + J_2)}{J_1J_2} \text{ so } \omega_n = \sqrt{\frac{K}{J_2} + \frac{K}{J_1}} \tag{6.8a}$$

The damping ratio is deduced from:

$$\frac{2\zeta}{\omega_n} = \frac{C}{K} \text{ so } \zeta = \frac{C}{2\sqrt{K}}\sqrt{\frac{1}{J_2} + \frac{1}{J_1}} \tag{6.8b}$$

If the damping ratio $\zeta < 1$, some transient oscillation will occur at a frequency

$$\omega = \omega_n\sqrt{1 - \zeta^2} = \omega_n\sqrt{1 - \frac{C^2}{4K}\left(\frac{1}{J_2} + \frac{1}{J_1}\right)} \text{ rad/s.}$$

The shaft acts as a torsional spring (see Table 6.1) in which the spring constant K (shaft stiffness) is given by:

$$K = \frac{\pi G r^4}{2l} \text{ Nm/rad} \tag{6.9}$$

where G is the shear modulus of the shaft material ($\approx 80\,\text{GN/m}^2$ for steel), l is the shaft length (m) and r is the shaft radius (m).

If C is zero or small, such that the oscillation is undamped or underdamped, a torsional damper (Table 6.1) could be added to the shaft. The damped oscillation frequency must have a value well away from any torque-pulsation frequency arising either in the load, e.g. if it is a compressor, or from the motor, if it is supplied through certain types of power-electronic circuit. These factors are of concern for the drives specialist. For present purposes, it will be assumed that the shaft is stiff enough to transmit the torque without twisting. This means that the combined inertia of the whole drive-system can be referred through to the motor shaft as say J kg m^2, by summing the total stored energy in the moving parts, $\Sigma\frac{1}{2}(J_n\omega_n{}^2 + M_m v_m{}^2)$, where the speeds of the elements ω_n and v_m correspond to a particular speed ω_m at the motor shaft (as determined by the gear ratios, see p. 191), and dividing this energy by $\frac{1}{2}\omega_m{}^2$. In a similar way by equating the power transmitted through the various forces and torques at their appropriate speeds, these can be referred to a motor torque giving the same power at the motor shaft running at speed ω_m. Allowing for the transmission, efficiency will increase these referred values.

For the stiff shaft, the last two terms of eqn (6.6) can be replaced by the right-hand side of eqn (6.7) with J_1 plus J_2 combined as J to give:

$$T_e = T_{loss} + J\frac{d\omega_m}{dt} + T_{load} = T_m + \frac{J\,d\omega_m}{dt} \tag{6.10}$$

since $p\theta_{m1} = \omega_m$ and $\theta_{m2} = \theta_{m1}$. On steady state when the speed has settled down, T_e is balanced by the total mechanical torque $T_m = T_{load} + T_{loss}$. The electromagnetic torque T_e is a function of the motor currents or may be expressed as a function of speed through the speed/torque curve. If the $T_m(\omega_m)$ characteristic is of simple form, eqn (6.10) is of first order and easily solved. In any case:

$$J\frac{d\omega_m}{dt} = T_e - T_m \quad \text{so} \quad \frac{d\omega_m}{dt} = \frac{T_e - T_m}{J}.$$

If the torques are some known function of ω_m, the last expression is the f(θ) (where θ here is ω_m) required for the solution of eqn (6.3).

Example of electromechanical system in transient state

Because the d.c. machine equations are relatively simple, this machine provides an easy introduction to the application of the mechanical equations just described. For the present, the armature inductance will be neglected so that the armature current at any instantaneous speed ω_m, and e.m.f. e, with supply voltage V, will be given by $(V-e)/R$ and e itself will be $k_\phi \omega_m$. Hence, during a speed transient, the electromagnetic torque T_e will be:

$$k_\phi i_a = k_\phi \frac{(V - k_\phi \omega_m)}{R} = \frac{k_\phi V}{R} - \frac{k_\phi^2 \omega_m}{R}.$$

This must be balanced against $T_m + J\,d\omega_m/dt$ from eqn (6.10) giving:

$$\frac{k_\phi V}{R} - \frac{k_\phi^2 \omega_m}{R} = T_m + J\frac{d\omega_m}{dt}.$$

To get this into the standard form of eqn (6.1), the coefficient of the variable must be brought to unity which means dividing throughout by k_ϕ^2/R and rearranging:

$$\frac{JR}{k_\phi^2}\frac{d\omega_m}{dt} + \omega_m = \frac{V}{k_\phi} - \frac{RT_m}{k_\phi^2} (= \omega_{mf}). \tag{6.11}$$

The solution given by eqn (6.2) can now be applied for sudden step changes of V, R or T_m. Step changes of flux are not practicable because of the relatively slow field time-constant. It is assumed that T_m is not a function of ω_m; i.e. is constant. Under these circumstances, the speed time-constant is JR/k_ϕ^2. Since this is a function of mechanical, electrical and magnetic parameters it is sometimes referred to as τ_m, the electromechanical time-constant. If T_m was a function of speed, say $T_m = k_2\omega_m$, then the value of τ_m would be $JR/(k_\phi^2 + k_2 R)$. Note that this equation covers all modes of operation with V taking various values, positive, negative and zero, see Section 3.5. It has only one more term than the general speed/torque equation for a d.c. machine. When steady-state speed has been reached, $d\omega_m/dt = 0$ and the equation is then identical with eqn (3.8).

Invoking the solution given by eqn (6.2b), we can write:

$$\omega_m = \omega_{m0} + (\omega_{mf} - \omega_{m0})(1 - e^{-t/\tau_m})$$

and Figure 6.4 shows three transients:

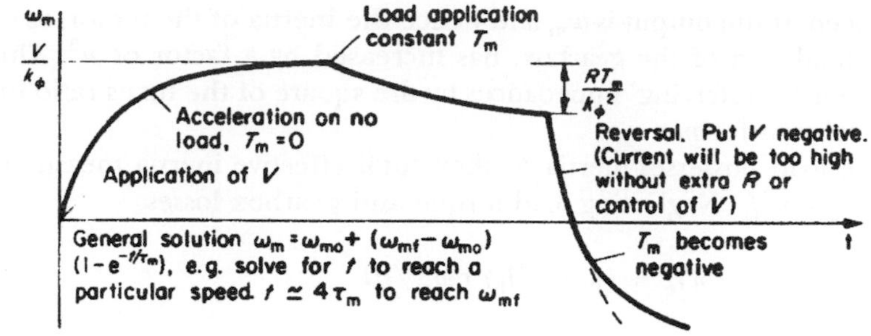

Figure 6.4 *Electromechanical transient on d.c. machine.*

(a) Acceleration on no load, V change from 0 to V.
(b) Sudden application of load T_m after reaching no-load speed.
(c) Reversal of V, usually with extra limiting resistance. Note that beyond zero speed, T_m, if a passive load, would also become negative as well as V and the speed response would become slower. τ_m governs all response times, with the appropriate value of resistance inserted.

Gearing

Although many electrical drives involve direct coupling to the mechanical load, especially in larger-power units, it is often necessary to interpose gearing, usually to perform a speed reduction. This can occur on certain small-power units and when a fairly high torque at low speed is required, though the power is small. As the physical size (and price) of a motor depends on its torque (power/speed), it is advantageous to have a small, low-torque, high-speed motor. (Very small motors can be built to run up to 5×10^5 rev/min.) A gearbox is then used as the mechanical equivalent of an electrical transformer; the power on each side is torque × angular speed, neglecting loss.

Let the gear ratio be n:1 and the output shaft speed be ω_m. The motor speed is therefore $n\omega_m$. If the motor torque is T, the torque on the output shaft (neglecting losses) will be nT.

If we measure the kinetic energy of the rotor, it will be: kinetic energy of motor = $\frac{1}{2}$ × inertia × (angular speed)^2

$$\text{K.E.} = \tfrac{1}{2} J_1 (n\omega_m)^2$$

By conservation of energy, this could equally well be measured at the output shaft side of the gearbox:

$$\text{K.E.} = \tfrac{1}{2}(n^2 J_1)\omega_m^2$$

The speed at the output is ω_m and hence the inertia of the motor, as seen at the load end of the gearbox, has increased by a factor of n^2. This is analogous to 'referring' impedances by the square of the turns ratio in an electrical transformer.

If the load possesses inertia J_2, then total effective inertia measured at load $= J_2 + n^2J_1$. Neglecting load torque and gearbox losses,

$$nT_e = (J_2 + n^2J_1)\, d\omega_m/dt$$

$$\therefore \text{Motor torque} = T_e = (J_2/n + nJ_1)\, d\omega_m/dt$$

In a robot, for example, a motor for operation of one axis has to be carried around on another axis. Hence the motor should be as small and light as possible. To maximise the ratio of acceleration $d\omega_m/dt$ to torque T_e available from the motor, we need to minimise $(J_2/n + nJ_1)$:

$$\therefore d/dn(J_2/n + nJ_1) = -J_2/n^2 + J_1 = 0 \text{ for minimum.}$$

$$\therefore n = \sqrt{(J_2/J_1)} = \text{optimum gear ratio.}$$

6.4 Transfer function for a drive with linear load

The transfer function derived here is for a d.c. machine with constant flux, stiff shaft and no gearing, the load being first represented by an inertia and torsional (viscous) damping.

At speed ω_m, generated e.m.f. $= e = k_\phi \omega_m$, where k_ϕ is a constant, and torque $T_e = k_\phi i$,

$$V = e + i_a R + L\, di_a/dt$$

Using the Laplace transform, p = d/dt, and considering that quantities have zero initial value,

$$V(p) = e(p) + Ri_a(p) + pLi_a(p) = e(p) + i_a(p)[R + pL] \qquad (6.12)$$

$$i_a(p) = \frac{V(p) - e(p)}{R + pL} = \frac{V(p) - k_\phi \omega_m(p)}{R + pL}$$

Consider the simple case in which the mechanical load consists only of a damping torque proportional to speed ($T_m = k_2\omega_m$). This allows a transfer function representing the motor impedance to be developed. Linearisation of load characteristics can be carried out using eqn (6.3).

Acceleration torque is acceleration multiplied by inertia (as for force = mass × acceleration)

$$T_e = J\,d\omega_m/dt + T_m = J\,d\omega_m/dt + k_2\omega_m$$

This is a first-order equation, similar to eqn (6.1). The 'mechanical time-constant' is J/k_2. Taking the Laplace transform,

$$T_e(p) = Jp\omega_m(p) + k_2\omega_m(p) = \omega_m(p)[Jp + k_2] = k_\phi i_a(p)$$

$$\text{Hence, } \omega_m(p) = \frac{k_\phi i_a(p)}{Jp + k_2} \text{ and } e(p) = \frac{k_\phi^2 i_a(p)}{Jp + k_2}$$

From eqn (6.12),

$$i_a(p)[R + pL] = V(p) - e(p) = V(p) - \frac{k_\phi^2 i_a(p)}{Jp + k_2}$$

$$\therefore i_a(p)\left[R + pL + \frac{k_\phi^2}{Jp + k_2}\right] = V(p)$$

The transfer function i_a/V within the drive system is:

$$\frac{i_a(p)}{V(p)} = \frac{1}{R + pL + k_\phi^2/(Jp + k_2)}$$

$$= \frac{Jp + k_2}{(R + pL)(Jp + k_2) + k_\phi^2}$$

$$= \frac{Jp + k_2}{p^2 LJ + p(k_2 L + JR) + (k_\phi^2 + Rk_2)} \qquad (6.13)$$

$$= \frac{k_2(1 + pJ/k_2)/(k_\phi^2 + Rk_2)}{p^2(L/R)JR/(k_\phi^2 + Rk_2) + p(k_2 L + JR)/(k_\phi^2 + Rk_2) + 1}$$

Note the presence of the time-constant $JR/(k_\phi^2 + Rk_2)$ referred to in the discussion following eqn (6.11). L/R is the armature (electrical) time-constant. If numerical values are substituted, the denominator can be factorised to give a transfer function:

$$\frac{i_a(p)}{V(p)} = \frac{k'(1 + p\tau_1)}{(1 + p\tau_2)(1 + p\tau_3)}$$

where k' is a constant $(k_2/(k_\phi^2 + Rk_2))$, τ_1 is the mechanical time-constant J/k_2 and τ_2 and τ_3 are time-constants which may be found by factorising the denominator, once numerical values have been substituted.

$T_e(p)/V(p)$ can be found by multiplying the transfer function in eqn (6.13) by k_ϕ, i.e.

$$\frac{T_e(p)}{V(p)} = \frac{k_\phi(Jp + k_2)}{p^2LJ + p(k_2L + JR) + k_\phi^2 + Rk_2}$$

Also, since $T_e(p) = \omega_m(p)[Jp + k_2]$, $\dfrac{\omega_m(p)}{T_e(p)} = \dfrac{1}{Jp + k_2}$

$$\frac{\omega_m(p)}{V(p)} = \frac{T_e(p)/V(p)}{Jp + k_2} = \frac{k_\phi}{p^2LJ + p(k_2L + JR) + k_\phi^2 + Rk_2} \quad (6.14)$$

Returning to eqn (6.13); if there is no torsional damping, i.e. $k_2 = 0$, so the mechanical load is zero. Putting $k_2 = 0$ in eqn (6.14):

$$\frac{\omega_m(p)}{V(p)} = \frac{k_\phi}{p^2LJ + pJR + k_\phi^2}$$

$$= \frac{1/k_\phi}{p^2LJ/k_\phi^2 + pJR/k_\phi^2 + 1}$$

$$= \frac{1/k_\phi}{p^2LJR/Rk_\phi^2 + pJR/k_\phi^2 + 1}$$

$$= \frac{1/k_\phi}{p^2\tau_m\tau_e + p\tau_m + 1} \quad (6.15)$$

where τ_e is the electrical time-constant L/R and τ_m is the 'electro-mechanical time-constant' JR/k_ϕ^2 mentioned earlier.

If a load torque T_m, is subtracted from T_e as shown in Figure 6.5, the presence of T_m will affect ω_m, which will, in turn, affect T_e. The effect on speed may be found by superposition, as this is a linear system, i.e. by finding the contribution to $\omega_m(p)$ due to T_m when the input voltage V is

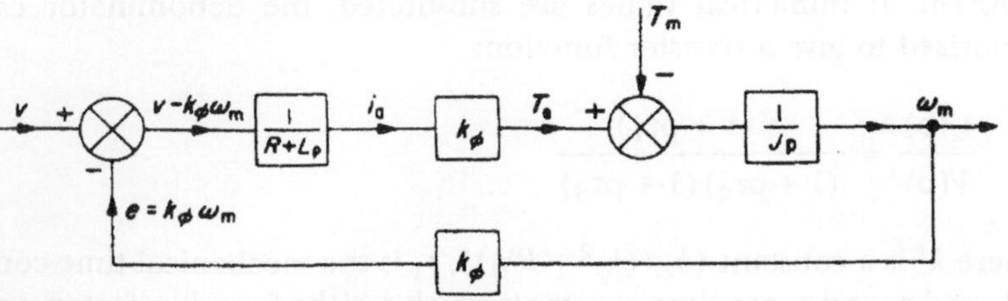

Figure 6.5 *System block diagram of d.c. motor.*

zero. With $V = 0$ in Figure 6.5, taking the product of the 'boxes' round the loop,

$$T_e(p) = \frac{-k_\phi^2 \omega_m(p)}{R + Lp}$$

and $\omega_m(p) = \{T_e(p) - T_m(p)\}/Jp$

Substituting for $T_e(p)$:

$$\omega_m(p) = \frac{-k_\phi^2 \omega_m(p)}{Jp(R + Lp)} - \frac{T_m(p)}{Jp}$$

$$\therefore \quad \omega_m(p)\{Jp(R + Lp) + k_\phi^2\} = -T_m(p)(R + Lp)$$

$$\therefore \quad \omega_m(p) = \frac{-T_m(p)(R + Lp)}{Jp(R + Lp) + k_\phi^2}$$

$$= \frac{-T_m(p)(R + Lp)}{k_\phi^2(JLp^2/k_\phi^2 + JRp/k_\phi^2 + 1)}$$

$$= \frac{-T_m(p)(1 + p\tau_e)R}{k_\phi^2(p^2\tau_m\tau_e + p\tau_m + 1)}$$

The resulting speed in the presence of load torque T_m is then, from eqn (6.15):

$$\omega_m(p) = \frac{1/k_\phi}{p^2\tau_m\tau_e + p\tau_m + 1}\left\{V(p) - \frac{T_m(p)(1 + p\tau_e)R}{k_\phi}\right\}$$

If the load torque is constant (k_1), $T_m(p) = k_1/p$. When steady state is reached, the term in brackets reduces to the steady-state e.m.f., since $T_m/k_\phi = I_a$ and p is zero. On multiplying by $1/k_\phi$, it becomes the final steady-state speed E/k_ϕ.

If the load torque is a linear function of ω_m ($k_2\omega_m$), then:

$$T_m(p) = k_2\omega_m(p)$$

and the result is the same as in eqn (6.14).

If the load torque is a quadratic function involving ω_m^2, the above equations can only be used if the torque is linearised at the operating value, e.g. as in Example 6.14.

By comparing the denominator term $(p^2\tau_m\tau_e + p\tau_m + 1)$ with eqn (6.5), it can be shown that the system will be oscillatory ($\zeta < 1$) if $\tau_m^2 - 4\tau_m\tau_e < 1$, i.e. if

$\tau_m < 4\tau_e$. In this case, the undamped natural frequency is $\omega_n = \sqrt{(1/(\tau_m\tau_e))}$, the damping ratio ζ is $\tau_m/2\omega_m = \frac{1}{2}\sqrt{(\tau_m/\tau_e)}$ and the actual oscillation frequency is:

$$\omega = \omega_n\sqrt{(1-\zeta^2)} = \sqrt{(1/\tau_m\tau_e)}\sqrt{(1-\tfrac{1}{4}(\tau_m/\tau_e))}$$

$$= \sqrt{\frac{1}{\tau_m\tau_e} - \frac{1}{4\tau_e^2}}$$

6.5 Linear control system analysis methods

The principal requirements for any control system are speed (e.g. decay rate, settling time), stability (e.g. damping ratio) and accuracy (which may be determined from the final value of the error signal). Various graphical methods (root loci, Bode plot, Nyquist plot, Nichols chart) are available to obtain information on these from the transfer function of a linear system and can be found in control engineering texts. Similar results may be obtained from other representations such as state-space.

Although the input signal can be of any form, two types of test signal are frequently used in control system analysis, as described below.

Step inputs or impulses are used in time response. Although a realistic input will consist of several components, e.g. step, ramp, parabola, dead time, etc., an idealised step or impulse is used as a test signal for analysis. The equations for first- and second-order systems and the examples in this chapter illustrate exponential and damped sinusoidal responses when a step change is applied to the input. In systems with higher-order transfer functions, it is not feasible to solve the differential equations to plot the response but equivalent information can be found graphically from a root-locus plot, for example.

A sinusoidally varying input over a range of frequency is used in frequency-response analysis. Note that these signals represent sinusoidal variations of the control signals for test purposes only and do not represent the frequency of a.c. applied for power input. If the control reference input of a linear system is subjected to a sinusoidal perturbation, the system's output will be a sine wave of the same frequency ω but altered in magnitude and phase angle. As the frequency of the input wave is altered, the magnitude and phase of the output change and a plot of the variation provides useful information on the system. p is replaced in the transfer function G(p) by $j\omega$. The most popular frequency response plot is the Bode plot. This consists of two graphs, one for magnitude (plotted in dB) and one for phase shift, both plotted against frequency on a logarithmic scale.

For analysis and design work, the Bode plot of the open-loop transfer function G(jω) is normally used. Assuming unity feedback, the closed-loop transfer function is

$$\frac{G(j\omega)}{1 + G(j\omega)}.$$

If $G(j\omega) = -1$, the closed-loop transfer function becomes indeterminate, i.e. will have no steady-state output for a finite input; in other words, the system has lost stability. To determine the degree of stability for a system, the closeness of G(jω) to −1 is assessed. −1 can be represented as $|1|\,\underline{/180°}$ and the 'phase margin' is defined as the margin by which the phase shift introduced $\underline{/G(j\omega)}$ is less than 180°, at the frequency at which $|G(j\omega)| = 1$. For second-order system phase-margins up to 60°, the phase margin (in degrees) is approximately one hundred times the damping ratio ζ. A damping ratio of one represents a phase margin of around 75° for a second-order system. The 'gain margin' is defined as the margin by which $|G(j\omega)|$ is below 1 at the frequency at which $\underline{/G(j\omega)} = 180°$. The bandwidth, which is the frequency at which a system's closed-loop frequency response passes through −3 dB, is often used as a measure of the speed of a system.

PI and PID controllers

In many cases, assembly of standard units creates a control system which does not meet its specification in terms of speed, stability and accuracy. An additional controller may be added in any control loop to modify the response. Usually, the controller acts on the error signal before it is converted to a higher voltage or other physical quantity. The most popular general-purpose controllers are the proportional + integral (PI) and proportional + integral + derivative (PID) controllers.

$$\text{PID: } K(1 + 1/p\tau_i + p\tau_d) = \frac{K(1 + p\tau_i + p^2\tau_i\tau_d)}{p\tau_i}$$

$$\text{PI: } K(1 + 1/p\tau_i) = \frac{K(1 + p\tau_i)}{p\tau_i}$$

Figure 6.6 shows the three terms of the PID controller: the first is proportional to the 'error' signal, which is the difference between reference (demanded) input and feedback of the measured output; the second and third are the integral and derivative of the 'error' signal respectively.

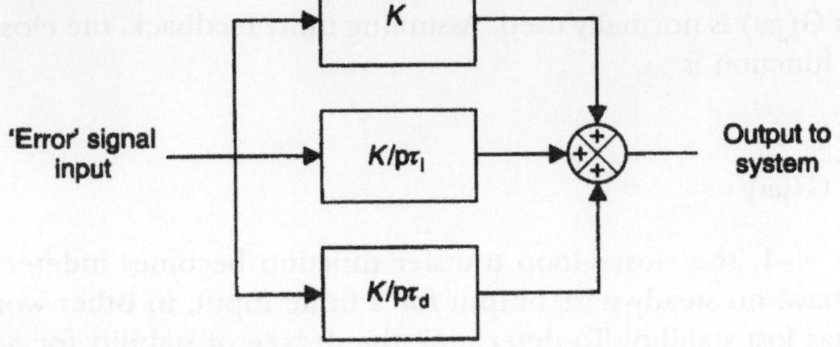

Figure 6.6 *PID controller.*

The addition of the integral of the error is particularly useful in a drive system, since it removes steady-state error. If a speed control system had reached its demanded speed, the speed error would be zero. If the motor voltage was proportional to the speed error, however, there would then be no voltage, so there has to be an error to provide voltage to keep the motor running. An integrator will give a constant output when its input is zero ($\int 0\ dt$ = constant) to maintain voltage when the error is zero. If the input changes, an error voltage will exist and the integrator will charge or discharge and steady state will eventually be reached with zero error and an altered integrator output voltage.

The derivative term provides a negative term when the error is rapidly decreasing and thus helps to prevent overshoot of the final value. In most applications, the derivative term is unnecessary.

Design methods for PI and PID controllers can be found in control engineering textbooks. A well-established 'rule of thumb' for initial selection of K, τ_i and τ_d for systems of higher than second order is as follows:

(i) With no integral or derivative action ($\tau_i = \infty$, $\tau_d = 0$), increase K to K_m, at which gain the system *just* oscillates continually at f Hz.

(ii) Set $K = 0.6K_m$, $\tau_i = \frac{1}{2}f$, $\tau_d = 8f$

For PI action only, set $K = 0.45K_m$, $\tau_i = 1/1.2f$.

Many industrial motor controllers of the types discussed in Chapter 7 incorporate a closed-loop speed control option including user-settable PI or PID parameters.

Example 6.1

A machine winding of resistance 0.5 Ω and inductance 0.25 H is subjected to a 125 Hz square wave of 10 volts amplitude and equal on and off periods. Calculate the maximum current which will be reached in the winding, assuming that a current path exists even when the applied voltage is zero, e.g. if a 'freewheel' path is provided, as in Figure 7.3.

Let the current be i and assume that voltage V is applied at zero time. The circuit equation can be expressed as

$$\frac{L\,\mathrm{d}i}{\mathrm{d}t} + Ri = V$$

$$\therefore \frac{L\,\mathrm{d}i}{R\,\mathrm{d}t} + i = V/R$$

$$\therefore \frac{\tau\,\mathrm{d}i}{\mathrm{d}t} + i = I_f,$$

where τ (= L/R) is the time-constant, as defined in Section 6.1, and I_f is the final steady-state value of i.

When the step change in voltage, say, is applied, the solution follows from eqn (6.2a).

$$i = I_f(1 - e^{-t/\tau}) + i_0 e^{-t/\tau}$$

The time-constant τ is $L/R = 0.5$ and the final current I_f is $V/R = 20$ A. The circuit is of first order and, after a time interval, it will have settled to a steady variation between fixed maximum and minimum current levels as the voltage is switched on and off. Let the initial current when the voltage is applied be i_1.

$$i = 20(1 - e^{-2t}) + i_1 e^{-2t}$$

i_1 will be the 'trough' of the current waveshape. At 125 Hz, the ON and OFF periods of voltage each last for 0.004 seconds. If the current rises exponentially to reach i_2 at the end of the ON period,

$$i_2 = 20(1 - e^{-0.008}) + i_1 e^{-0.008}$$

$$= 0.159 + 0.992 i_1$$

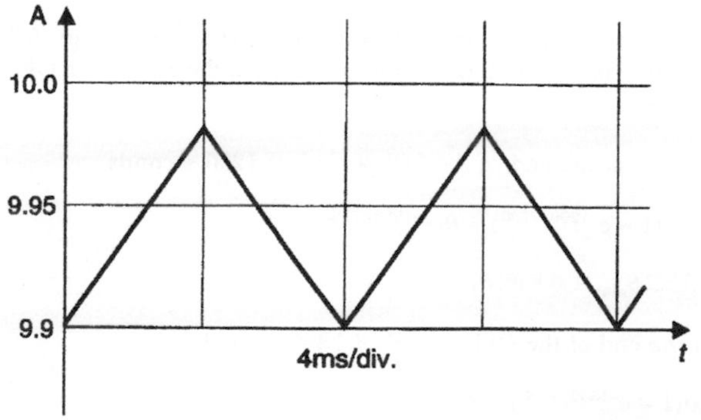

Figure E.6.1

During the OFF period, $I_f = 0$, as the current would eventually reach zero if the voltage was not reapplied.

$$\therefore i = i_2 e^{-2t}$$

After 0.004 seconds, the current returns to i_1.

$$\therefore i_1 = i_2 e^{-0.008} = 0.992 i_2$$

$$\therefore i_2 = 0.159 + (0.992)^2 i_2$$

$$\therefore i_2 = \underline{9.98\,\text{A}} \text{ and } i_1 = 9.9\,\text{A}$$

The graph of current against time for steady variation of current is shown in Figure E6.1.

Example 6.2

The temperature rise $\theta(t)$ of a particular motor can be assumed to follow a first-order differential equation, as in Example 6.1. The derivation of the first-order transient equation corresponding to eqn (6.1), from the relevant physical variables, is the subject of Tutorial Example T6.1, p. 351.

The time-constant is 2 hours, i.e.

$$7200 \frac{d\theta(t)}{dt} + \theta(t) = \theta_{max}$$

where θ_{max} is the steady-state temperature which would be reached if the machine runs continuously.

The machine operates on a duty cycle in which it is clutched to its load for 20 minutes and then declutched to run on no load for 30 minutes. This cycle is repeated continually. The steady temperature rise when running on no load continuously is 10°C and, when operating on the above duty cycle, the maximum temperature rise at the end of an ON period is 50°C. In the event of a timing failure, a thermostat is set at 60°C and shuts down the drive. Calculate:

(a) the minimum temperature rise above ambient when operating the above duty cycle;
(b) the maximum temperature rise if both the timing circuit – which sets the ON and OFF periods – and the thermostat fail to protect the system.

This example can be tackled in a similar way to Example 6.1, using eqn (6.2a) directly. The minimum temperature is 10°C and θ_2, the temperature at the end of the ON period, is 50°C, both above ambient.

(a) From eqn (6.2a), at the end of the ON period of 1200 seconds,

$$\theta_2 = \theta_{max}(1 - e^{-1200/7200}) + \theta_1 e^{-1200/7200}$$

$$\therefore \quad 50 = 0.1535\theta_{max} + 0.8465\theta_1$$

Similarly, at the end of the OFF period of 1800 seconds,

$$\theta_1 = 10(1 - e^{-1800/7200}) + 50e^{-1800/7200}$$

$$= \underline{41.15°\text{C}}$$

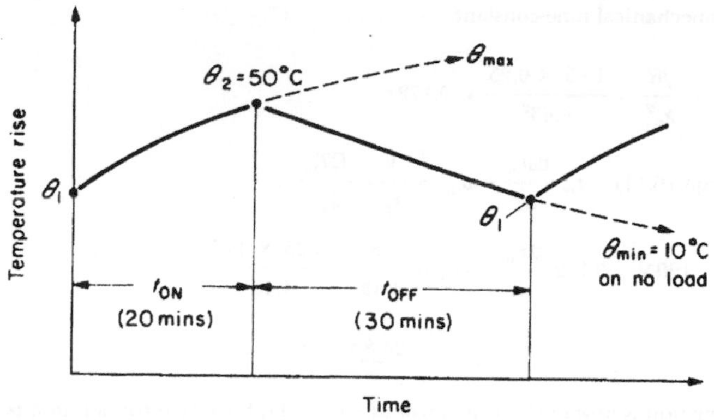

Figure E.6.2

(b) Substituting in the previous equation,

$$50 = 0.1535\theta_{max} + 0.8465 \times 41.15$$

$$\therefore \theta_{max} = \underline{98.8°C}$$

The graph of temperature vs. time is shown in Figure E.6.2.

Example 6.3

The machine of Examples 3.2 and 3.3 is to run as a motor with 220 V applied to the shunt field winding and with the series winding unconnected. The total coupled inertia is 13.5 kgm^2 and the motor runs against a constant, total mechanical torque of 177 Nm. Armature inductance can be neglected.

(a) Calculate the steady-state speed ω_{mo} rad/s when the armature voltage controller is adjusted to 100 V.

(b) The controller output is now increased in a step from 100 to 120 V. Calculate the electromechanical time constant τ_m and find an expression for the speed transient. What are the initial and the final (steady-state) armature currents and the final speed?

(c) Assuming the mechanical load had instead been a damping torque ($k_2\omega_m$) of the same value (177 Nm) at the speed ω_{mo}, determine k_2 and repeat the calculation of part (b) with this new condition.

(a) Referring to Examples 3.2 and 3.3, 220 V across the shunt field winding would give 2 A field current and a value of k_ϕ = 4.43 Nm/A. Armature resistance R_a = 0.25 Ω. Armature current for 177 Nm = T_e/k_ϕ = 177/4.43 = 39.95 A

$$\text{Speed at 100 V} = \frac{E}{k_\phi} = \omega_{mo}$$

$$= \frac{100 - 0.25 \times 39.95}{4.43} = \underline{20.32 \text{ rad/s}}$$

(b) Electromechanical time-constant

$$\tau_m = \frac{JR}{k_\phi^2} = \frac{13.5 \times 0.25}{4.43^2} = 0.172\,\text{s}$$

$$\text{From eqn (6.11)}, \quad \tau_m \frac{d\omega_m}{dt} + \omega_m = \frac{V}{k_\phi} - \frac{RT_m}{k_\phi^2}$$

$$\text{For } V = 120\,\text{V}, \quad 0.172 \frac{d\omega_m}{dt} + \omega_m = \frac{120}{4.43} - \frac{0.25 \times 177}{4.43^2}$$

$$= \underline{24.83\,\text{rad/s}} = \omega_{mf}$$

The equation is now in the same form as eqn (6.1), for which the solution is eqn (6.2a), i.e.

$$\omega_m = \omega_{mf}(1 - e^{-t/\tau_m}) + \omega_{mo}e^{-t/\tau_m}$$

$$= \underline{24.83(1 - e^{-t/0.172}) + 20.32e^{-t/0.172}}$$

or alternatively in the form of eqn (6.2b):

$$\omega_m = 20.32 + 4.51(1 - e^{-t/0.172})$$

Initial current, neglecting inductance

$$= \frac{V - k_\phi\omega_{mo}}{R} = \frac{120 - 4.43 \times 20.32}{0.25} = \underline{120\,\text{A}}$$

This value is three times the rated current and semiconductor equipment, which has a short thermal time constant, would have to be rated to withstand the overload. However, circuit inductance would slow down the rise of current and lower the peak value slightly; see Example 6.10.

$$\text{Final steady-state current} = \frac{120 - 4.43 \times 24.83}{0.25} = \underline{40\,\text{A}}$$

which is the same as at 100 V (neglecting calculator round-off errors), since it is a constant-torque load.

(c) For a damping load-torque, $T_m = k_2\omega_{mo}$,

$$\text{from which} \quad k_2 = \frac{177}{20.32} = \underline{8.7\,\text{Nm/rad}}$$

By substituting $k_2\omega_m$ for T_m in eqn (6.11),

$$\frac{JR}{k_\phi^2}\frac{d\omega_m}{dt} + \omega_m = \frac{V}{k_\phi} - \frac{Rk_2\omega_m}{k_\phi^2}$$

Hence,

$$\frac{JR}{k_\phi^2}\frac{d\omega_m}{dt} + \omega_m\left(1 + \frac{Rk_2}{k_\phi^2}\right) = \frac{V}{k_\phi}$$

Referring to eqn (6.1), the time-constant is the coefficient of $d\omega_m/dt$ when the coefficient of ω_m is 1. In this case, the time-constant is:

$$\frac{JR}{k_\phi^2}\,\frac{1}{(1 + Rk_2/k_\phi^2)}, \quad \text{i.e. } \tau = JR/(k_\phi^2 + Rk_2)$$

$$\therefore \quad \tau = 13.5 \times 0.25/\{(4.43)^2 + (0.25 \times 8.7)\} = 0.155 \text{ seconds}$$

Also, when the steady-state speed ω_{mf} is reached ($d\omega_m/dt = 0$),

$$\omega_{mf}\left(1 + \frac{Rk_2}{k_\phi^2}\right) = \frac{V}{k_\phi}$$

$$\therefore \quad \omega_{mf}\left(1 + \frac{0.25 \times 8.7}{(4.43)^2}\right) = \frac{120}{4.43}$$

$$\therefore \quad \omega_{mf} = 24.4 \text{ rad/s}$$

As above,

$$\omega_m = \omega_{mf}(1 - e^{-t/\tau_m}) + \omega_{mo}e^{-t/\tau_m}$$

$$= 24.4(1 - e^{-t/0.155}) + 20.32e^{-t/0.155}$$

The final steady-state speed is 24.4 rad/s.

The initial current would be the same as in part (a) but the final current would be

$$T_m/k_\phi = k_2\omega_m/k_\phi = 8.7 \times 24.4/4.43 = 47.9 \text{ A.}$$

Compared with part (b), the torque has increased and the speed is reduced.

In the discussion of eqn (6.13), the time-constant of the mechanical components (J/k_2) was defined. It would generally be expected to be the longest time-constant and dominate the response of the drive. In this example, $J/k_2 = 13.5/8.7 = 1.55$ seconds. However, the time-constant found above is a tenth of this value. In discussion of transfer functions, it was shown that a closed-loop system generally has a smaller time-constant than its open-loop equivalent. An electric motor is effectively a closed-loop system, as a reduction in speed caused by increase in load leads to a reduction in e.m.f. which, in turn, produces an increase in current and hence an increase in torque to compensate for the speed drop. This can be seen very clearly by considering the block diagram of the d.c. motor in Figure 6.5.

Example 6.4

A flywheel is added on to the shaft of the motor for which particulars are given in Example 6.3. With the field again fully excited from 220 V, the armature is then connected to the same

supply but through a limiting resistor which will set the starting current at twice the full-load value – 2 × 40 A. The constant load-torque is coupled during starting. The speed reaches 200 rev/min in 10 seconds. Estimate the additional inertia contributed by the flywheel.

$$\text{Total armature circuit resistance} = \frac{220}{80} = 2.75\,\Omega$$

From eqn (6.11):

$$\tau_m \frac{d\omega_m}{dt} + \omega_m = \frac{V}{k_\phi} - \frac{RT_m}{k_\phi^2} = \frac{220}{4.43} - \frac{2.75 \times 177}{4.43^2}$$

$$= 24.86\,\text{rad/s}$$

The initial speed is zero, so the solution of the equation, from eqn (6.2a), gives:

$$\omega_m = 24.86\,(1 - e^{-t/\tau_m})$$

Speed reaches 200 rev/min = 20.94 rad/s in 10 seconds.

Substituting: $20.94 = 24.86\,(1 - e^{-10/\tau_m})$

from which: $e^{10/\tau_m} = \dfrac{1}{1 - 20.94/24.86}$

taking logarithms:

$$10/\tau_m = 1.85, \text{ giving } \tau_m = 5.4 = JR/k_\phi^2 \text{ seconds}$$

Hence total inertia $J = 5.4 \times 4.43^2/2.75 = 38.5\,\text{kgm}^2$

The flywheel inertia is therefore $38.5 - 13.5 = \underline{25\,\text{kgm}^2}$

Note that the electromechanical time-constant is much larger than in Example 6.3 because of the higher inertia and circuit resistance.

Retardation tests

The above example represents one method of measuring the inertia by a transient test, applying a step voltage through a suitable limiting resistor. An alternative would be to take a speed/time curve as the machine slows down under the action of a known torque, e.g. the mechanical loss. From eqn (6.10) $J = -T_m/(d\omega_m/dt)$ since T_e would be zero. This could be checked at various speeds during retardation, allowing for non-linearity if present in both numerator and denominator. This retardation test is sometimes used, with known inertia, to determine the (unknown) losses.

Example 6.5

A 500-V d.c. series motor drives a fan, the total mechanical load torque being given by the expression: $T_m = 10 + (\omega_m/4.2)^2$ Nm. An external 7.5 Ω resistor is added at starting to limit the current when full voltage is applied, and the motor is allowed to run up to the balancing speed corresponding to this circuit condition. The resistor is then cut out and again the speed is allowed to rise to the new balance condition in this single-step starting procedure.

(a) Calculate the starting current.
(b) Calculate the two balancing speeds, noting that the machine resistance itself is 0.8 Ω.
(c) Estimate the currents at the two balancing speeds, and on changeover.
(d) Estimate the time to accelerate from 0 to 100 rev/min if total inertia is 14.5 kgm².

A magnetisation curve at 550 rev/min gave the following information:

Field current I_f	20	30	40	50	60	70	A
Generated e.m.f. (E_{test})	309	406	468	509	534	560	V

Calculation of ω_m/T_e curves proceeds in a similar manner to Examples 3.16 and 3.17.

$k_\phi = E_{test}/(550 \times 2\pi/60)$	5.36	7.05	8.12	8.84	9.27	9.72	Nm/A
$T_e = k_\phi I_a = k_\phi I_f$	107.2	211.5	324.8	442	556.2	680.4	Nm
For high resis. $E = 500 - (0.8 + 7.5)I_f$	334	251	168	85	2	−81	V
$\omega_m = E/k_\phi$	62.3	35.6	20.7	9.6	0.2	−8.3	rad/s
For low resis. $E = 500 - 0.8I_f$	484	476	468	460	452	444	V
$\omega_m = E/k_\phi$	90.3	67.5	57.6	52	48.8	45.6	rad/s
T_m (use high resis. ω_m) $10 + \left(\dfrac{\omega_m}{4.2}\right)^2$	230	81.8	34.3	15.2	10		

(a) Starting current, ($\omega_m = 0$) $= 500/(0.8 + 7.5) =$ 60.2 A.

(b) From the above tabular calculations the two ω_m/T_e curves and the ω_m/T_m curve are plotted on the figure. The two balancing speeds at the intersections are 50 and 64.5 rad/s = 477 and 616 rev/min.

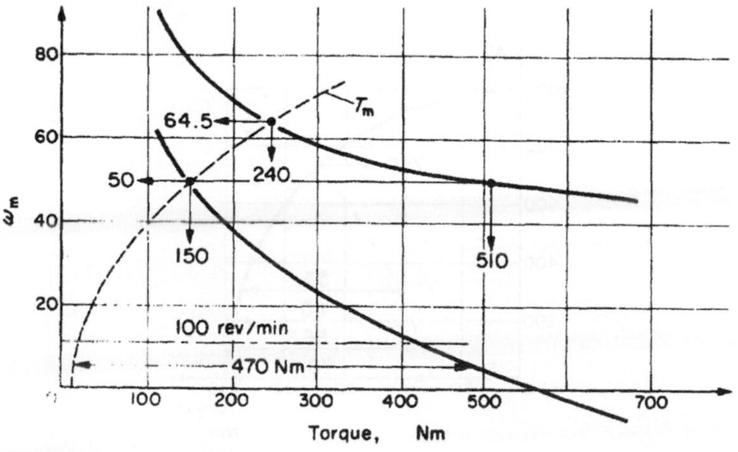

Figure E.6.5

(c) The corresponding torques are 150 and 240 Nm. Since $T_e = k_\varphi I_f$ there is a unique relationship between torque and I_f and one can interpolate between the torque/I_f points in the table to estimate the corresponding currents at 25 and 33 A. At changeover, the torque rises to 510 Nm and the current will be approximately 56 A, a little less than at the first step of starting.

(d) Between 0 and 100 rev/min (10.47 rad/s), the mean accelerating torque $T_e - T_m$, by measurement from the curve, is 470 Nm and since from eqns (6.10) and (6.3):

$$T_e - T_m = J\frac{\Delta\omega}{\Delta t}, \quad \Delta t = \frac{14.5 \times 10.47}{470} = \underline{0.323 \text{ sec.}}$$

Example 6.6

A d.c. shunt motor has its supply voltage so controlled that it produces a speed/torque characteristic following the law:

$$\text{Rev/min} = 1000\sqrt{1 - (0.01T_e)^2}$$

where T_e is in Nm. The total mechanical load, including machine loss-torque, has the following components: Coulomb friction 30 Nm; Viscous friction (α speed) 30 Nm at 1000 rev/min; and fan-load torque [α(speed)^2], 30 Nm at 1000 rev/min. The total coupled inertia is 4 kgm^2. Determine the balancing speed and also calculate the time to reach 98% of this speed, starting from rest.

From the given laws of the speed/torque relationships, the curves are calculated below and plotted on the figure.

T_e	20	40	60	70	80	90	100		Nm
$N = 1000\sqrt{1 - (0.01T_e)^2}$	980	917	800	714	600	436	0		rev/min
N	200	400	500	600	700	800	900	1000	rev/min
$T_m = 30\left[1 + \left(\frac{N}{1000}\right) + \left(\frac{N}{1000}\right)^2\right]$	37.2	46.8	52.5	58.8	65.7	73.2	81.3	90	Nm

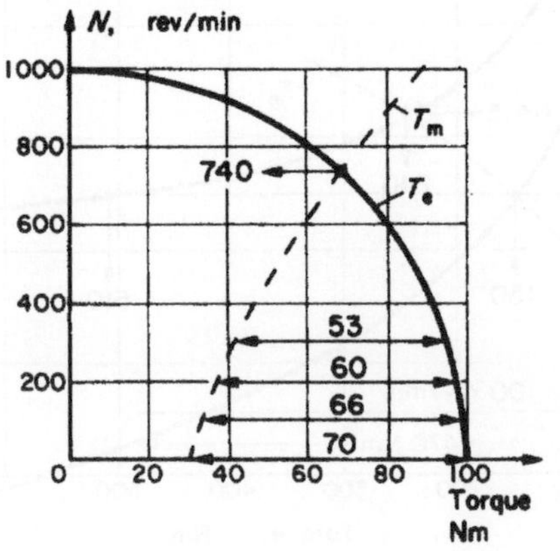

Figure E.6.6

For speed/time calculations, we require the accelerating torque $T_e - T_m = J(\Delta\omega_m/\Delta t)$, over a series of intervals from zero to the top speed. Accuracy falls off when $T_e - T_m$ becomes small because Δt is inversely proportional to this. Extra intervals are taken therefore near the final speed.

From the intersection of the curves, the balancing speed is 740 rev/min and 98% of this is 725 rev/min. The following table is completed with the aid of readings from the graph.

N	0	100	200	300	400	500	600	650	690	725
$T_{acc} = T_e - T_m$	70	66	60	53	44.5	34	22	14.5	8	3
ω_m	0	10.5	20.9	31.4	41.9	52.4	62.8	68.1	72.3	75.9

Mean T_{acc}	68	63	56.5	48.8	39.3	28	18.3	11.3	5.5
$\Delta\omega_m$	10.5	10.5	10.5	10.5	10.5	10.5	5.3	4.2	3.6
$\Delta t = \dfrac{4\Delta\omega_m}{\text{mean } T_{acc}}$	0.62	0.66	0.74	0.86	1.07	1.5	1.14	1.48	2.67

$t = \Sigma\Delta t$	0	0.62	1.28	2.02	2.88	3.95	5.45	6.59	8.07	10.74

Speed/time coordinates can now be read from the above table and a curve plotted if desired. The time to 98% of the balancing speed is 10.74 seconds. Note that the longest times apply to the final build-up intervals and accuracy here is relatively poor.

Example 6.7

The induction motor of Example 4.10 is to be braked to standstill by reversing the phase sequence of the supply to the stator. The mechanical load remains coupled and the total drive inertia is 0.05 kgm². An additional speed-torque coordinate will be required to construct the reverse sequence characteristic and this may be taken as (∓ 1500 rev/min; ± 3 Nm). Make an approximate estimate of the time to zero speed.

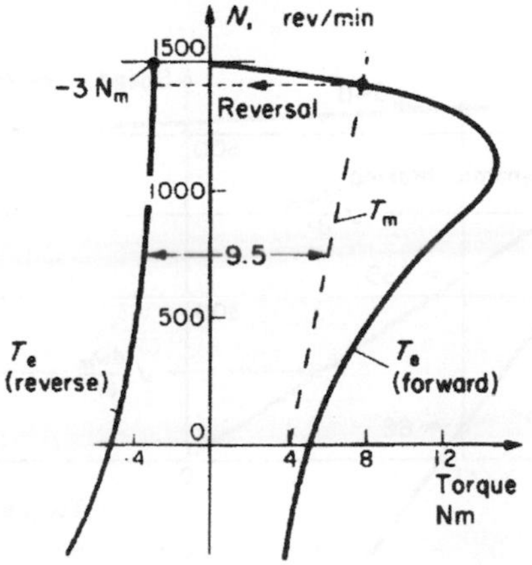

Figure E.6.7

The reverse-sequence characteristic is a mirror image of the forward-sequence characteristic but only the portion in the top left-hand quadrant is required. The figure shows the two ω_m/T_e curves over the required range and the load characteristic has determined the normal speed as 1420 rev/min. It is also seen to be approximately parallel to the 'tail end' of the reverse characteristic and a mean value of the decelerating torque down to zero speed is measured as 9.5 Nm.

$\Delta\omega_m$, taking just one interval, is: $1420 \times 2\pi/60 = 149.75$ rad/s.

$$\text{Hence} \quad \Delta t = \frac{J\,\Delta\omega_m}{T_e - T_m} = \frac{0.05 \times 149.75}{9.5} = \underline{0.79 \text{ sec.}}$$

Note: if the supply is not switched off at zero speed, the machine will run up in the reverse direction. This problem has illustrated plugging braking.

Example 6.8

A 500-V, 60-hp, 600-rev/min, d.c. shunt motor has a full-load efficiency of 90%. The field-circuit resistance is 200 Ω and the armature resistance is 0.2 Ω. Calculate the rated armature current and hence find the speed under each of the following conditions at which the machine will develop rated electromagnetic torque.

(a) Regenerative braking; no limiting resistance;
(b) Plugging (reverse current) braking – external limiting resistor of 5.5 Ω inserted;
(c) Dynamic braking – external limiting resistor of 2.6 Ω inserted.

Rated field-current is maintained and armature reaction and brush drop may be neglected.

The machine is to be braked from full-load motoring using the circuit configurations of (b) and (c). What time would it take in each case to bring the machine to rest? The inertia of the machine and coupled load is 4.6 kg m² and the load, which is coulomb friction, is maintained.

Using the same equation as in Example 3.6 but noting the absence of external field resistance:

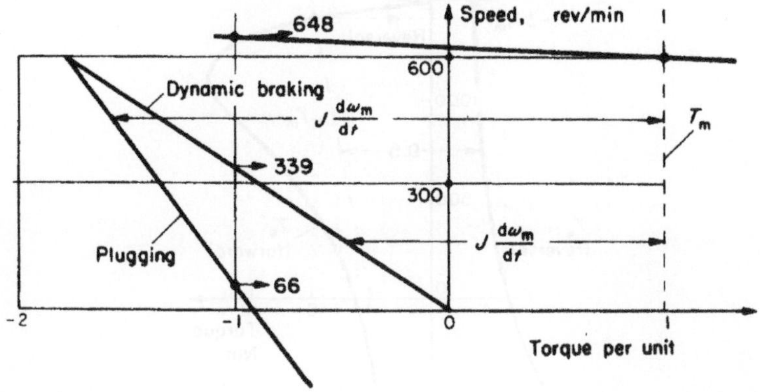

Figure E.6.8

$$\eta_R = \frac{60 \times 746}{500 \times I_{aR} + (500)^2/200} = \frac{90}{100}, \text{ from which } I_{aR} = \underline{97\,A}.$$

$$\text{Rated flux } k_{\phi R} = \frac{500 - 0.2 \times 97}{600 \times 2\pi/60} = 7.65\,\text{Nm/A, which is maintained constant.}$$

The braking-circuit calculations are similar to those for Example 3.20 and the basic equation is:

$$\omega_m = \frac{V}{k_\phi} - \frac{RT_e}{k_\phi^{\,2}}.$$

The value of rated torque is $k_{\phi R} I_{aR} = 7.65 \times 97 = 742\,\text{Nm}$.

(a) For regeneration, $T_e = -742$. $\omega_m = \dfrac{500}{7.65} - \dfrac{0.2(-742)}{7.65^2} = 67.9 = \underline{648\,\text{rev/min}}.$

(b) For plugging, $T_e = -742$,

$V = -500$ and $R = 5.5 + 0.2$. $\omega_m = \dfrac{-500}{7.65} - \dfrac{5.7(-742)}{7.65^2} = 6.91 = \underline{66\,\text{rev/min}}.$

(c) For dynamic braking $T_e = -742$,

$V = 0$ and $R = 2.6 + 0.2$. $\omega_m = 0 - \dfrac{2.8(-742)}{7.65^2} = 35.5 = \underline{339\,\text{rev/min}}.$

To solve the final part of the question, eqn (6.11) could be used directly but instead will be built up from the data given. The electromagnetic torque $T_e = k_\phi i_a$ has to be balanced against $T_m + J\,d\omega_m/dt = 742 + 4.6\,d\omega_m/dt$ and:

$$\text{for (b): } T_e \text{ at any speed } \omega_m = k_\phi \frac{(V - k_\phi \omega_m)}{R} = 7.65\,\frac{(-500 - 7.65\omega_m)}{5.5 + 0.2}$$

$$= -671 - 10.27\omega_m,$$

$$\text{for (c): } T_e \text{ at any speed } \omega_m = k_\phi \frac{(-k_\phi \omega_m)}{R} = 7.65\,\frac{(-7.65\omega_m)}{2.6 + 0.2} = -20.9\omega_m,$$

where the appropriate values of V and R have been inserted. In both cases, the initial speed starts off from the full-load value which is $600 \times 2\pi/60 = 62.8\,\text{rad/s} = \omega_{mo}$, in the solution given by eqn (6.2b), namely $\omega_m = \omega_{mo} + (\omega_{mf} - \omega_{mo})(1 - e^{-t/\tau_m})$.

For (b) the differential equation is therefore: $-671 - 10.27\omega_m = 742 + 4.6\,d\omega_m/dt$, which can be rearranged in standard form as: $0.448\,d\omega_m/dt + \omega_m = -137.6$, and the standard solution is: $\omega_m = 62.8 + (-137.6 - 62.8)(1 - e^{-t/0.448})$.

The question asks for the stopping time; i.e. when $\omega_m = 0$. Substituting and rearranging:

$$\frac{-62.8}{-200.4} = 1 - e^{-t/0.448},$$

$$e^{t/0.448} = \frac{1}{1 - 62.8/200.4} = 1.456.$$

from which t, the time to stop, is $0.448 \times 0.376 = \underline{0.168 \text{ sec.}}$

For (c) the differential equation is: $-20.9\omega_m = 742 + 4.6\, d\omega_m/dt$,

which can be rearranged in standard form as: $0.22\, d\omega_m/dt + \omega_m = -35.5$

and the standard solution is: $\omega_m = 62.8 + (-35.5 - 62.8)(1 - e^{-t/0.22})$

For the stopping time, putting $\omega_m = 0$:

$$\frac{-62.8}{-98.3} = 1 - e^{-t/0.22},$$

$$e^{t/0.22} = \frac{1}{1 - 62.8/98.3} = 2.769,$$

from which the time t to stop is $0.22 \times 1.018 = \underline{0.224 \text{ sec.}}$

Dynamic braking gives a slower stopping time, even though the peak armature current is about the same in (b) and (c) (as could be checked); and the load torque, which in this case is a major braking force, is the same. A study of the figure will show why this is so. For plugging, (b), the machine will run up in reverse rotation after stopping unless it is switched off.

Example 6.9

A 250-V d.c. shunt motor has an armature resistance of 0.15 Ω. It is permanently coupled to a constant-torque load of such magnitude that the motor takes an armature current of 120 A when running at rated speed of 600 rev/min. For emergency, provision must be made to stop the motor from this speed in a time not greater than 0.5 seconds. The peak braking current must be limited to twice the rated value and dynamic braking is to be employed with the field excited to give rated flux. Determine the maximum permissible inertia of the motor and its coupled load, which will allow braking to standstill in the specified time. Calculate also the number of revolutions made by the motor from the initiation of braking, down to standstill.

If after designing the drive as above, it was found that the stopping time was too long and had to be reduced to 0.4 seconds, determine the reduced value of resistance necessary to achieve this, and calculate the increased value of peak current.

At the rated condition, e.m.f. $E = 250 - 0.15 \times 120 = 232$ V. So:

$$k_{\phi R} = \frac{232}{600 \times 2\pi/60} = \frac{232}{62.8} = 3.69 \text{ Nm/A and } T_{eR} = 3.69 \times 120 = 443 \text{ Nm}.$$

The limiting resistance must keep the current to 2 × 120 = 240 A and since on dynamic braking the current is E/R in magnitude = 232/R on changeover; R = 232/240 = 0.967 Ω, an extra 0.967 − 0.15 = 0.817 Ω.

As in the previous question, $T_e = k_\phi \dfrac{(-k_\phi \omega_m)}{R} = \dfrac{-3.69^2}{0.967}\omega_m = -14.1\omega_m$.

Forming the mechanical balance equation: $T_e = T_m + J\,d\omega_m/dt$

$$-14.1\omega_m = 443 + J\,d\omega_m/dt,$$

and rearranging: $\dfrac{J}{14.1}\dfrac{d\omega_m}{dt} + \omega_m = -31.4.$

Standard solution is, from eqn (6.2b): $\omega_m = \omega_{mo} + (\omega_{mf} - \omega_{mo})(1 - e^{-t/\tau_m})$

$$= 62.8 + (-31.4 - 62.8)\ (1 - e^{-0.5 \times 14.1/J})$$

For zero speed: $0 = -31.4 + 94.2e^{-7.05/J},$

from which: $e^{7.05/J} = \dfrac{1}{31.4/94.2}$ and $J = 6.42\,\text{kgm}^2$.

Substituting this value of J gives the general expression for speed under these conditions:

$$\omega_m = 62.8 + (-31.4 - 62.8)(1 - e^{-t \times 14.1/6.42})$$

$$= -31.4 + 94.2e^{-2.196t}.$$

To find the number of revolutions turned through we require to integrate

$$\int \frac{d\theta}{dt}\,dt = \int d\theta.$$

$$\text{Hence } \theta = \int_0^{0.5} \omega_m\,dt = \left[-31.4t + \frac{94.2}{-2.196}e^{-2.196t}\right]_0^{0.5}$$

$$= [-15.7 - 42.9(0.3335 - 1)] = 12.9\ \text{radians}.$$

∴ Number of revolutions to stop = 12.9/2π = 2.05.

If the stopping time is to be changed to 0.4 seconds and R is unknown,

$$T_e = \frac{-3.69^2}{R}\omega_m.$$

The mechanical balance equation is now:

$$\frac{-13.63}{R}\omega_m = 443 + 6.42\,d\omega_m/dt.$$

Rearranging: $0.471R\dfrac{d\omega_m}{dt} + \omega_m = -32.5R$.

The solution is: $\omega_m = 62.8 + (-32.5R - 62.8)(1 - e^{-t/0.471R})$.

With t = 0.4 seconds: $= -32.5R + (32.5R + 62.8)e^{-0.849/R}$,

$$1 = \left(1 + \frac{1.93}{R}\right)e^{-0.849/R}.$$

An explicit solution for R is not possible, but by trying a few values of R somewhat below the previous value of 0.967 Ω a solution close to the correct answer is quickly found. For $R = 0.6\ \Omega$ the R.H.S. of the equation is 1.02 so the additional series resistor must be reduced to about 0.6 – 0.15 = 0.45 Ω.

The peak current on changeover would be 232/0.6 = 387 A = 3.22 *per unit* which is a considerable increase on the previous value of 2 *per unit* for 0.5 seconds stopping time. The machine designer would have to be consulted to approve this increase.

Example 6.10

A small permanent magnet, 100-V d.c. motor drives a constant-torque load at 1000 rev/min and requires an input of 250 watts. The armature resistance is 10 Ω. The motor is to be reversed by a solid-state contactor which can be assumed to apply full reverse-voltage instantaneously. The inertia of the motor and drive is 0.05 kgm². Calculate the time:

(a) to reach zero speed
(b) to reach within 2% of the final reverse speed.

The armature inductance can be neglected, but assuming it is 1 henry, make an estimate of the actual peak current during reversal.

$$I_a = P/V = 250/100 = 2.5\text{ A},$$

$$k_\phi = \frac{E}{\omega_m} = \frac{100 - 10 \times 2.5}{1000 \times 2\pi/60} = 0.7162\text{ Nm/A}.$$

Rated torque = $k_\phi I_a = 0.7162 \times 2.5 = 1.79$ Nm.

(a) During the transient:

$$T_e = T_m + J\,d\omega_m/dt \text{ and } V = -100$$

$$0.7162\left(\frac{-100 - 0.7162\omega_m}{10}\right) = 1.79 + 0.05\,d\omega_m/dt.$$

Rearranging: $0.9748\, d\omega_m/dt + \omega_m = -174.5$.

Solution is: $$\omega_m = \omega_{mo} + (\omega_{mf} - \omega_{mo})(1 - e^{-t/\tau_m}).$$

Substituting $\omega_m = 0$: $$0 = 104.7 + (-174.5 - 104.7)\,(1 - e^{-t/0.9748})$$

$$= -174.5 + 279.2e^{-t/0.9748}.$$

from which: time to stop, t = 0.458 sec.

Acceleration in the reverse direction will also be exponential of time-constant τ_m, and to 98% of final speed, time will be $4\tau_m = 3.9$ seconds. Total time = 4.36 seconds.

(b) If L is neglected, peak current on changeover $= \dfrac{-100 - 75}{10} = -17.5$ A,

current at zero speed $= \dfrac{-100}{10} = -10$ A,

giving a current waveform as shown on Figure E.6.10 over the period 0.458 seconds.

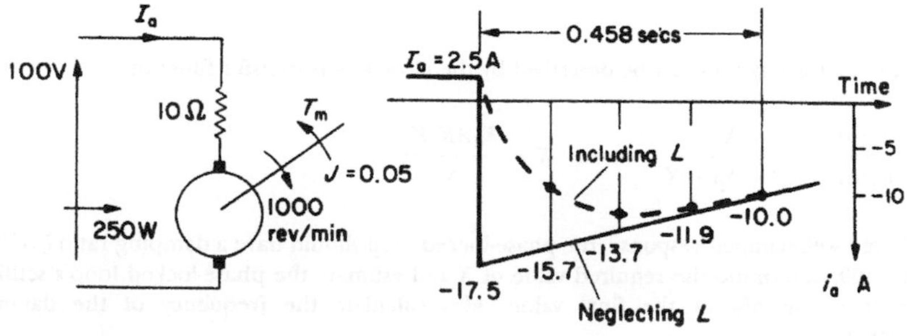

Figure E.6.10

Electrical time-constant $\tau_e = L/R = 1/10 = 0.1$ seconds which is approximately 1/4 of stopping time. Dividing this part of the current wave into four intervals, an estimate of the current actually reached can be based on the exponential response of Figure 6.1.

After 1st interval, current will be approximately $0.632 \times 15.7 = 9.9$ A.

After 2nd interval, current will be approximately $0.864 \times 13.7 = 11.8$ A.

After 3rd interval, current will be approximately $0.95 \times 11.9 = 11.3$ A.

So the actual current peak will be about 12 A, not the 17.5 A calculated with inductance neglected. See Example 6.14 for equations including effect of inductance.

Example 6.11

A phase-locked loop is a frequency control system and is frequently used for synchronising power-electronic controllers in electrical drive applications to external sources, such as a mains supply. Figure E.6.11 shows a typical phase-locked loop arrangement. ω_i is the input (reference) frequency, ω_0 is the frequency delivered by the system, which is higher than ω_i but is synchronised to it, and ω_f is the feedback frequency. A difference between ω_i and ω_f is shown by a varying phase shift which is sensed by the phase detector. The voltage-controlled oscillator generates the new frequency which has to be synchronised to ω_i and the counter divides it to the same units as ω_i.

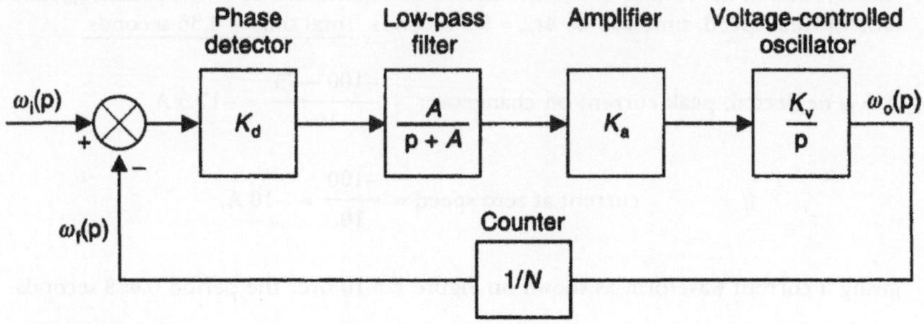

Figure E.6.11

Show that the system can be described by the closed-loop transfer function

$$\frac{\omega_f(p)}{\omega_i(p)} = \frac{X}{p^2 + Ap + X} \quad \text{where} \quad X = \frac{K_d A K_a K_v}{N}$$

To have a well-damped response, the phase-locked loop should have a damping ratio ζ of 0.7. If $A = 100$, determine the required value of X and estimate the phase-locked loop's settling time to within 5% of the final value. Also calculate the frequency of the damped oscillation.

$\omega_f(p)$ has been deliberately defined as the output signal, as it is directly comparable with ω_i and is a measure of ω_0. The open-loop transfer function is then $\omega_f(p)/\{\omega_i(p) - \omega_f(p)\}$, which can be found by multiplying the blocks of the diagram together, i.e.

$$\frac{\omega_f(p)}{\omega_i(p) - \omega_f(p)} = K_d \times \frac{A}{(p + A)} \times K_a \times \frac{K_v}{p} \times \frac{1}{N} = \frac{X}{p(p + A)} = G(p)$$

The closed-loop transfer function is:

$$\frac{\omega_f(p)}{\omega_i(p)} = \frac{G(p)}{1 + G(p)} = \frac{X/(p^2 + Ap)}{1 + X/(p^2 + Ap)} = \frac{X}{p^2 + Ap + X}$$

With $A = 100$, $\dfrac{\omega_f(p)}{\omega_i(p)} = \dfrac{X}{p^2 + 100p + X}$

Comparing the denominator with the standard second-order transfer function in eqn (6.5),

$$2\zeta\omega_n = 100.$$

If $\zeta = 0.7$, then $\omega_n = 100/1.4 = 71.4$.

$$\therefore \quad X = \omega_n^2 = \underline{5102}$$

The decay rate $= \sigma = \zeta\omega_n = 0.7 \times 71.4 = 50\,\text{s}^{-1}$.

Hence the settling time to within 5% of the final value is:

$$3/\sigma = 3/50 = \underline{60\,\text{ms}}$$

The frequency of damped oscillation is:

$$\omega_n\sqrt{(1-\zeta^2)} = 71.4\sqrt{(1-0.7^2)} = 51\,\text{rad/s} = \underline{8.1\,\text{Hz}}$$

Example 6.12

Figure E.6.12 shows a control system in which the angle of a shaft is controlled in response to a reference voltage V_{ref}. The control voltage V formed by (V_{ref} – position feedback voltage V_0 – speed-proportional voltage $k_v\omega_m$), where ω_m is the shaft speed, is the input to a drive system of transfer function $K/(1 + 0.02p)$. The torque output T from the drive system is the input to a speed-reduction gearbox.

If the motor inertia is $4 \times 10^{-4}\,\text{kgm}^2$ and the load inertia is $4\,\text{kgm}^2$, find the optimum gear ratio.

With the gear ratio found above and $K = 0.44\,\text{Nm/V}$, $K_v = 0.8\,\text{V/rad/s}$, show that:

$$\frac{V_0(p)}{V_e(p)} = \frac{275}{p(p^2 + 50p + 220)}$$

By approximating the system to a second-order system, estimate the natural frequency and damping ratio of the system.

The optimum gear ratio is:

$$\sqrt{(\text{load inertia/motor inertia})} = \sqrt{(4/4 \times 10^{-4})} = \underline{100}$$

The total inertia is then $(J_2 + n^2J_1) = (4 + 4) = 8\,\text{kgm}^2$.

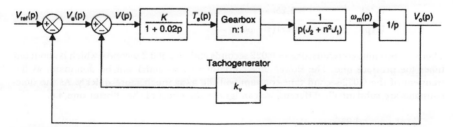

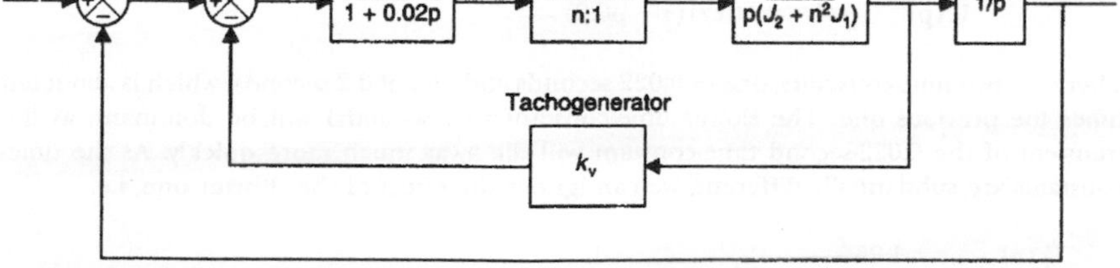

Figure E.6.12

Inserting numerical values for the inertia and gear ratio,

$$\frac{\omega_m(p)}{T(p)} = \frac{n}{p(J_2 + n^2 J_1)} = \frac{100}{8p}$$

With $K = 0.44$,

$$\frac{T(p)}{V(p)} = \frac{0.44}{1 + 0.02p}$$

The transfer function G(p) relating $V(p)$ and $\omega(p)$ is:

$$G(p) = \frac{\omega_m(p)}{T(p)} \times \frac{T(p)}{V(p)} = \frac{\omega_m(p)}{V(p)} = \frac{44}{8p(1 + 0.02p)}$$

Since $K_v = 0.8$, $V(p) = V_e(p) - 0.8\omega_m(p)$, where V_e is the error voltage.

$$\therefore \quad \omega_m(p) = G(p)\{V_e(p) - 0.8\omega_m(p)\}$$

$$\therefore \quad \frac{\omega_m(p)}{V_e(p)} = \frac{G(p)}{1 + 0.8G(p)}$$

and $\quad V_0(p) = \omega_m(p)/p$

Hence the open-loop transfer function will be:

$$\frac{V_0(p)}{V_e(p)} = \frac{G(p)}{p(1 + 0.8G(p))} = \frac{44/(8p + 0.16p^2)}{p[1 + 35.2/(8p + 0.16p^2)]}$$

$$= \frac{44}{p(0.16p^2 + 8p + 35.2)}$$

$$= \frac{275}{p(p^2 + 50p + 220)}$$

Factorising, $\dfrac{V_0(p)}{V_e(p)} = \dfrac{275}{p(p + 44.9)(p + 4.9)}$

Dividing throughout by 4.9×44.9,

$$\frac{V_0(p)}{V_e(p)} = \frac{1.25}{p(1 + p0.022)(1 + p0.2)}$$

There are two time-constants, one of 0.022 seconds and one of 0.2 seconds, which is about ten times the previous one. The slower time-constant (0.2 seconds) will be dominant, as the transient of the 0.022-second time-constant will die away much more quickly. As the time-constants are substantially different, we can ignore the effect of the shorter one, i.e.

$$\frac{V_0(p)}{V_e(p)} \approx \frac{1.25}{p(1 + p0.2)} = G'(p), \text{ say}$$

The relationship between the open- and closed-loop transfer functions is the same as in Figure 6.2b, i.e.

$$V_{ref}(p) - V_0(p) = V_e(p) = V_0(p)/G'(p)$$

$$\therefore \quad V_{ref}(p) = V_0(p)\{1 + 1/G'(p)\}$$

Hence the closed-loop transfer function is:

$$\frac{V_0(p)}{V_{ref}(p)} = \frac{G'(p)}{1 + G'(p)} = \frac{1.25/(p + 0.2p^2)}{1 + 1.25/(p + 0.2p^2)}$$

$$= \frac{1.25}{0.2p^2 + p + 1.25} = \frac{6.25}{p^2 + 5p + 6.25}$$

By comparison with the standard form of the second-order transfer function in eqn (6.5),

$$\omega_n{}^2 = 6.25. \quad \therefore \omega_n = \underline{2.5\,\text{rad/s.}}$$

$$2\zeta\omega_n = 5. \quad \therefore \zeta = 5/5 = \underline{1.}$$

A damping ratio of 1 is known as 'critical damping', as a lower damping ratio implies the presence of some oscillation when a step input is applied. For position control applications, overshoot in position is undesirable and $\zeta \approx 1$ is normally considered appropriate.

Example 6.13

An induction motor with rotor inertia 5 kgm^2 drives a fan load of inertia 500 kgm^2 via a 2-metre steel shaft of 0.1 m diameter and 10:1 gearbox. The motor has a torque pulsation of around 400 Hz and the fan has a pulsating load of 16 Hz. Determine the natural frequency of the shaft's torsional vibration and check whether this is close to either of the system frequencies. Find the torsional damping constant required to give a damping ratio of 0.7. The steel shaft has a shear modulus $G = 80\,\text{GN/m}^2$.

From eqn (6.9),

$$K = \frac{\pi G r^4}{2l} = \frac{\pi \times 8 \times 10^{10} \times (0.05)^4}{4} = 392\,700\ \text{Nm/radian}$$

For a shaft in torsion without damping, from eqn (6.8a):

$$\omega_n = \sqrt{\frac{K}{J_2} + \frac{K}{J_1}}$$

where $J_1 = 5$ and J_2 (referred to the motor shaft) $= 500/n^2 = 5$

$$\therefore \quad \omega_n = \sqrt{\frac{392\,700}{5} + \frac{392\,700}{5}} = 396\,\text{rad/s} = 63\,\text{Hz}$$

This lies between the resonance frequencies but is almost a multiple of 16 Hz. From eqn (6.8b),

$$\zeta = 0.7 = \frac{C}{2\sqrt{K}}\sqrt{\frac{1}{J_1}+\frac{1}{J_2}}$$

$$\therefore \quad C = \frac{1.4\sqrt{(392\,700)}}{\sqrt{(0.4)}} = 1387\,\text{Nm per rad/s}$$

and $\omega = 63\sqrt{(1-0.7^2)} = \underline{45\,\text{Hz}}$, which is clear of load and motor pulsation frequencies.

Example 6.14

A permanent magnet d.c. motor with $k_\varphi = 0.54$ is controlled in speed by a closed-loop system, as shown in Figure E.6.14, including a power-electronic control amplifier which has a gain of 100. The motor's armature has resistance 0.7 Ω and inductance 0.1 H. The armature inertia is 0.05 kgm^2 and frictional damping is negligible. The motor is connected through a 10:1 gearbox to a load which has inertia 5 kgm^2 and a torque which is proportional to (speed)^2; load torque = $0.01 \times \omega_m^2$.

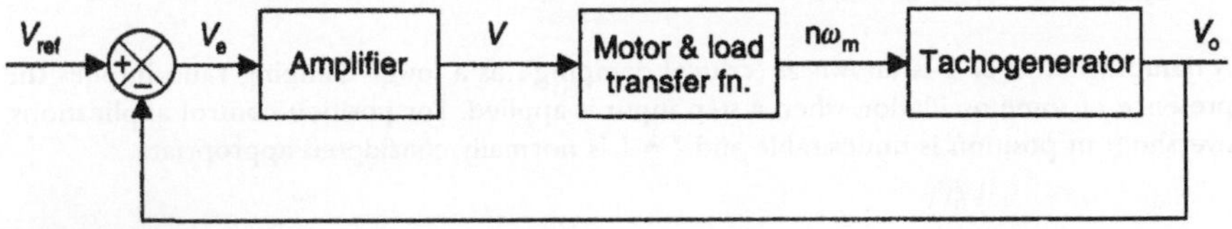

Figure E.6.14

The motor shaft speed is sensed by a tachogenerator, which is a precision d.c. permanent-magnet generator; it produces 9.5 V/1000 rev/min and the 'error' signal which is the input to the power-electronic amplifier is derived by subtracting the tachogenerator output from a reference voltage.

By approximating the load torque characteristic about an operating point of 100 rad/s, derive a closed-loop transfer function for the system and calculate its natural frequency and damping ratio.

$$\text{Load torque} = T_m = 0.01(\omega_m)^2$$

From eqn (6.3),

$$dT_m/d\omega_m = 0.02\omega_m \approx \Delta T/\Delta\omega_m$$

Around an operating point of $\omega_m = 100\,\text{rad/s}$,

$$\Delta T_m/\Delta\omega_m = k_2 = 2\,\text{Nm per rad/s}$$

The motor inertia referred to the output shaft is

$$(10)^2 \times 0.05 = 5\,\text{kgm}^2$$

The total inertia referred to the output shaft is then 10 kgm^2.

The motor and load transfer function $\omega_m(p)/V(p)$ can be found using eqn (6.14) and inserting system constants $LJ = 1$, $k_2L = 0.2$, $RJ = 7$, $Rk_2 = 1.4$, $k_\phi{}^2 = 0.29$:

$$\frac{\omega_m(p)}{V(p)} = \frac{k_\phi}{p^2LJ + p(k_2L + JR) + Rk_2 + k_\phi{}^2}$$

$$= \frac{0.54}{p^2 + 7.2p + 1.69}$$

The tachogenerator constant is in the units normally stated on data sheets.

Converting, 9.5 V/1000 rev/min = 0.091 V per rad/s.

However, the tachogenerator is coupled to the motor shaft which rotates at ten times the speed of the output shaft, i.e. $V_0(p)/n\omega_m(p) = 0.091$.

Since the amplifier has a gain of 100, $V(p)/V_e(p) = 100$.

The open-loop transfer function is

$$\frac{V_0(p)}{V_e(p)} = \frac{V_0(p)}{n\omega_m(p)} \times \frac{n\omega_m(p)}{V(p)} \times \frac{V(p)}{V_e(p)}$$

$$= \frac{0.091 \times 10 \times 0.54 \times 100}{p^2 + 7.2p + 1.69}$$

$$= \frac{49.14}{p^2 + 7.2p + 1.69} = G(p), \text{ say.}$$

The closed-loop transfer function is

$$\frac{V_0(p)}{V_{ref}(p)} = \frac{G(p)}{1 + G(p)} = \frac{49.14}{p^2 + 7.2p + 50.83}$$

Comparing this with the coefficients of first (p) and second (p^2) derivative terms in eqn (6.4),

$$\omega_n = \sqrt{(50.83)} = \underline{7.13 \text{ rad/s}}$$

and $2\zeta\omega_n = 7.2$.

Hence $\zeta = \underline{0.5}$, which is a reasonable value for a speed control system.

Example 6.15

The motor and load of Example 6.14 are to be used in a position control system in which a position sensor on the output shaft produces an output of 0.05 V per degree of rotation. Sensors are briefly discussed in Chapter 7. Find the system's closed-loop transfer function and show that the system is unstable. Suggest a means of providing additional damping to stabilise the system.

The position sensor's constant can be expressed as 2.86 V/rad. Figure E.6.15a shows the block diagram of this system. V_0 is a voltage representing the shaft position and V_{ref} is a voltage representing the required shaft angle.

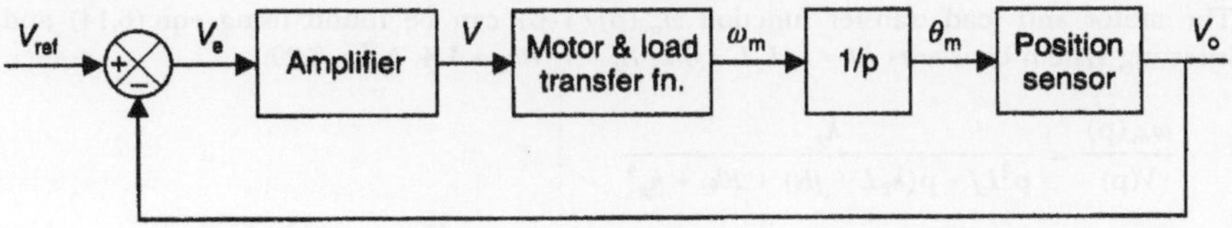

(a)

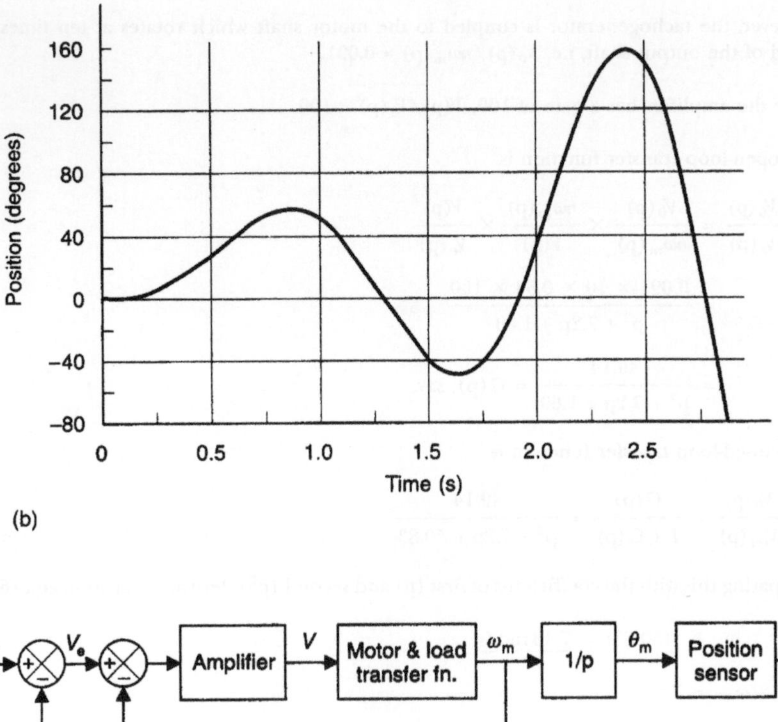

(b)

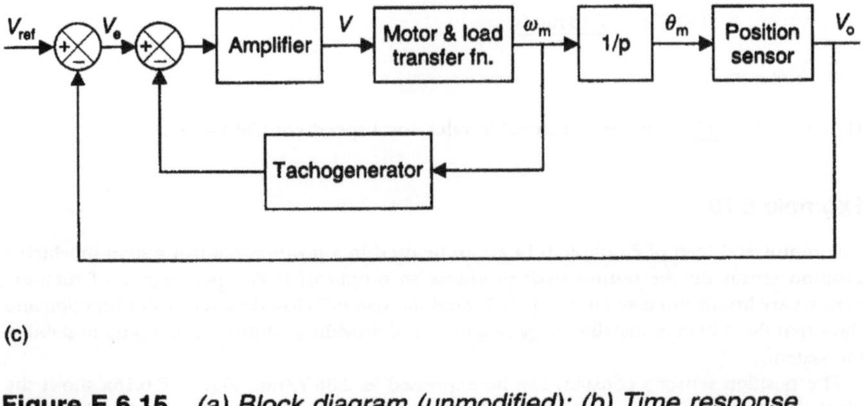

(c)

Figure E.6.15 *(a) Block diagram (unmodified); (b) Time response (unmodified); (c) Block diagram (with rate feedback)*

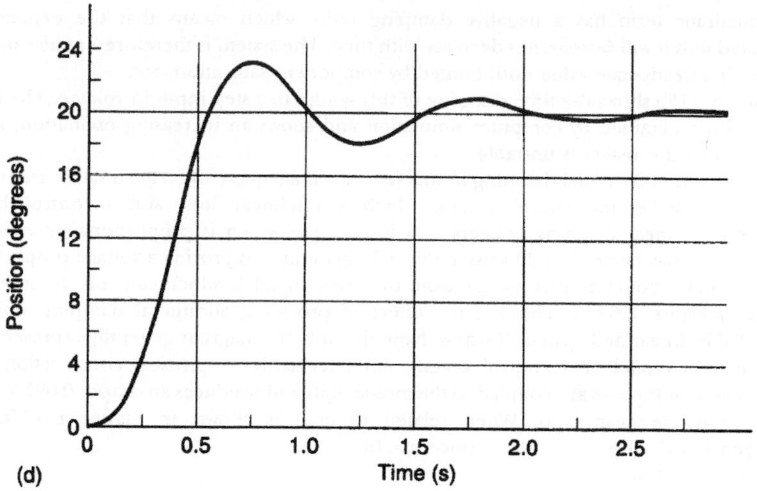

Figure E.6.15 *(d) Time response (with rate feedback).*

The motor and load transfer function is the same as in Example 6.14, i.e.

$$\frac{\omega_m(p)}{V(p)} = \frac{0.54}{p^2 + 7.2p + 1.69}$$

For the amplifier, $V(p)/V_e(p) = 100$.

For the position sensor, $V_0(p)/\theta(p) = 2.86$.

Also, the shaft angle θ is the integral of the shaft speed ω_m, i.e. $\theta(p)/\omega_m(p) = 1/p$.

Hence the open-loop transfer function is

$$\frac{V_0(p)}{V_e(p)} = \frac{V_0(p)}{\theta(p)} \times \frac{\theta(p)}{\omega_m(p)} \times \frac{\omega_m(p)}{V(p)} \times \frac{V(p)}{V_e(p)} = \frac{2.86 \times 1 \times 0.54 \times 100}{p(p^2 + 7.2p + 1.69)}$$

$$= \frac{154.4}{p(p^2 + 7.2p + 1.69)}$$

The closed-loop transfer function is found as in Example 6.14.

$$\frac{V_0(p)}{V_{ref}(p)} = \frac{154.4}{p^3 + 7.2p^2 + 1.69p + 154.4}$$

This can be factorised by trying out values of p to give:

$$\frac{V_0(p)}{V_{ref}(p)} = \frac{154.4}{(p^2 - 1.74p + 17.27)(p + 8.94)}$$

The quadratic term has a negative damping ratio, which means that the exponential associated with it will *increase*, not decrease with time. The system is therefore *unstable* and will not reach a steady-state value until limited by component saturation, etc.

Figure E.6.15b shows the time response of this system to a step input in voltage. The result is a response obtained by computer simulation and shows an increasing oscillation, which confirms that the system is unstable.

The amount of frictional damping in this system is inadequate to create a stable system but addition of further mechanical friction, which is non-linear, lossy and uncontrollable, is undesirable. Linear damping appears as a loss torque which is proportional to speed. A similar effect may be produced by using the tachogenerator to provide a voltage proportional to speed and subtracting that voltage from the error signal V_e which controls the motor, as shown in Figure E.6.15c. This has the effect of providing additional damping which is controllable, linear and virtually lossless. Note that a block diagram generally represents the mathematical model of a control system, not necessarily its physical construction. The tachogenerator is physically coupled to the motor shaft and produces an output $0.091 \times n\omega_m$, where n is the gear ratio. When related to ω_m, as shown in Figure E.6.15c, the tachogenerator's output is $0.91\omega_m$, since $n = 10$.

From Figure E.6.15c,

$$\omega_m(p) = \frac{100 \times 0.54}{p^2 + 7.2p + 1.69}\{V_e(p) - 0.91\omega_m(p)\}$$

$$\therefore \quad \omega_m(p)\left\{1 + \frac{49.14}{p^2 + 7.2p + 1.69}\right\} = \frac{54V_e(p)}{p^2 + 7.2p + 1.69}$$

$$\therefore \quad \omega_m(p)(p^2 + 7.2p + 50.83) = 54V_e(p)$$

$$\therefore \quad \frac{V_0(p)}{V_e(p)} = \frac{2.86\omega_m(p)}{pV_e(p)} = \frac{154.4}{p(p^2 + 7.2p + 50.83)}$$

$$\frac{V_0(p)}{V_{ref}(p)} = \frac{154.4}{p^3 + 7.2p^2 + 50.83p + 154.4}$$

This can be factorised by trying out values of p to give:

$$\frac{V_0(p)}{V_{ref}(p)} = \frac{154.4}{(p^2 + 3.14p + 38.1)(p + 4.05)}$$

In this case, the quadratic term has positive damping. Note that the damping ratio of 0.25 obtained by equating 38.1 to ω_n^2 and 3.14 to $2\zeta\omega_n$ is not a true measure of the stability of the system, as the system is of third order. However, it gives a useful guide as to the rate at which oscillations decay. The closed-loop step response shown in Figure E.6.15d was generated by computer simulation. In this case, the time response has a moderate overshoot but is stable, as expected. The amount of damping is inadequate for a position control application, in which no overshoot would normally be desired and additional damping is obtainable by amplifying the tachogenerator signal prior to its subtraction from V_e.

6.6 Duty cycles and ratings

Many drives are required to supply a varying load, following a cycle which is repeated continually. Correctly chosen equipment would only need to have a steady (continuous) rating somewhat less than the maximum power. However, if this is relatively high as on a 'peaky' load, it might be the major factor in determining the appropriate machine rating if the overload (maximum/continous rating) is beyond about 2. For the continuous rating, it is the r.m.s. value of the machine current which must be withstood without excessive temperature rise, not the lesser 'average' value corresponding to the mean power and actual energy consumption. Consider the straightforward case of a d.c. machine operating at constant flux. The torque is proportional to current, so if the time-varying torque is multiplied by the maximum speed to convert to power, an r.m.s. rating could be obtained from the power/time curve. It would be equivalent to r.m.s. torque (proportional to r.m.s. current) times maximum speed. This procedure is not strictly correct if the motor speed is varying, which in turn results in a different machine-loss pattern and inferior cooling capacity at low speeds. The situation is even less precise if considering the determination of r.m.s. power for a.c. machine drives, with the additional complication of power factor and less simple current/torque relationships. Various examples to follow will illustrate some of these points and the general philosophy of estimating machine ratings. But in practice, each type of drive requires specialist treatment for greater accuracy. Reference 3 discusses the issues in more detail.

If the approximation of using the power/time curve for r.m.s. power is employed, the power variation is often further simplified to a series of straight-line segments and the squared areas under these as summed to give the r.m.s. power as:

$$\sqrt{\frac{\int P^2 t\,\mathrm{d}t}{\Sigma t}}$$

where Σt must include any times during the cycle when the process is stopped. Consider a section of such a power/time curve which in general will have a trapezoidal shape, Figure 6.7. The area under the squared curve will be:

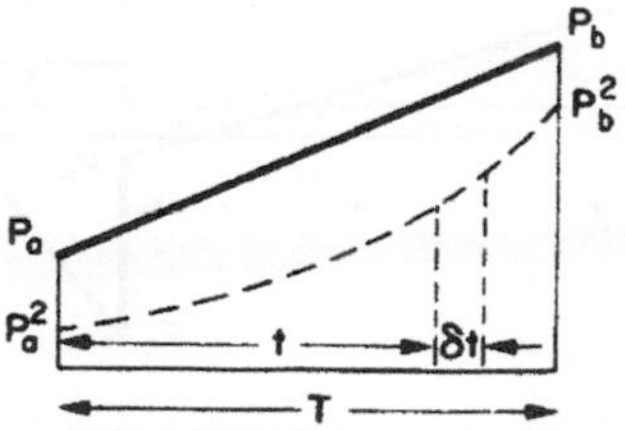

Figure 6.7 *Area under squared curve.*

$$\int_0^T \left[P_a + (P_b - P_a)\frac{t}{T} \right]^2 dt = \int_0^T \left[P_a^2 + 2(P_a P_b - P_a^2)\frac{t}{T} + (P_b^2 - 2P_a P_b + P_a^2)\frac{t^2}{T^2} \right] dt$$

which integrates finally to:

$$\frac{T}{3}(P_a^2 + P_a P_b + P_b^2). \tag{6.16}$$

The overall power/time cycle can be divided into a series of such generalised trapezia, e.g. $P_a = P_b$ for a rectangle and P_a or $P_b = 0$ for a triangle.

As an example of a duty cycle consider a mine hoist for which speed is accelerated from zero to a constant value for a period and is then decelerated back to zero, with a rest period before the next cycle. An overload torque is required for acceleration, which if constant as shown, requires a constant-torque component. During deceleration a similar, reverse torque is required and will give some regeneration. Figure 6.8

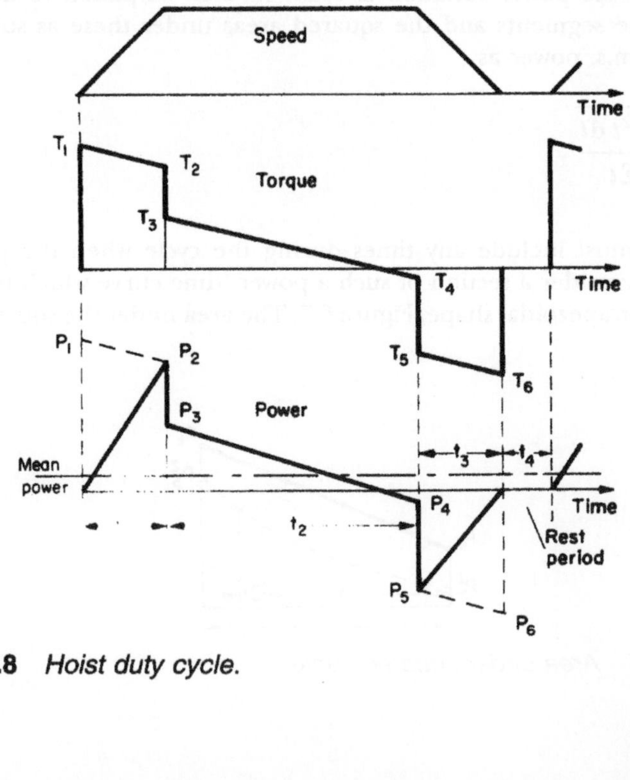

Figure 6.8 *Hoist duty cycle.*

shows such a cycle for a winder in which the static rope-torque is not balanced out, though this is more usual. The changing, unbalanced weight of the rope causes the slope in the torque curve. The power curve is approximated by straight lines as shown, though this is only strictly correct during the constant-speed region. The r.m.s. power can now be calculated by summing the three squared areas, using eqn (6.16) on each. Although the power starts from zero, the torque is often multiplied by the constant speed throughout – which is the torque/time curve to another scale – to allow for the inferior cooling at low speeds. The mean power is obtained from the area under the power curve as shown, part of which is negative. The peak power is P_2, but the machine must be capable of delivering the peak torque T_1 at the beginning of the cycle. Tutorial Example T6.11 refers to this figure, which is typical of the pattern for lift drives generally, though regeneration is not usually an economical proposition for lifts in low buildings.

Example 6.16

An 8-pole, 50-Hz induction motor coupled to a flywheel drives a load which requires a torque $T_0 = 110\,\text{Nm}$ when running light. For an intermittent period of 8 seconds, a pulse load rising instantaneously to 550 Nm is to be supplied. What must be the combined inertia of the system to ensure that the peak motor torque does not exceed 400 Nm? The motor characteristic may be taken as linear and giving a torque of 350 Nm at a slip of 5%.

If the coupled inertia was to be changed to 200 kgm^2, what would then be the peak motor torque with the same duty?

Intermittent loads of this kind are not an uncommon requirement. To rate the motor for the full peak would be wasteful. Instead, the stored energy in the inertia, $\frac{1}{2}J\omega_m^2$, supplemented by a flywheel if necessary, is partly extracted during the pulse, by designing the motor with a steep enough speed-regulation to release the required amount. Figure E.6.16 shows the duty pulse and the motor characteristic from the given data. The law of the motor characteristic is linearised as

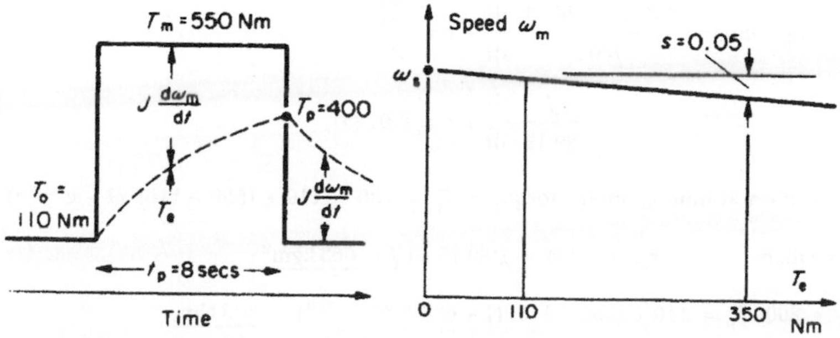

Figure E.6.16

$$T_e = k\frac{(\omega_s - \omega_m)}{\omega_s} = ks.$$

Substituting the point given: $350 = k \times 0.05$ so $k = 7000$.

It is not essential of course that the motor should be of the induction type and the characteristic is being simplified here by assuming it is a straight line.

From the mechanical balance equation: $T_e = T_m + J\,d\omega_m/dt$

substituting for ω_m and T_e: $ks = T_m + J\,d\omega_s(1-s)/dt,$

$$= T_m - J\omega_s\,ds/dt,$$

rearranging: $$\frac{J\omega_s}{k}\frac{ds}{dt} + s = \frac{T_m}{k},$$

which again is in standard form and the standard solution of eqn (6.2b) can be invoked.

$$s_p = s_0 + \left(\frac{T_m}{k} - s_0\right)(1 - e^{-(t_p/J\omega_s/k)}),$$

multiply by k: $ks = T_e = T_0 + (T_m - T_0)(1 - e^{-kt_p/J\omega_s})$.

There are five variables and the equation can be solved for any one of them, with all the others specified; i.e. minimum inertia (as in this question), maximum torque pulse, maximum pulse time t_p, steepness of speed regulation k and motor peak torque T_p (as in the second part of this question). Notice that the speed time-constant $\tau_m = J\omega_s/k$ is a different expression from those met in earlier examples in this chapter.

Substituting values: $$\frac{J \times 2\pi \times 50/4}{7000}\frac{ds}{dt} + s = \frac{550}{7000}$$

$$\frac{J}{89.13}\frac{ds}{dt} + s = 0.07857.$$

In the solution at time t_p, motor torque $= T_p = 400 = 110 + (550 - 110)\,(1 - e^{-8/\tau_m})$

from which: $\tau_m = 7.434 = J/89.13$ so $\underline{J = 663\ \text{kgm}^2}$.

With $J = 200$: $T_p = 110 + (550 - 110)\,(1 - e^{-(8 \times 89.13)/200}) = \underline{538\ \text{Nm}}$,

which is almost the same as the pulse torque because the inertia is too small.

Example 6.17

A high-power d.c. magnet of resistance 0.1 Ω is pulsed occasionally with 2000 A maintained constant for a period of 8 seconds. A 6-pole 50-Hz induction motor drives the supply generator, the speed regulation being set to give a speed of 800 rev/min at the end of the load pulse. The induction-motor torque at 800 rev/min is 1500 Nm. For the purposes of estimating the required flywheel effect, make the following assumptions:

(a) negligible light-load torque;
(b) d.c. generator efficiency 92%;
(c) induction-motor torque proportional to slip;
(d) the pulse is of rectangular shape having a magnitude corresponding to the generator coupling torque which occurs at the mean speed (900 rev/min).

Calculate this torque and the required flywheel effect of the motor-generator set.
Estimate how frequently the pulse could be applied while maintaining its magnitude and duration.

Generator input power = Magnet power/efficiency

$$= I^2R/\eta = \frac{2000^2 \times 0.1}{0.92}$$

Coupling torque at mean speed

$$= \frac{P}{\omega_m} = \frac{2000^2 \times 0.1}{0.92 \times 900 \times 2\pi/60} = \underline{4613\,\text{Nm}}$$

For the motor characteristic; at 800 rev/min, slip = (1000 − 800)/1000 = 0.2
The peak value, T_p (= T_e) = 1500 = ks = $k \times 0.2$ from which k = 7500.

Setting up the mechanical equation: $7500s = 4613 + J\,d\omega_m/dt = 4613 - J\omega_s\,ds/dt$.

Rearranging and substituting $\omega_s = 2\pi \times 50/3 = 104.7$ rad/s:

$$\frac{J \times 104.7}{7500}\frac{ds}{dt} + s = \frac{4613}{7500}$$

$$\frac{J}{71.63}\frac{ds}{dt} + s = 0.615 = s_f$$

The solution is: $$T_e = T_0 + (T_m - T_0)\,(1 - e^{-t_p/\tau_m})$$

Substituting at the end of the pulse: $1500 = 0 + (4613 - 0)\,(1 - e^{-8/\tau_m})$

from which: $$e^{8/\tau_m} = \frac{1}{1 - 1500/4613}$$

and: $$\tau_m = 20.34 = J/71.63$$

hence required inertia $$= \underline{1457\,\text{kgm}^2}$$

After the pulse has been removed, the motor speed will rise and recharge the inertia. The only change in the basic mechanical equation is that the load pulse will be replaced by the light-load torque which in this case is taken to be zero. The factors affecting the time-constant are not changed. Consequently, the slip will have returned to within 2% of its no-load value in a time equal to four time-constants.

Hence rest time must be at least $4\tau_m = 4 \times 20.34 = 81$ seconds.

The frequency of pulse application should not therefore exceed *one every* $1\frac{1}{2}$ *minutes.* If the frequency must be greater than this, then the analysis will have to be different since the minimum torque will no longer be T_0. The method used in Example 6.2 is applicable.

Example 6.18

A d.c. rolling-mill motor operates on a speed-reversing duty cycle. For a particular duty, the field current is maintained constant and the speed is reversed linearly from + 100 to −100 rev/min in a time of 3 seconds. During the constant-speed period, the first pass requires a steady rolling torque of 30 000 lbf ft for 6 seconds and the second reverse pass requires 25 000 lbf ft for 8 seconds. At the beginning and end of each pass there is a 1 second period of no load running at full speed. If the total inertia J referred to the motor shaft is 12 500 kgm², find the r.m.s. torque, r.m.s. power and peak motor power.

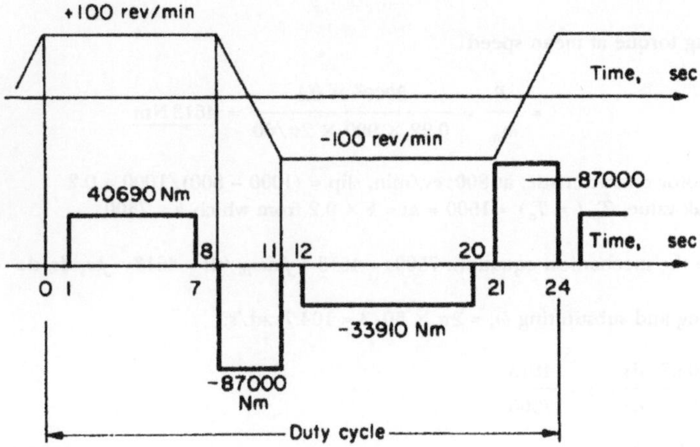

Figure E.6.18

This example, as an introduction to the effect of duty cycles on motor ratings, is somewhat simplified, but sufficiently realistic to illustrate the general method used when the duty cycle is not an intermittent pulse as in the previous two examples.

The figures are drawn from the data and should help to interpret these data. The torques are converted to Nm on multiplication by 746/550. During reversal, the inertia torque is far greater than any no-load torque, which has been neglected, giving:

$$\text{Reversing torque} = J\frac{d\omega_m}{dt} = 12\,500\,\frac{[100-(-100)]\times 2\pi/60}{3} = 87\,000\,\text{Nm}.$$

Since the motor is on fixed field current, the torque is proportional to armature current. Therefore if the r.m.s. value of the torque duty cycle is obtained, it will be proportional to the r.m.s. current of the motor, which determines the motor rating. Since the torque/time curves are all simple rectangles, the r.m.s. value is:

$$T_{rms} = 10^4 \sqrt{\frac{4.069^2 \times 6 + (8.7^2 \times 3) \times 2 + 3.391^2 \times 8}{24}} = \underline{51\,860\,\text{Nm}}.$$

Hence the motor power-rating =

$$\omega_m T_{rms} = \frac{2\pi \times 100}{60} \times 51\,860 = \underline{543\,\text{kW}}.$$

The motor must also be designed to withstand the peak torque and current corresponding to 87 000 Nm. The peak power is

$$\frac{2\pi \times 100}{60} \times 87\,000 = \underline{911\,\text{kW}}.$$

Example 6.19

The motor coach on a 4-car suburban train weighs 50 tonnes (1 tonne = 1000 kg) and the remaining coaches with passengers weigh a further 100 tonnes. All four axles are driven by d.c. series motors supplied through a.c./d.c. rectifiers. If the maximum coefficient of friction before wheel-slip is 0.2, calculate the maximum possible acceleration on level track and also up a gradient of 1 in 120. Allow an extra 10% on the weight for the stored energy of the machinery in rotary motion.

For a particular duty, the train is on level track and travels under the following conditions:

Mean starting current per motor 400 A, maintained by tap changing on the rectifier transformer to increase the voltage uniformly to 750 V on each motor. Acceleration up to 50 km/hour, whereupon the power is shut off and coasting commences.

Braking is imposed at the rate of 3 km/hour/sec.

Resistance to motion is 4.5 kgf/tonne normally but is 6.5 kgf/tonne when coasting.

Calculate:

(a) distance travelled in 2 minutes;
(b) average speed;
(c) energy consumption in watt hours;
(d) specific energy consumption in watt hours/tonne/km.

The characteristics of each series motor are as follows, in terms of motor current, torque converted to kilogram force at the motor-coach couplings, and linear speed in km/hour.

Current	400	320	240	160	120	80	A
Force per motor	2500	1860	1270	650	390	180	kgf
Speed	29	31.5	36.9	44.6	53	72.4	km/h

In the case of linear motion, the force equation is $F = M\,dv/dt$ newtons, where M is the mass in kg and velocity v is in metres/sec. If acceleration a is in the usual units of km/h per second and the effective mass for both translational and rotational stored energy is M_e tonnes:

$$F = (M_e \times 1000) \times \left(a \times \frac{1000}{60^2} \right) \text{ newtons}$$

in kgf: $$F = \frac{1}{9.81} \times \frac{10^6}{3600} \times M_e a = \underline{28.3\ M_e a \text{ kgf.}} \quad (6.17)$$

In the example: $M_e = (50 + 100) \times 1.1 = 165$ tonnes.

Maximum force before slipping at the driving axles $= (50 \times 0.2) \times 1000 = 10\,000$ kgf.

Train resistance $= (50 + 100) \times 4.5 = 675$ kgf.

Downwards force due to gravity on a gradient of 1 in 120 $= \dfrac{(50 + 100)}{120} \times 1000 = 1250$ kgf.

The motor-torque-pulsations due to the rectifier supply will be neglected.

From eqn (6.13), on level track: $a = \dfrac{10\,000 - 675}{28.3 \times 165} = \underline{2 \text{ km/h/sec}}$ (1.25 mph/sec);

on gradient: $a = \dfrac{10\,000 - 675 - 1250}{28.3 \times 165} = \underline{1.73 \text{ km/h/sec.}}$

Calculation of the speed/time curve will also use eqn (6.17) in the form:

Accelerating force $= 28.3 M_e \dfrac{\Delta v}{\Delta t}$ from which $\Delta t = \dfrac{28.3 M_e \times \Delta v}{\text{Accel. force}}$

The method is similar to Example 6.6 for rotary motion.
The motor characteristics are repeated here in terms of the total force.

Current	400	320	240	160	120	80	A
Total force	10 000	7440	5080	2600	1560	720	kgf
Speed v	29	31.5	36.9	44.6	53	72.4	km/h
Accel. force (–675)	9325	6765	4405	1925	885	45	kgf

Mean acc. force	9325	8045	5385	3165	1405	465	kgf
Δv	29	2.5	5.4	7.7	8.4	19.4	km/h
$\Delta t = \dfrac{28.3 M_e \Delta v}{\text{mean force}}$	14.5	1.45	4.5	11.36	27.9	194.8	s

Time t from start	14.5	15.95	20.45	31.8	69.7	264.5	s

(a) From these results, the speed/time and current/time curves can be plotted as on the figure. At 50 km/h, power is shut off and coasting commences at a decelerating rate of:

$$\frac{\text{Resistance force}}{28.3M_e} = \frac{6.5 \times 150}{28.3 \times 165} = 0.209\,\text{km/h/sec.}$$

From a time of 2 minutes, the braking line can be drawn in at the specified rate of 3 km/h/sec to complete the speed/time curve. The distance travelled is now the area under this curve from $v = dx/dt$ so $x = \int v\,dt$. By counting squares, or any other method, this area is 4700 km/h × seconds and dividing by 60^2 gives distance as 1.306 km (0.81 mile).

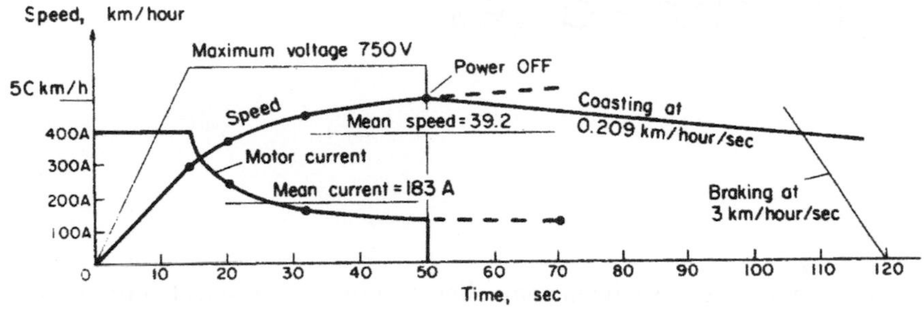

Figure E.6.19

(b) Mean speed $= \dfrac{1.306}{120/3600} = \underline{39.2\,\text{km/h}}$ (24.3 mph).

(c) Energy in first accelerating period is (mean voltage) × 400 A × $\Delta t = \dfrac{750}{2} \times 400 \times 14.5$

and in the second period is: 750 V × (mean current) × $\Delta t = 750 \times 183 \times (50 - 14.5)$.

For the four motors, in watt hours:

$$\text{energy consumed} = \frac{4}{3600}(375 \times 400 \times 14.5 + 750 \times 183 \times 35.5) = \underline{7830\ \text{watt hours}}$$

the mean current during the second period being obtained from the area under the current/time curve.

(d) This performance figure is $\dfrac{7830}{150 \times 1.306} = \underline{40\ \text{watt hours/tonne km}}$ (63.4 wh/ton mile).

7 Power-electronic/electrical machine drives

Electrical machines have been controlled through power-electronic circuits since the early days of mercury-arc-type rectifiers. The development of compact power semiconductors has increased such applications enormously. Only a limited amount of power can be dissipated by such a small device and so it must be operated as a switch, either open – rated at maximum circuit voltage and virtually zero current; or closed – rated at maximum circuit current and the relatively small forward-voltage-drop. Power-electronic circuits therefore involve ON/OFF switching and waveforms which are neither pure d.c. nor pure sinusoidal a.c., but the average d.c. voltage or the average a.c. voltage and fundamental frequency can be controlled as desired. It is no longer possible to calculate the complete performance so simply as for the ideal waveforms because of the harmonics introduced, but even if these are neglected the errors in the general electromechanical performance-calculations are not usually serious.

An electrical drive consists of an electrical motor and control gear. In the case of a variable-speed application, it will usually involve power-electronic converter(s) and control equipment, which may include a microcomputer. Although many applications require only a fixed-speed motor, variable-speed drives are common throughout industry and in the domestic market. The objective of this chapter is to introduce power-electronic equipment and the equipment arrangements and control schemes used in variable-speed drives. Both a.c. and d.c. drives are considered; a.c. drives form the majority of large drives currently being installed.

The power-electronic devices briefly described below are the semiconductor switches most commonly used in providing efficient and compact power-control of electric motors. For production of specified speed, position and torque, d.c. machine voltage and current require control; in addition, a.c. drives often require control of frequency. At high power levels, efficiency-related quantities such as harmonic content and power-factor must also be considered. For all drives, electromagnetic interference and compatibility must now be considered more carefully.

There are two principal reasons for employment of variable-speed drives:

- To drive a mechanical load at variable speed.
- To allow a motor to be accelerated in an efficient and controlled way.

In addition, reversal and recovery of stored kinetic energy in the load during braking may be necessary for overall reduction of energy consumption and associated losses.

7.1 Power-electronic devices

Symbols for the power-electronic devices most commonly used in variable-speed drives are shown in Figure 7.1.

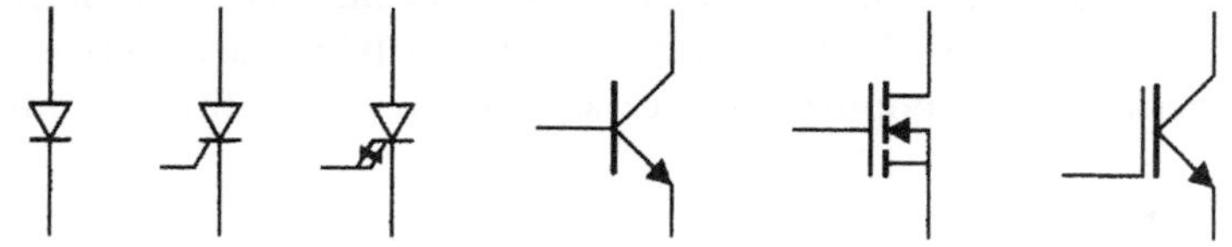

Figure 7.1 *Power-electronic devices.*

Power diodes

Familiarity with the semiconductor P-N junction is assumed. Important parameters for a diode are its blocking voltage rating, its forward current voltage drop (which define its thermal capability and power handling) and its reverse recovery time. For faster switching, removal of stored charge must be rapid, and this is achieved by gold or platinum doping. Fast switching is obtained at the penalty of greater voltage drop and higher leakage current. At lower voltages, the fast-recovery epitaxial diode (FRED) is available. Other 'fast' devices such as Schottky diodes exist but conduction losses prevent their use at other than low power level.

Power bipolar transistor

The characteristics of the NPN transistor should be familiar. However, at high power levels, power loss in the device prohibits its use in the linear amplifying region, where collector current is proportional to the controlling (base) current. The device is therefore operated as a solid-state switch, in either cut-off or saturated condition.

Power MOSFET

The MOSFET (metal-oxide silicon field-effect transistor) is an attractive device for power use, as it has a very fast switching speed (giving minimal switching loss) and no second breakdown problem. As a voltage-driven device, its drive power is small enough to be provided by an integrated circuit. However, the forward voltage drop is larger than the saturation voltage of a bipolar transistor and MOSFETs are not available in the same power range as bipolar transistors.

IGBT

The IGBT (insulated-gate bipolar transistor) is a combination of a MOSFET input stage driving a bipolar output stage. IGBT ratings are similar to those available for bipolar transistors but driving requirements are those of a MOSFET. The ratings of available IGBTs are increasing and the devices are now popular for d.c./a.c. inverters into the MW range.

SCR

The SCR (silicon controlled-rectifier), commonly referred to as the thyristor, is only one member of the thyristor family. It is a 4-layer device (PNPN) and hence contains a blocking (NP) junction. This junction may be broken down (avalanched) by high temperature or voltage, but the normal method of 'firing' the SCR to initiate conduction is by injection of carriers from the gate. Once in avalanche, this junction cannot regain its blocking ability until restored to the 'off' state by removal of anode current, i.e. the device will not turn itself off in the presence of unidirectional current. 'Inverter-grade' SCRs were once common for high-power drives but have now been replaced by newer devices. However, the rugged 'rectifer-grade' SCR is still applied to the largest variable-speed drives. The relatively small amount of gate power can trigger a large anode current and blocking capability extends to several kV. To increase the gate power and thereby speed turn-on, larger devices use a 'pilot' SCR to trigger the main device.

GTO

The GTO (gate-turn-off thyristor) is a device which overcomes the problem of turning off an SCR without the use of separate circuitry to extinguish the anode–cathode current. Abstraction of gate current allows the device to be turned off. This does, however, require a substantial quantity of energy to be transferred from the gate.

Transducers and sensors

The closed-loop control schemes referred to in Chapters 6 and 7 require sensing of physical quantities to generate a useable feedback signal which may be compared with the input reference. The control signal is often a ±10 volt d.c. signal or is converted to digital form if required. A few sensors provide direct digital output.

Current may be measured by the voltage drop across a small resistor but is more commonly sensed using a Hall-effect transducer, which produces a

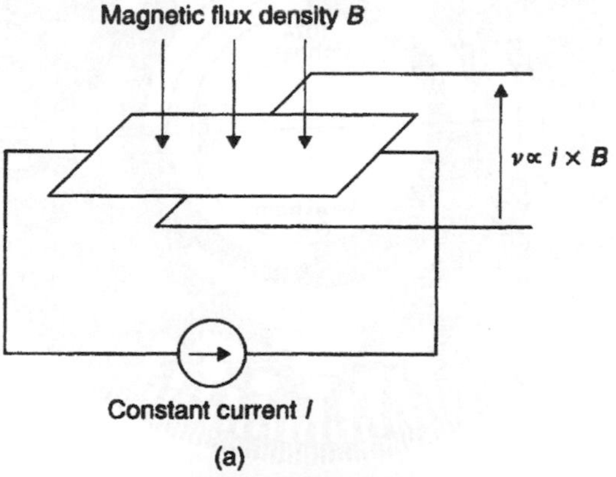

Figure 7.2 *Sensors: (a) Hall-effect sensor; (b) Incremental encoder disc (courtesy of Hohner (UK) Ltd)*

voltage proportional to the electromagnetic field (of flux density B) created by the current being measured (Figure 7.2a).

Speed may be measured using a d.c. tachogenerator, as in Example 6.12, but digital devices are now more common. The output is a train of pulses produced when poles or optical lines pass a detector. The count of pulses or time between pulses provides a direct digital measure of shaft speed.

A similar form of sensor is used for digital position-sensing. In an 'incremental' encoder, the passage of a transparent disc (such as that

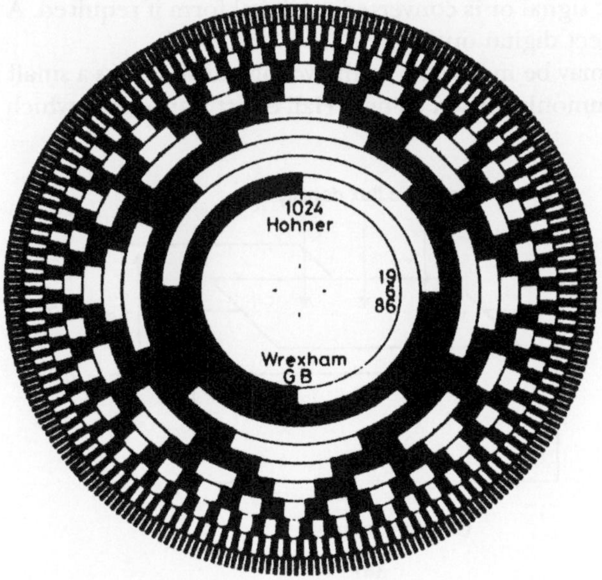

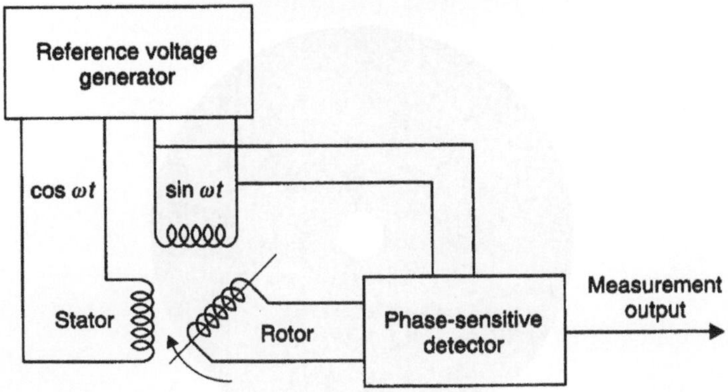

(d)

Figure 7.2 *(c) Absolute encoder disc (courtesy of Hohner (UK) Ltd); (d) Resolver.*

shown in Figure 7.2b) with lines on its periphery causes a series of pulses to be received by an optoelectronic sensor which detects light through the disc from an infra-red or laser source. A 'start' marker is usually provided on a second channel. Position from the reference is indicated by incrementing a counter with the sensor output.

An 'absolute' encoder has multiple channels, each of which reads one binary digit; a typical 10-line encoder disc is shown in Figure 7.2c. This will generate a 10-bit binary number (hexadecimal numbers 0000 to 3FFF) and one revolution will be divided into 2^{10} (1024) portions, each giving a unique code and indicating shaft position in 0.35° increments. When all digits are 'ones' (hexadecimal 3FFF), allowance must be made by the digital processor to ensure that the following transition to 0000 is registered as only a 1-bit change.

The same problem would be encountered in a potentiometer, which is the simplest device for providing an output voltage proportional to angle of rotation, at the point of transition from maximum to minimum voltage. Instrumentation potentiometers can be rotated continuously and have an active length of around 355°.

The resolver (Figure 7.2d) is a device which can be rotated without a sudden transition in output, but at the penalty of an analogue output which is linear only around the reference position. It is basically a high-frequency transformer with a rotatable reference and two stationary windings; a synchro is a similar device with three windings. The resolver provides theoretically stepless position information and linear readout of position is obtainable via a synchro-to-digital converter integrated circuit, which could typically offer up to 16-bit resolution.

Motor torque is less frequently sensed, but the torque at the coupling can be measured by load cells on the motor frame, or by the effect of shaft torsion. In either case, the signal requires processing or amplification before it can be used.

7.2 Chopper controlled d.c. machine

This is one of the simplest power-electronic/machine circuits. With a battery, it is currently the most common electric road vehicle controller; the 'chopper' is also used for some d.c. rail traction applications. The technique is similar to that used in low-power switched amplifiers and isolating d.c./d.c. converters. For motoring, power is switched ON and OFF rapidly by a power-electronic switch (an IGBT in Figure 7.3a). For fixed field excitation (current or permanent magnet) the motor speed is almost proportional to the modulation factor 'δ', since this determines the average d.c. voltage applied. Its value is:

$$\delta = \frac{t_{ON}}{t_{ON} + t_{OFF}}$$

When the IGBT receives gate drive, it appears as a low-resistance switch and thus the supply voltage, minus the back-e.m.f. of the motor at that time, appears across the resistance and inductance of the armature. As in

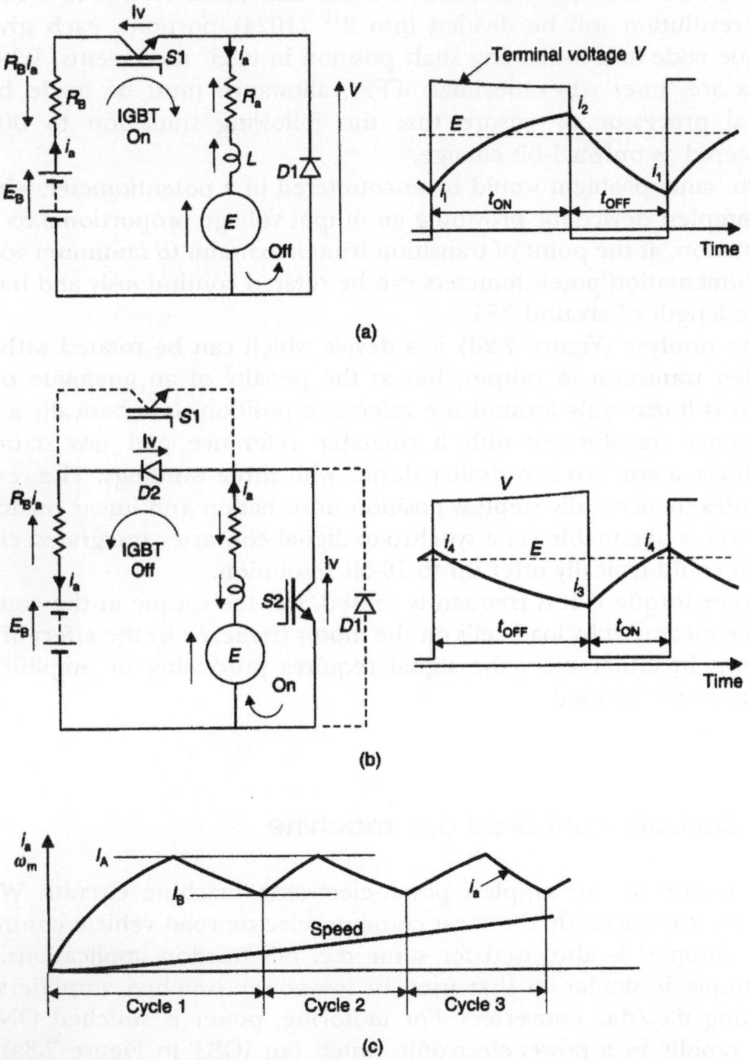

Figure 7.3 *Chopper-fed d.c. machine: (a)* MOTORING *(Motoring conventions); (b)* GENERATING *(Generating conventions); (c) Acceleration between limits.*

Example 6.1, the current rises exponentially, as shown in Figure 7.3a. At the end of t_{ON}, the IGBT turns off and the current path from the supply is removed. As the armature inductance will produce a voltage to oppose the decay of current, a freewheel diode is provided across the motor to clamp this voltage and prevent the breakdown voltage of the IGBT being exceeded. During t_{OFF}, the terminal voltage becomes that of the diode voltage drop, which is also equal to the machine e.m.f. plus the voltages in the resistance and inductance, as long as current continues to flow. The voltage across the inductance will have changed sign as the current is decreasing. If the current does fall to zero at low values of δ (i.e. discontinuous current), the terminal voltage rises to that of the machine e.m.f. during this short part of the cycle.

For constant flux, the e.m.f. may be assumed constant during the cycle. If the motor is series excited, then the e.m.f. does vary and can be expressed as $k_{fs} i_a$ where k_{fs} is the mean slope of the magnetisation curve in volts per amp at the appropriate speed, and over the small current-pulsation range from i_1 to i_2.

The volt drops of the semiconductors can be included as constant voltages of say 1 V and the battery resistance R_B is in circuit during the ON period. The equations are:

	IGBT/ON	IGBT/OFF
	$E_B = E + (R_a + R_B) i_a + Lpi_a + 1$ (7.1a)	$0 = E + R_a i_a + Lpi_a + 1$ (7.1b)
Rearranging:	$\frac{L}{R} pi_a + i_a = \frac{E_B - E - 1}{R}$	$\frac{L}{R_a} pi_a + i_a = \frac{-E - 1}{R_a}$
	$= I_{max}$	$= I_{min}$
Solutions: [Eqn (6.2b)]	$i_2 = i_1 + (I_{max} - i_1)(1 - e^{-t_{ON}/\tau})$	$i_1 = i_2 + (I_{min} - i_2)(1 - e^{-t_{OFF}/\tau})$

Note in the above that E_B is the battery e.m.f., $R = (R_a + R_B)$ and the two time-constants are different because in the OFF condition the resistance is just R_a.

For the series motor, the e.m.f. = $k_{fs} i_a$ so the equations become:

	$E_B = k_{fs} i_a + R i_a + Lpi_a + 1$ (7.2a)	$0 = k_{fs} i_a + R_a i_a + Lpi_a + 1$ (7.2b)
Rearranging:	$\frac{L}{R + k_{fs}} pi_a + i_a = \frac{E_B - 1}{R + k_{fs}}$	$\frac{L}{R_a + k_{fs}} pi_a + i_a = \frac{-1}{R_a + k_{fs}}$

The equations are still of first order but the values are different and the response faster as shown by the reduced time-constant. Increase of current causes increase of flux.

From the viewpoint of energy conservation, it is important to consider regenerative braking where possible and the chopper circuit lends itself simply to this.

The circuit inductance serves as an energy store and the motor is to be braked by extracting mechanical energy from the load and returning it to the supply, but the e.m.f. is smaller than the supply voltage. Semiconductor switch S1 and diode D1 carry out the motoring function as outlined above and now shown (connected by broken lines) in the circuit of Figure 7.3b. When braking is to occur, S2 is turned on, short-circuiting the motor and causing the current to build up rapidly. When a maximum current is reached, S2 is turned off, forcing the current to decay. di_a/dt becomes negative, so $L\,di_a/dt$ reverses to supplement the e.m.f., allowing the motor terminal voltage to exceed that of the supply. Generated current then flows via D2 to the supply until the level of braking power falls below a set limit. S2 is then turned on again and the cycle repeats, as shown in Figure 7.3b. The equations are:

	IGBT ON	IGBT OFF
	$0 = E - R_a i_a - Lpi_a - 1$ (7.3a)	$E_B = E - (R_a + R_B) i_a - Lpi_a - 1$ (7.3b)
Rearranging:	$\frac{L}{R_a} pi_a + i_a = \frac{E - 1}{R_a} = I_{max}$	$\frac{L}{R} pi_a + i_a = \frac{E - E_B - 1}{R} = I_{min}$
Solutions:	$i_4 = i_3 + (I_{max} - i_3)(1 - e^{-t_{ON}/\tau})$	$i_3 = i_4 + (I_{min} - i_4)(1 - e^{-t_{OFF}/\tau})$

All the equations are first order and the current waveforms will therefore be exponential in form. For the series machine when regenerating, the exponential of the transient term is positive. This could be expected, since the self-excited generator is basically unstable and only restrained by the onset of saturation and demagnetising effects of armature m.m.f.

Acceleration and deceleration between current limits

The operating frequency of a chopper may range from 100 Hz at high power levels to tens of kHz at low powers. As the electrical-circuit time-constant is much less than mechanical ones, speed changes little during t_{ON} or t_{OFF}. Hence the e.m.f. will remain constant for a separately excited or permanent-magnet machine. The chopping frequency may be constant or may be arranged such that the 'hysteresis band' over which the current varies from maximum to minimum is maintained, so long as there is scope for further adjustment of the modulation factor. The mean torque is then 'constant' and the rate of change of speed will be constant. Thereafter the current will fall until steady-state speed is reached. Figure 7.3c shows the initial part of such a current/time, speed/time schedule.

Example 7.1

An electric train is powered by four 750-V d.c. series motors. The motor resistance and inductance are respectively 0.25 Ω and 6 mH. A total line inductance of 10 mH and resistance of 0.1 Ω is in series with the supply. The fixed voltage from the traction supply is regulated by the chopper of Figure 7.3a, operating at 200 Hz. When the machine is running at 800 rev/min, the generated e.m.f. per ampere can be taken as an average value of 0.79 V/A. If the modulation factor δ is 80%, calculate the maximum and minimum currents, allowing 2 V for semiconductor loss, and find the average torque of each motor.

The mass of the fully loaded train is 120 tonnes and its resistance to motion on level track is 500 N. Each motor is geared to the wheel of the motor coach by a 3:1 ratio and the coach wheel tread diameter is 1.0 m. Estimate the train's rate of acceleration under the above conditions.

When the IGBT is turned on, the 750 V supply voltage is connected to the motor circuit, which has e.m.f. $0.79i_a$, total circuit resistance 0.35 Ω, total circuit inductance 0.016H and 2 V semiconductor voltage drop.

From the equations in Example 6.1 with the e.m.f. added, or from eqns (7.2a) and (7.2b) in the tables above,

$$V = E + Ri_a + L\,di_a/dt + 2$$

$$750 = 0.79i_a + 0.35i_a + 0.016\,di_a/dt + 2$$

Rearranging: $656 = i_a + 0.014\,di_a/dt$

When the current reaches a theoretical maximum, i.e. $di_a/dt = 0$, $i_a = 656$ A. From Example 6.1, with $I_f = 656$, $\tau = 0.014$,

$$i_a = 656(1 - e^{-t/0.014}) + i_1 e^{-t/0.014}$$

where i_1 is the current at the start of t_{ON}.

At 200 Hz chopping frequency with $\delta = 0.8$, $t_{ON} = 4$ ms.

$$e^{-0.004/0.014} = 0.751$$

$$\therefore \quad i_2 = 656(1 - 0.751) + 0.751 i_1 = 163 + 0.751 i_1$$

where i_2 is the value of i_a at the end of t_{ON}, i.e. the start of t_{OFF}.

During t_{OFF}, the total circuit resistance and inductance are 0.25 Ω and 6 mH, respectively, and the semiconductor loss (2 V) is that of the freewheel diode. The supply voltage is switched OFF.

$$\therefore \quad 0 = 0.79 i_a + 0.25 i_a + 0.006\, di_a/dt + 2$$

$$\therefore -1.92 = i_a + 0.0058\, di_a/dt$$

Assuming that this path were connected permanently, i_a would eventually decay to –1.92 A, assuming that the motor maintained constant speed and the diode could continue to conduct, even when the current is negative. From Example 6.1 again,

$$i_a = -1.92(1 - e^{-t/0.0058}) + i_2 e^{-t/0.0058}$$

since i_2 is the current at the start of t_{OFF}. Assume that the current has gained a steady variation. t_{OFF} at 200 Hz, $\delta = 0.8$, lasts for 1 ms, after which i_a has reached i_1 again.

$$e^{-0.001/0.0058} = 0.842$$

$$\therefore \quad i_1 = -1.92(1 - 0.842) + 0.842 i_2 = -0.303 + 0.842 i_2$$

Eliminating i_1, $\quad i_2 = 163 + 0.751(-0.303 + 0.842 i_2)$

$$\therefore i_2(1 - 0.632) = 163 - 0.228$$

$$\therefore \quad i_2 = \underline{442\text{ A}}$$

$$\therefore \quad i_1 = -0.303 + 0.842 \times 442 = \underline{372\text{ A}}$$

For a series motor, torque $= k_\phi \times i_a = \dfrac{0.79 i_a \times i_a}{\omega_m}$

In this case, the average torque will be proportional to the average value of $(i_a)^2$, i.e. the mean-square current. As the electrical L/R time constant is considerably larger than t_{ON} or t_{OFF}, it is reasonable to assume that the graph of current against time roughly follows straight lines.

From eqn (6.16), the area under the current-squared curve for t_{ON} is:

$$0.004(372^2 + 372 \times 442 + 442^2)/3 = 664.2$$

and the area under the current-squared curve for t_{OFF} is:

$$0.001(372^2 + 372 \times 442 + 442^2)/3 = 166.1$$

The mean current-squared is then $\dfrac{664.2 + 166.1}{0.005} = 166\,060$

$$\therefore \text{Torque} = \frac{0.79 \times 166\,060}{800 \times 2\pi/60} = \underline{1566\,\text{Nm}}$$

$$\therefore \text{Tractive force at wheel} = \frac{\text{torque} \times \text{gear ratio}}{\text{wheel radius}} = \frac{1566 \times 3}{0.5} = 9.4\,\text{kN}$$

For four motors, total tractive force = 37.6 kN. With 500 N rolling resistance, the tractive force available for acceleration is 37.1 kN. The train mass is 120 tonnes.

$$\therefore \text{acceleration} = 37.1/120 = \underline{0.31\,\text{m/s}^2}$$

Example 7.2

An electric automobile is powered by a permanent magnet d.c. motor rated at 72 V, 200 A. The motor resistance and inductance are, respectively, 0.04 Ω and 6 mH. The chopper controller maintains the motor current between 200 and 220 A, as long as sufficient voltage is available, and k_ϕ is 0.233 V per rad/s. The battery resistance is 0.06 Ω and semiconductor voltage drops are 1.2 V and 1 V in the semiconductor switch and freewheel diode, respectively. The total mechanical load referred to the motor shaft at a motor speed of 1000 rev/min is 24 45 Nm and the total inertia referred to the motor shaft is 1.2 kgm^2.

Calculate the chopping frequency and modulation factor for the conditions given. Calculate the acceleration and deceleration rates in rad/s^2 and, assuming these rates are maintained, determine the time to accelerate from zero to 1000 rev/min and to decelerate to zero from 1000 rev/min.

At 1000 rev/min, 104.7 rad/s, the generated e.m.f. is $104.7k_\phi$ = 24.4 V. Total resistance when motor connected to supply = 0.1 Ω.

Acceleration:

For the ON period, with battery voltage,

$$V = E \quad + Ri_a \quad + L\,di_a/dt \qquad + \text{IGBT drop}$$

$$\therefore 72 = 24.4 + 0.1i_a + 0.006\,di_a/dt + 1.2$$

and rearranging in standard form of eqn (6.1):

$$464 = i_a + 0.06\,di_a/dt$$

The maximum current which would be reached if the IGBT were on continuously at this motor speed occurs when $di_a/dt = 0$, i.e. $I_f = 464\,\text{A}$.

The standard solution to this equation for an initial current $i_1 = 200\,\text{A}$ is, from eqn (6.2a) or Example 7.1:

$$i_a = 464(1 - e^{-t/0.06}) + 200e^{-t/0.06}$$

At the end of t_{ON}, the current is 220 A

$$\therefore\ 220 = 464(1 - e^{-t_{ON}/0.06}) + 200e^{-t_{ON}/0.06}$$

$$\therefore\ t_{ON} = 0.00473 \text{ seconds}$$

During the OFF period, the resistance is only that of the armature and the semiconductor voltage drop is that of the freewheel diode.

$$\therefore\ 0 = 24.4 + 0.04i_a + 0.006\ di_a/dt + 1$$

Rearranging: $-635 = i_a + 0.15\ di_a/dt$

From eqn (6.2a), $i_a = -635(1 - e^{-t/0.15}) + i_2 e^{-t/0.15}$

where i_2 is the current at the start of t_{OFF}, i.e. 220 A. At the end of t_{OFF}, $i_a(t) = 200$ A.

$$\therefore\ 200 = -635(1 - e^{-t_{OFF}/0.15}) + 220e^{-t_{OFF}/0.15}$$

$$\therefore\ t_{OFF} = 0.00355 \text{ seconds}$$

$$\therefore\quad \delta = \frac{0.00473}{0.00473 + 0.00355} = \underline{0.571}$$

and chopping frequency = $(0.00473 + 0.00355)^{-1}$ = $\underline{120.8 \text{ Hz}}$

Deceleration:

For regenerative braking, the circuit is as shown in Figure 7.3b with the suggestion to use a generating convention, namely current direction in the same sense as the machine e.m.f., giving eqns (7.3a) and (7.3b) as in the third table of equations.

Hence, for the ON period with the machine short-circuited,

$$0 = E \quad - Ri_a \quad - L\ di_a/dt \quad - 1.2$$

$$= 24.4 - 0.04i_a - 0.006\ di_a/dt - 1.2$$

Rearranging:

$$580 = i_a + 0.15\ di_a/dt$$

for which the solution is:

$$\therefore \quad i_a = 580(1 - e^{-t/0.15}) + i_3 e^{-t/0.15}$$

i_3 is the current at the start of t_{ON}, i.e. 200 A; $i_a = i_4 = 220$ A at the end of t_{ON} and would have reached a maximum of 580 A eventually.

$$\therefore \; 220 = 580(1 - e^{-t_{ON}/0.15}) + 200e^{-t_{ON}/0.15}$$

$$\therefore \; t_{ON} = 0.00811 \text{ seconds}$$

When the current is directed through the diode towards the supply in the OFF period, the semiconductor voltage drop is 1 V and the supply resistance is added to that of the armature.

From eqn (7.3b):

$$72 = 24.4 - 0.1i_a - 0.006 \, di_a/dt - 1$$

Rearranging:

$$-486 = i_a + 0.06 \, di_a/dt$$

for which the solution is:

$$i_a = -486(1 - e^{-t/0.06}) + i_4 e^{-t/0.06}$$

$i_4 = 220$ A is the current at the start of t_{OFF} and $i_3 = 200$ A at the end of t_{OFF}.
Substituting:

$$\therefore \; 200 = -486(1 - e^{-t_{OFF}/0.06}) + 220e^{-t_{OFF}/0.06}$$

$$\therefore \; t_{OFF} = 0.00172 \text{ seconds}$$

$$\therefore \quad \delta = \frac{0.00811}{0.00811 + 0.00172} = \underline{0.825}$$

and chopping frequency = $(0.0195 + 0.003895)^{-1}$ = 101.7 Hz

The average value of i_a is $\frac{1}{2}(220 + 200)$ = 210 A

giving a 'constant' mean torque T_e $= 210k_\phi$

$$= 210 \times 0.233 = \underline{48.9\,\text{Nm}}$$

During acceleration,

$$\mathrm{d}\omega_m/\mathrm{d}t = \frac{T_e - T_m}{J} = \frac{48.9 - 24.45}{1.2} = \underline{20.38\ \mathrm{rad/s^2}}$$

Accelerating time to 1000 rev/min (104.7 rad/s) = 104.7/20.38

$$= \underline{5.14\ \text{seconds}}$$

During deceleration, the torque is the same magnitude but reversed.

$$\text{Constant } \mathrm{d}\omega_m/\mathrm{d}t = \frac{T_e - T_m}{J} = \frac{-48.9 - 24.45}{1.2} = \underline{-61.13\ \mathrm{rad/s^2}}$$

Deceleration time = 104.7/61.13 = $\underline{1.71\ \text{seconds}}$

7.3 Thyristor-converter drives

Although manufacture of d.c. motors, particularly in larger sizes, has declined in favour of a.c. drives, the simplicity and low cost of the d.c. motor controlled from a controlled rectifier retains a market share for such systems. Rectifiers are also required to produce d.c. sources for inverters used in a.c. drive systems. Both single-phase and three-phase rectifiers are employed.

Single-phase rectifier

Single-phase rectifiers for producing d.c. supplies for inverters are usually uncontrolled, i.e. composed of diodes only; the variation of voltage and frequency is carried out by the output stage, usually by pulse width modulation (PWM – see Section 7.4). However, single-phase controlled rectifiers are still employed as variable voltage sources for d.c. servomotors used in control systems. Figure 7.4a shows the circuit of a single-phase fully controlled bridge rectifier. The average rectifier output from Appendix B (with $m = 2$ for single phase) is $2E_p(\cos\alpha)/\pi$, where E_p is the peak a.c. line voltage and α is the firing-delay angle. The e.m.f. of the motor is considered constant and the net voltage available to drive current through the impedances is the rectifier output voltage shown less the e.m.f. In Figure 7.4b, $\alpha < 90°$; it is assumed that the motor has sufficiently high inductance to maintain current over short intervals when the rectifier's output voltage is negative. In Figure 7.4c, $\alpha > 90°$ and the average output voltage of the rectifier is negative. Power flow is therefore from the load to the supply (inversion); this condition can only be maintained if the motor

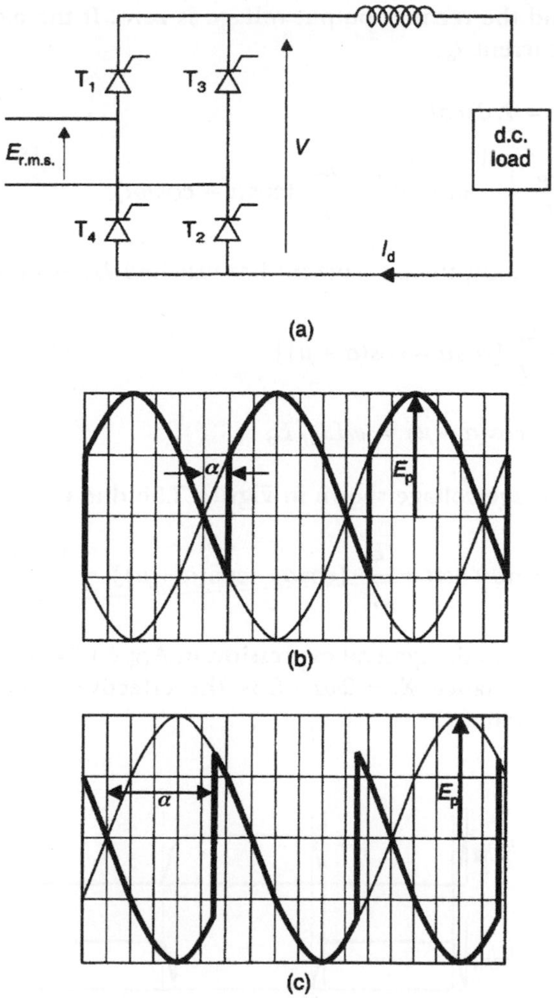

Figure 7.4 *Fully controlled single-phase bridge rectifier: (a) Circuit; (b) Output voltage, $\alpha < 90°$; (c) Output voltage, $\alpha > 90°$.*

e.m.f. is aiding the flow of current, i.e. generating. If inversion is not required, a simpler 'half-controlled' bridge formed of two SCRs and two diodes may be used. The waveforms in Figures. 7.4b and 7.4c assume that the supply current can reverse instantaneously when there is a change in the SCR conduction pattern. However, presence of inductance in the supply prevents sudden reversal and the result is that the supply voltage is committed for a short interval in supporting the $L\,di/dt$ in the supply inductance. During this 'overlap' interval, all four devices conduct (the load current rising from zero in one pair and decaying towards zero in the other pair) and the rectifier output voltage is zero. If the a.c. current is i and the d.c. current I_d,

$$E_p \sin \omega t = L\, di/dt$$

$$\therefore \quad i = \frac{E_p}{L}\int_{\alpha}^{\omega t} \sin \omega t dt = \frac{E_p}{\omega L}(\cos\alpha - \cos\omega t)$$

At angle $(\alpha + \mu)$, say, the a.c. current has reached I_d, as in Figure 7.5a.

$$\therefore \quad I_d = \frac{E_p}{\omega L}[\cos\alpha - \cos(\alpha + \mu)]$$

$$\therefore \quad \cos\alpha - \cos(\alpha + \mu) = \omega L I_d / E_p$$

The loss of average voltage shown in Figure 7.5b due to 'overlap' is

$$\frac{E_p}{\pi}\int_{\alpha}^{\alpha+\mu} \sin \omega t \, d\omega t = \frac{E_p}{\pi}[\cos\alpha - \cos(\alpha + \mu)] = \frac{\omega L I_d}{\pi}$$

This corresponds to the general expression in Appendix B, with $m = 2$ and commutating reactance $X_c = 2\omega L$. L is the effective supply inductance between a.c. terminals.

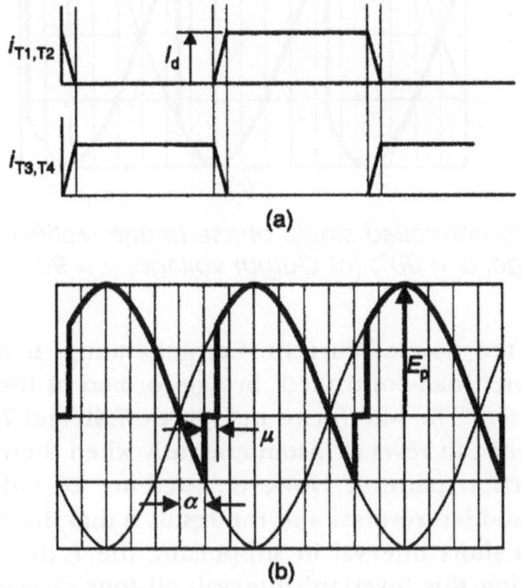

Figure 7.5 *Effect of overlap: (a) Device currents; (b) Output voltage.*

The a.c. current taken by the rectifier with a large inductance in the load consists of rectangular blocks of current. When a half-controlled rectifier bridge is used, the length of each block will vary with α. If the load is capacitive, e.g. for a constant-voltage supply for an inverter, the a.c. current will be in the form of large, short pulses. The level of supply harmonic-currents is now subject to international standards; electrical drives, which consume a high proportion of generated power, must have provision for reduction of current harmonics. At low power levels, switched mode methods based on standard integrated circuit controllers may be used. At high power levels, filtering may be used. A passive filter (Figure 7.6) adds

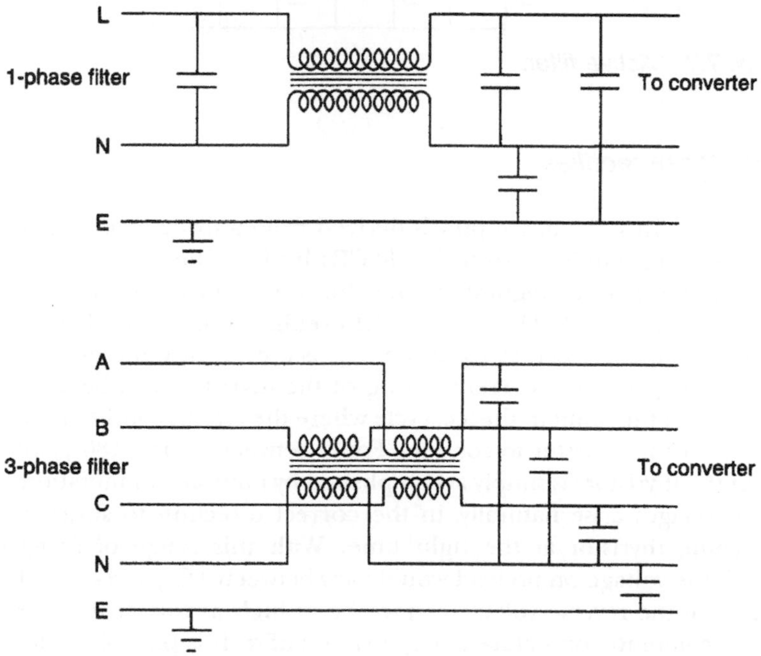

Figure 7.6 *Passive line filters.*

a.c. energy storage which allows the harmonics to be supplied from the filter. Active waveshaping is a more modern technique and is suitable for high-power drives. Two principal methods are emerging. The first uses pulse-width modulation, i.e. splitting of the current conduction period into short blocks of unequal length defined by a modulating sine wave (see Section 7.4). Turn-off switches are required, either in the rectifier or in the d.c. output, to carry out the necessary switching. The second method uses a separate, active line-power-conditioner (Figure 7.7). This is more expensive, but can be added to an existing system.

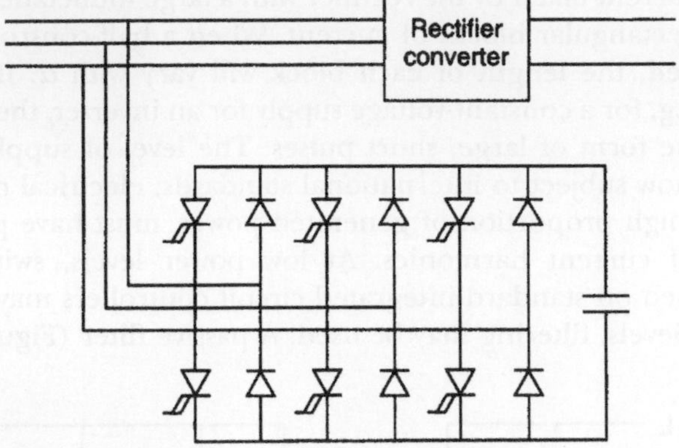

Figure 7.7 *Active filter.*

Three-phase rectifier

For larger drives, the d.c. supply is derived from a three-phase a.c. system, generally using a fully controlled (all-SCR) bridge. The circuit diagram for the 3-phase bridge configuration and the terminal-voltage waveforms are shown on Figure 7.8. The average 'Thevenin' e.m.f. behind the bridge circuit can be expressed as $E_{do} \cos \alpha$. E_{do} is the mean value of the bridge output voltage on zero current. Firing of the thyristors can be delayed by angle α from the point in the a.c. cycle where the circuit conditions are first suitable for the thyristor to conduct. For α between 0° and 180°, switching over of the thyristors is simply accomplished by natural commutation, since circuit voltages arise naturally, in the correct direction to switch off the conducting thyristor at the right time. With this range of firing-angle control, the voltage on no load would vary between $+E_{do}$ and $-E_{do}$, though in practice the reverse voltage cannot be as high as this, from considerations of commutation failure at large values of α. It is possible to linearise the transfer characteristic by suitable control in the firing circuits. The effect of supply and transformer-leakage inductance on the delay of current transfer (commutation) from thyristor to thyristor is to cause a voltage drop which is proportional to current, so that the d.c. terminal voltage can be expressed finally as $V = E_{do} \cos \alpha - kI$. See Appendix B for detailed explanation of these voltage relationships.

The same terminal voltage appears across the machine, for which it can be expressed as $V = E_m + RI$. Both of these expressions average out the harmonics, assume that the d.c. circuit inductance is high and that the current is continuous. They enable the steady-state behaviour to be calculated and can be represented graphically by straight lines on the two right-hand quadrants of a V/I, 4-quadrant diagram,[1] see Figure 7.9.

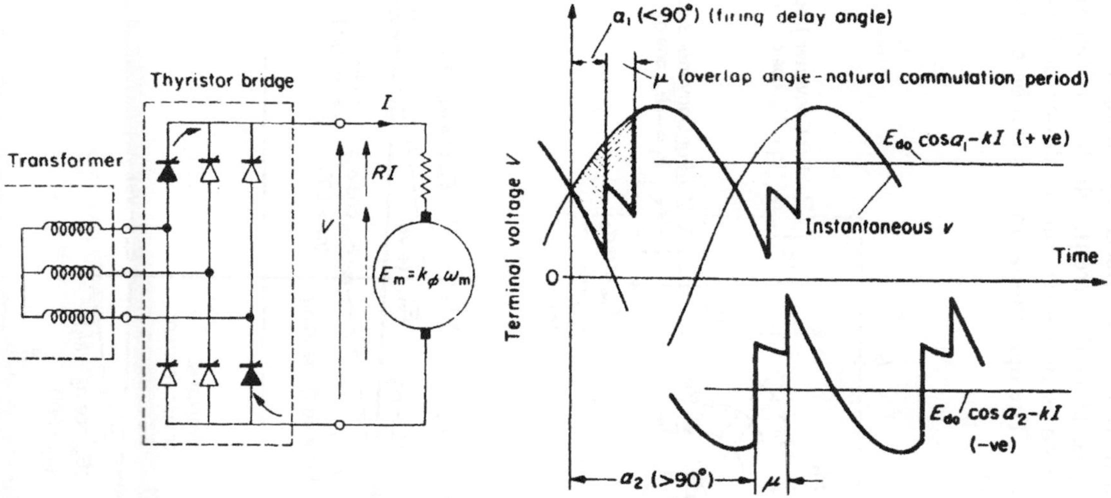

Figure 7.8 *3-Phase thyristor converter/d.c. machine system.*

The current in the circuit is unilateral, but the voltage polarity of either element can be reversed and hence cause reverse power flow. For the power-electronic circuit this means making α greater than 90°, when the rectifying action is changed to inverting action, from d.c. to a.c. For the machine, the polarity can be reversed by changeover of armature connections; by reversal of field current, which is rather slower, or by reversing rotation which is of course much slower and not usually a practicable proposition. The operational modes can be determined from the power expressions:

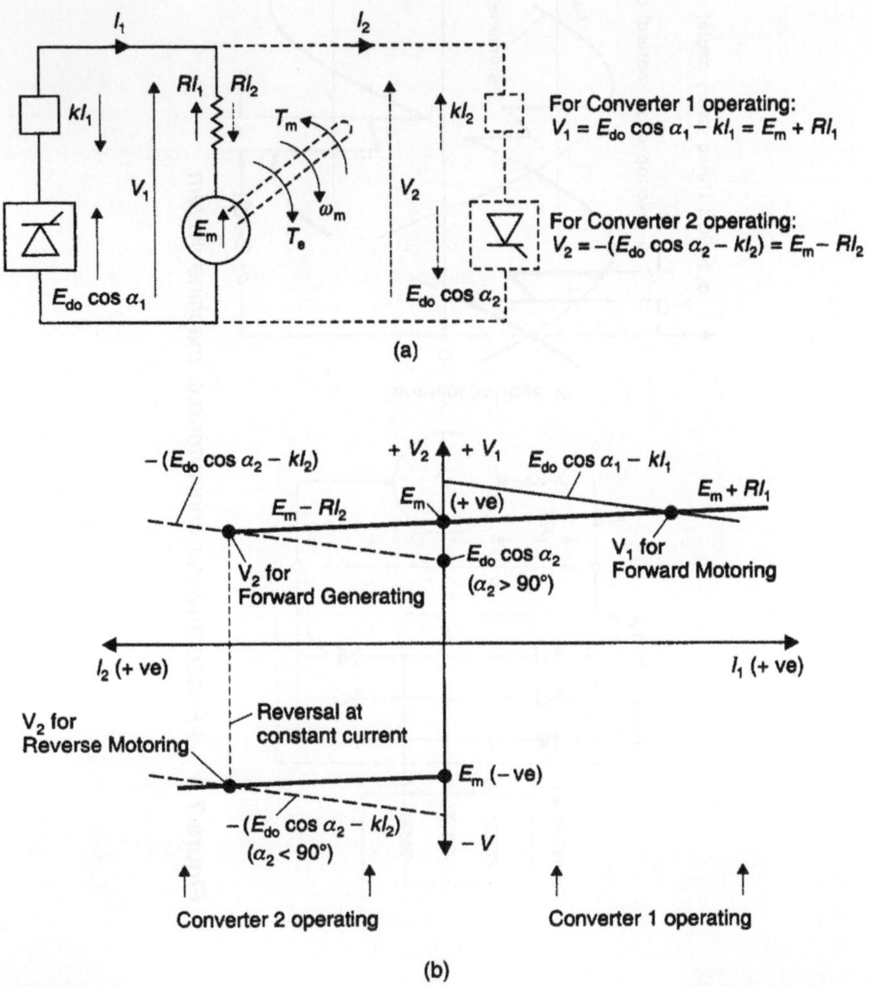

Figure 7.9 *Dual converter: (a) Circuit; (b) 4-quadrant diagram showing reversal.*

Machine motoring: $E_m I_a$ +ve; $(E_{do} \cos \alpha) I_a$ +ve;
P.E. circuit rectifying.

Machine generating: $E_m I_a$ –ve; $(E_{do} \cos \alpha) I_a$ –ve;
P.E. circuit inverting.

Machine plugging: $E_m I_a$ –ve; $(E_{do} \cos \alpha) I_a$ +ve;
P.E. circuit rectifying.

To cover the usual four quadrants of machine operation without a switching changeover, duplication of power-electronic equipment is used, with one bridge for one direction of motor current and the other bridge in inverse parallel for the opposite direction of current. The bridge controls are interlocked so that they cannot operate as short circuits on one another. The circuit is shown on Figure 7.9 together with an indication of how a speed reversal takes place. Converter 1 has its firing signals removed; i_1 falls to zero and after a few milliseconds delay (dead time), Converter 2 is fired.

On the circuit, positive voltages are shown by the arrowheads, though in the equations, these voltages may have negative values. The equations are:

Converter 1 operating:

$$V_1 = E_{do} \cos \alpha_1 - kI_1 = E_m + RI_1 \quad (7.4)$$

Converter 2 operating:

$$V_2 = -(E_{do} \cos \alpha_2 - kI_2) = E_m - RI_2 \quad (7.5)$$

These equations are shown as straight lines on the figure, the intersection of the machine and converter characteristics giving the operating points.

Typically, reversal is carried out with the control rapidly reversing current to maintain a constant 'mean' value and give uniform deceleration and acceleration. Figure 7.10 shows a computer simulation of a dual converter drive which is executing a duty cycle similar to that encountered in a reversing steel-mill drive. A steel billet is processed through the rolls at some nominally constant rolling torque and, when it emerges from the rolls, the torque falls to the no-load value. The rolls are then reversed as quickly as possible within a permissible short-time overload armature current. Steel enters the rolls again after they have been adjusted to give the appropriate pressure on the billet for this reverse 'pass'; see also Example 6.18.

The top graph records the d.c. machine terminal voltage V, the machine e.m.f. e, which is proportional to speed since flux is set constant, and the

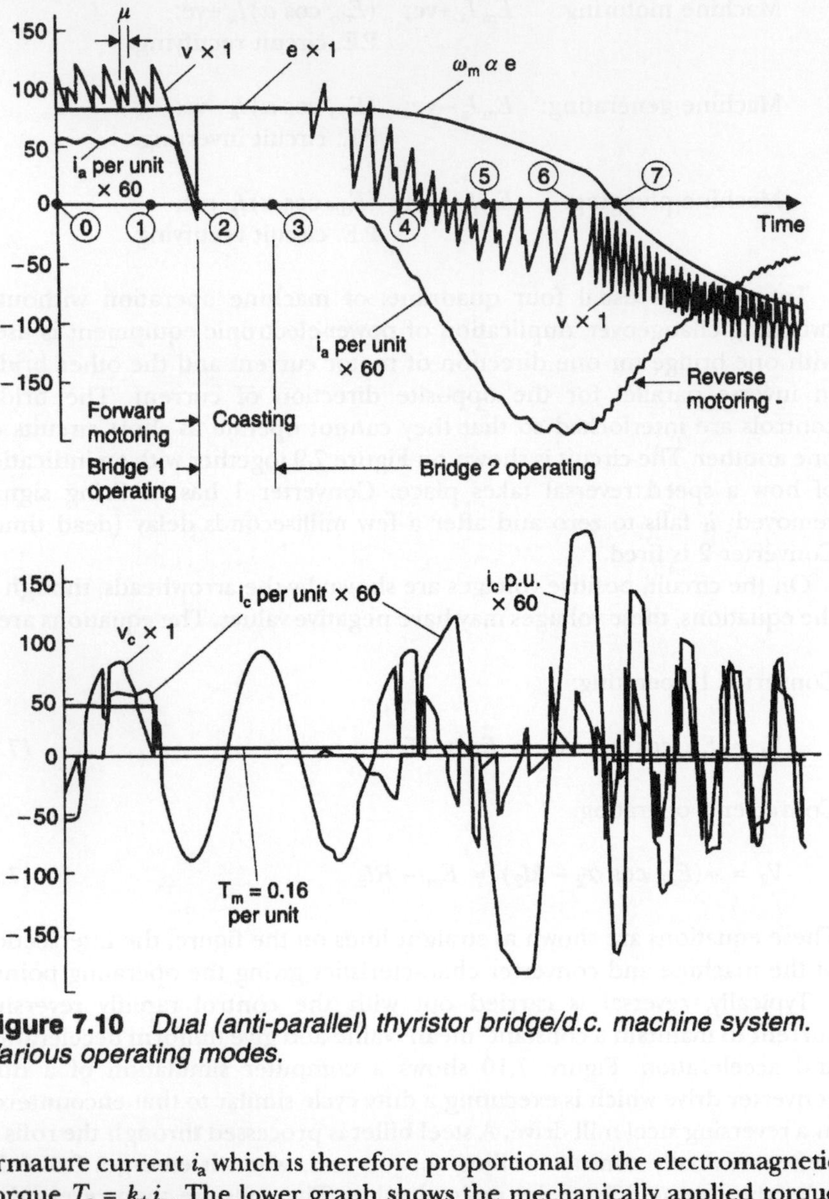

Figure 7.10 *Dual (anti-parallel) thyristor bridge/d.c. machine system. Various operating modes.*

armature current i_a which is therefore proportional to the electromagnetic torque $T_e = k_\phi i_a$. The lower graph shows the mechanically applied torque with the C-phase current and terminal-voltage waveforms. The notches in the phase-voltage waveform are caused by the 'effective' short circuit of the phases during overlap. The half-wave of phase current is therefore a little longer than 120° and follows the d.c. current waveform during most of this period. It is far removed from the idealised square wave assumed in approximate analysis of rectifier circuits; see Appendix B.

Points – with reference to base line on Figure. 7.10

0--1 Bridge 1 is conducting with $\alpha_A = 30°$. $T_m = 1$ *p.u.* The overlap angle μ is about 20°. Bridge 2 is inhibited. Currents are plotted in *per-unit* terms.

1 T_m reduced to 0.16 *p.u.* and Bridge 1 phased back to $\alpha_A = 150°$.

1--2 Current falls naturally to zero, a short time after terminal voltage v reverses, the delay being due to the d.c.-circuit inductance.

2--3 Coasting on no load for ≏ 20 ms (dead time), to ensure Bridge 1 is 'dead' and cannot short circuit Bridge 2 when this is fired. $T_e = 0$. The e.m.f. (and speed) rise slightly as the load is thrown off but fall thereafter, the terminal voltage then being the same as the e.m.f. since there is no current.

3 Bridge 1 inhibited and Bridge 2 fired at an initial firing-delay angle $\alpha_B = 150°$.

3--4 α_B phased forward at a linear rate of 1° per 4° of ωt change. Current eventually starts to flow as the high (reversed) bridge voltage falls below the machine e.m.f. Regeneration commences into the a.c. supply as can be seen by the phase relation between i_C and v_C which is greater than 90°. With this particular rate of α_B change, apart from the initial single-current pulse, the machine current increases at a rate of approximately 160 *per-unit*/sec, which is typical of a d.c. fast-response reversing drive.

4 Here, the mean d.c. terminal voltage is zero, so no average regenerated power is fed from the machine, though it is not consuming any power from the power-electronic system either.

4--6 The speed is falling so the mechanical energy stored is given up to provide the armature losses. The fundamental of i_C, though at low power-factor, is becoming less than 90° out of phase with v_C fundamental, so the a.c. system is providing power, consumed in transformer and thyristor losses. The system as a whole is therefore plugging until the speed is zero, the current reaching a value over 3 *p.u.* – approximately constant for a short time.

5 By the time this point is reached, α_B has become 30° and is fixed at this thereafter.

6 The time scale is changed here during the last stages of reversal.

7 Speed is now zero and the machine changes from plugging to motoring in reverse; note that the mechanical friction torque will reverse at this point also.

Example 7.3

The power amplifier which supplies the motor of Example 6.14 is a single-phase fully controlled rectifier bridge which has a voltage drop of 2.5 V. The input to the bridge is 115 V

a.c. r.m.s. at 50 Hz and the transformer has a leakage inductance of 2 mH. Calculate the bridge firing-delay angle which will produce a load acceleration of 5 rad/s² when the load speed is 15 rad/s. The load is now a frictional (constant) torque of 30 Nm but the total inertia referred to the load shaft is still 10 kgm².

For 5 rad/s², the accelerating torque is $5 \times 10 = 50\,\text{Nm}$

Load torque = $\underline{30}$ Nm

Total torque = 80 Nm

With a 10:1 gear ratio, the motor must supply 8 Nm at 150 rad/s.

Motor current = Torque/k_ϕ = 8/0.54 = 14.8 A

Motor e.m.f. at 150 rad/s = $k_\phi\,\omega_m = 0.54 \times 150 = 81\,\text{V}$

∴ motor voltage = e.m.f. + IR = $81 + 14.8 \times 0.7 = 91.37$

The voltage loss due to 'overlap' is

$$\omega LI/\pi = 2\pi \times 50 \times 0.002 \times 14.8/\pi = 2.96\,\text{V}$$

Hence the bridge output to overcome overlap and diode voltage drop must be:

$$91.37 + 2.96 + 2.5 = 96.83\,\text{V}$$

$$\therefore 96.83 = \frac{230\sqrt{2}\cos\alpha}{\pi}$$

$$\therefore \alpha = 20°.7$$

Example 7.4

A d.c. motor is supplied from a three-phase power system at 415 V r.m.s. line-to-line via a dual fully controlled bridge converter system which has 4 V device voltage drop. The motor armature resistance is 0.2 Ω and supply inductance may be neglected. Find the firing angles and d.c. machine e.m.f.s for the following conditions:

(a) Machine motoring from converter 1 at 100 A and a terminal voltage of 500 V.
(b) Machine regenerating through converter 2 at 100 A and the same terminal voltage.

(a) From eqn (7.4), $V_1 = E_{do}\cos\alpha - kI_1 = E_m + RI_1$

Here, $V_1 = 500\,\text{V}$, $E_{do} = \dfrac{3 \times 415 \times \sqrt{2}}{\pi} = 560.4\,\text{V}$ and $kI_1 = 4\,\text{V}$

$$\therefore\quad 500 = 560.4\cos\alpha - 4$$

$$\therefore \cos\alpha = 504/560.4 = 0.899 \quad \therefore \alpha = \underline{25°.9}$$

and $E_m = 500 - 0.2 \times 100 = \underline{480\,\text{V}}$

(b) From eqn (7.5), $V_2 = -(E_{do} \cos \alpha - kI_2) = E_m - RI_2$

Here, $V_2 = 500\,\text{V},\ E_{do} = 560.4\,\text{V},\ kI_2 = 4\,\text{V}$

$$\therefore\ 500 = -(560.4 \cos\alpha - 4)$$

$$\therefore \cos\alpha = 496/(-560.4) = -0.885 \quad \therefore \alpha = \underline{152.3°}$$

$$E_m = V + RI = 500 + 20 = \underline{520\,\text{V}}$$

Example 7.5

Determine, for the following conditions, the appropriate firing angles and d.c. machine e.m.f.s for a d.c. machine/thyristor-bridge system for which $E_{do} = 300\,\text{V}$, the bridge circuits absorb 15 V, including overlap voltage drop at rated motor current and the machine has a *per-unit* resistance of 0.05 based on its rated voltage of 250 V.

(a) Machine motoring at rated load current and with its terminal voltage at 250 V.
(b) Machine generating at rated load current and with its terminal voltage at 250 V.
(c) Machine plugging at rated load current and with its terminal voltage at 250 V.

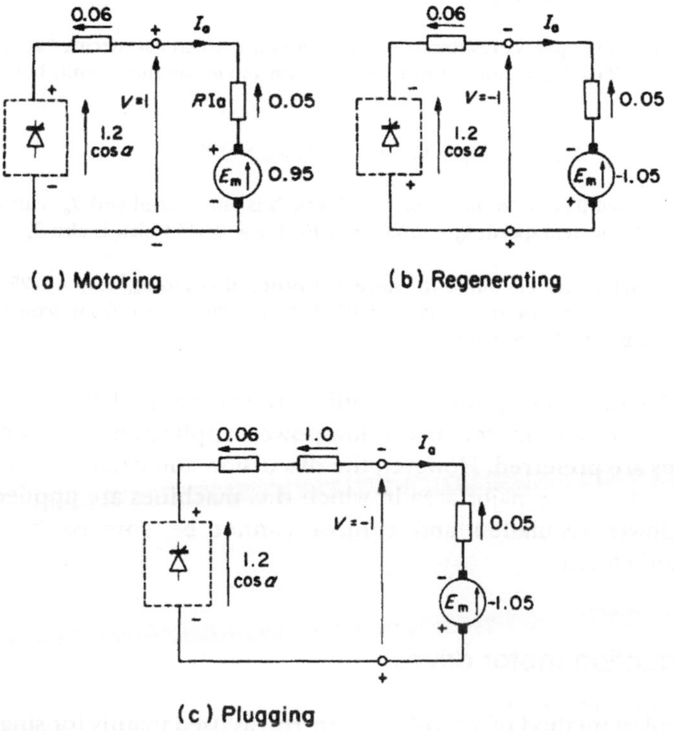

Figure E.7.5

(d) For condition (a), what would be the torque and speed if:
(i) flux is at rated value?
(ii) speed is 1.5 *per unit*?
(e) If the motor load for condition (a) is such that the torque is proportional to (speed)^2, what firing angle would be necessary to have the motor running at half speed with rated flux?

Conditions (a), (b) and (c) are shown on the figure. The question will be worked out in *per unit* for convenience, but can be worked out in actual values and checked against the answers. (Take rated current as 100 A say, and let this be 1 *per-unit* current.) The overall equations are:

$$E_{do}\cos\alpha - kI_a = V = E_m + RI_a = k_\phi\omega_m + RT_e/k_\phi. \qquad (7.6)$$

E_{do} = 300/250 = 1.2 *per unit* k = 15/250 = 0.06 *per unit*.

1 *per-unit* speed = $V_R/k_{\phi R}$ = 1/1 say 1000 rev/min.

∴ rated speed = $E_R/k_{\phi R}$ = 0.95/1 = 950 rev/min.

(a) $V = 1 = 1.2\cos\alpha - 0.06\times 1.$ $\quad\therefore \cos\alpha = 1.06/1.2;\ \alpha = 27°.96.$

(b) $V = -1 = 1.2\cos\alpha - 0.06\times 1.$ $\quad\therefore \cos\alpha = -0.94/1.2;\ \alpha = 141°.6.$

(c) $V = -1 = 1.2\cos\alpha - 0.06\times 1 + 1\times 1;$ $\quad\therefore \cos\alpha = 0.06/1.2;\ \alpha = 87°.1.$

Condition (c) requires a resistor to absorb 1 *per-unit* V, which was specified at the machine terminals. Plugging is not a normal steady-state mode for this circuit, but see Figure 7.10.

(d) $V = 1 = E_m + RI_a = k_\phi\cdot\omega_m + 0.05\times 1$, so $k_\phi\omega_m = 0.95$.

(i) For rated flux (1 *per unit*), ω_m = 0.95 which is rated speed and T_e is also 1 *p.u.*
(ii) For 1.5 *per-unit* speed, $k_\phi = E/\omega_m$ = 0.95/1.5 = 0.633 which is also T_e.

(e) With rated flux, current will be $(\frac{1}{2})^2$ since T_e is proportional to $\omega_m{}^2$ ∴ I_a = 0.25. Substitute in equation: 1.2 cos α = (0.06 + 0.05)0.25 + 0.95 × 0.5 from which cos α = 0.5025/1.2 = 0.4188 $\underline{\alpha = 65.2}$.

Other electronic and power-electronic circuits are used for d.c. machine control, especially in the many low-power applications for which d.c. machines are preferred. However, the discussion and examples in Sections 7.2 and 7.3 cover a major area in which d.c. machines are applied and in which power regulation and control cannot be covered by a single integrated circuit.

7.4 Induction motor drives

The simplest method of a.c. voltage control, as used mainly for single-phase a.c./d.c. 'universal' commutator motors in domestic appliances, is the triac (or back-to-back SCR pair) in the a.c. line. This method, however, produces low motor efficiency and a waveform containing substantial harmonic content. Examples 4.9 and 4.10 illustrate also the limitations to the range of speed control by this method. As seen in Chapter 4, the three-phase induction machine, which is the most common motor installed in industry, can only operate at fixed flux when both voltage and frequency are controllable. This can be seen in Examples 4.6 and 4.11. In a commercial inverter, the ratio of voltage to frequency can normally be specified by adjustment of a user parameter in the control circuits. Where current regulation is required, either a current sensor and feedback or a current estimator may be provided for closed-loop current control.

Voltage-source inverter

Most a.c. variable speed drives employ three-phase 'voltage-source' inverters consisting of six semiconductor power switches, as in Figure 7.11. The source is a constant d.c. voltage, usually obtained by rectifying single-

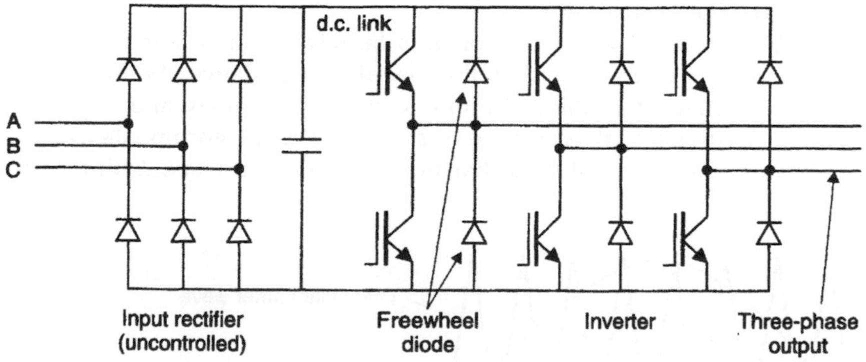

Figure 7.11 *Voltage-source inverter.*

or three-phase a.c. and smoothing the d.c. with a large capacitance. As in the chopper (Section 7.2), freewheel diodes are provided in reverse parallel with the power switching devices of the inverter to allow energy stored in machine inductance to decay when the relevant power switch is turned off; current will circulate until the stored energy has been dissipated. The freewheel diodes also allow current to flow back to the supply during regeneration. However, the capacitance has a limited ability to receive energy and the input rectifier would have to be fully controlled and have a large inductance to maintain current if inversion back to the supply system were to take place.

If each switch conducts for its maximum conduction period (approximately 180°), the line voltage output between inverter terminals consists of 'quasi-square' waveforms of 120° and the phase voltage in the star-connected load is stepped, as shown in Figure 7.13. A small time interval must be allowed for one device to fully turn off before the device in the same leg turns on. Varying the overall cycle time of the switching patterns alters the frequency of the output. Varying the conduction period will affect the r.m.s. voltage produced and will also vary the harmonic content of the voltage. Machines are fairly reactive and will have a filtering action, i.e. the current harmonics will generally be lower than the values imposed by the voltage. Harmonic reduction coupled with voltage control is conventionally applied by pulse width modulation (PWM), in which the voltage waveform is split into a number of separate pulses of pre-defined length and starting angle. A number of PWM methods exist, e.g. sinusoidal, harmonic elimination, symmetrical etc. The sinusoidal method shown in Figure. 7.12 gives a good overall reduction of significant harmonics and is well supported by integrated circuit waveform generators. More sophisticated or custom harmonic elimination methods would require PWM generation by a microcontroller. PWM is implemented only on voltage source inverters.

The PWM voltage waveform contains largely higher order harmonics which are more easily filtered by the machine impedance. As the voltage waveform contains harmonics, harmonic currents will exist and harmonic torques, additional to those caused by slot ripple and rotor imbalance, will be produced as rotor and stator harmonic currents interact. Additionally,

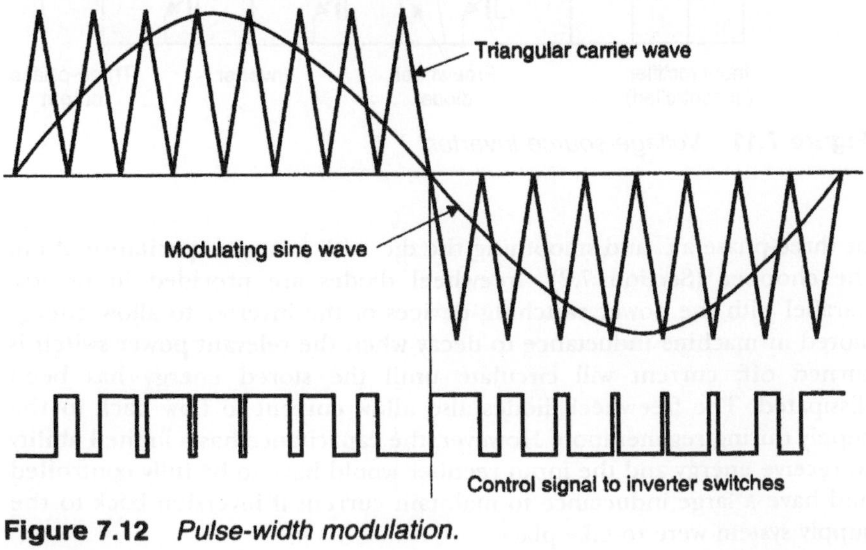

Figure 7.12 *Pulse-width modulation.*

saturation may cause space harmonics which will lead to further torque pulsations. The calculation of harmonic torques is therefore complex and unique in any situation. Fortunately, there is a fair degree of cancellation of random effects and PWM ensures that voltage harmonics remain modest up to higher voltages at which the need for a large r.m.s. voltage requires longer PWM pulses, and eventually half-cycle device conduction, as in Figure 7.13, where the line-to-line voltage includes all odd voltage harmonics except those which are multiples of three. However, frequency will also be higher and the filtering action of machine reactances is greater than at low frequency.

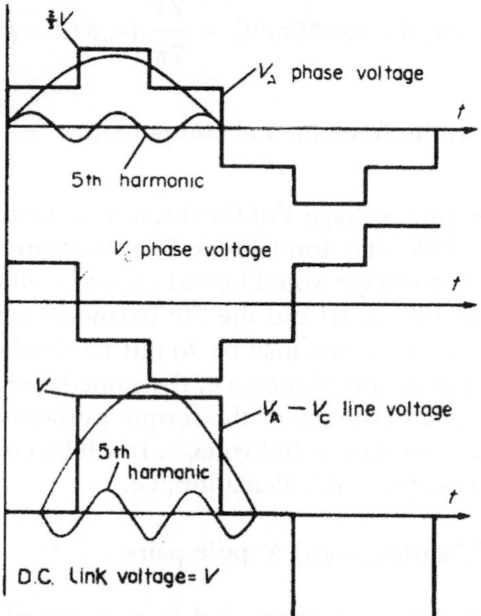

Figure 7.13 *Quasi-square inverter waveforms.*

At maximum voltage, as shown in Figure 7.13, the fundamental and harmonic components of the line voltage may be found by using Fourier series. If the origin of the waveform is set at the commencement of the fundamental of the line voltage in Figure 7.13, the analysis will contain only sine components and odd harmonics, i.e. the Fourier components a_0 and b_n are zero. The remaining Fourier coefficients are:

$$a_n = \frac{2}{\pi}\int_{\pi/6}^{5\pi/6} V \sin n\omega t \, d\omega t = \frac{2V}{\pi}\left[-\cos n\omega t\right]_{\pi/6}^{5\pi/6}$$

where V is the voltage of the d.c. input to the inverter.

$$a_1 = \frac{2V}{\pi}[\cos \pi/6 - \cos 5\pi/6] = \frac{2V}{\pi}(\sqrt{3}/2 + \sqrt{3}/2) = \frac{2\sqrt{3}V}{\pi}$$

$$a_3 = \frac{2V}{3\pi}[\cos 3\pi/6 - \cos 15\pi/6] = \frac{2V}{3\pi}(\cos \pi/2 - \cos \pi/2) = 0$$

i.e. the waveform contains no third, ninth etc. harmonic components.

$$a_5 = \frac{2V}{5\pi}[\cos 5\pi/6 - \cos 25\pi/6] = \frac{2V}{5\pi}(-\sqrt{3}/2 - \sqrt{3}/2) = \frac{-2\sqrt{3}V}{5\pi}$$

$$a_7 = \frac{2V}{7\pi}[\cos 7\pi/6 - \cos 35\pi/6] = \frac{2V}{7\pi}(-\sqrt{3}/2 - \sqrt{3}/2) = \frac{-2\sqrt{3}V}{7\pi}$$

e.g. if $a_1 = 2\sqrt{3}V/\pi$, the fundamental component $= \dfrac{2\sqrt{3}V \sin \omega t}{\pi}$

For example, if the input voltage V of the d.c. link between the rectifier and the inverter was 600 V, the fundamental component of the inverter's maximum output line voltage would be 661 sin ωt, i.e. 468 V r.m.s. The 5th harmonic would be 132 sin ωt and the 7th harmonic would be 95 sin ωt. Steady-state harmonic currents may be found by dividing the harmonic voltage by the harmonic impedance, i.e. the impedance at frequency $n\omega$.

We can now make an estimate of the torque pulsation for the inverter-fed induction motor working at full voltage. It will be convenient to recall eqn (5.15) for the purposes of calculation, i.e.

$$T_e = 3M' I_m I_2' \sin(\varphi_m - \varphi_2) \times \text{pole pairs} \qquad (7.7)$$

If the machine windings are connected in star, the phase-voltage waveforms are quite different from the line voltage, as indicated on Figure 7.13 for phases A and C. These are subtracted to give v_{AC}, which is the quasi-square line-voltage waveform of Figure 7.4. For all waveforms, 3rd and triplen harmonics are not present, but the other harmonics combine to give the usual 30° shift between phase and line waves. When this is allowed for, on the harmonic scale, the expression for the phase voltage waveform can be deduced as:

$$\frac{1}{\sqrt{3}} \times \frac{2\sqrt{3}}{\pi} \times V\left(\sin \omega t + \frac{\sin 5\omega t}{5} + \frac{\sin 7\omega t}{7} \ldots\right)$$

$$= \sqrt{2}V_1\left(\sin \omega t + \frac{\sin 5\omega t}{5} + \frac{\sin 7\omega t}{7} \ldots\right),$$

where $V_1 = \sqrt{2}V/\pi$ is the r.m.s. value of the fundamental component of the phase voltage. The 5th harmonic waveform is sketched in on Figure 7.5 to show why the harmonic signs have changed from the line-voltage expression.

Considering each voltage harmonic separately, applied to its own equivalent circuit with appropriate correction for frequency-increased reactances, and voltages reduced to V_1/n, we could calculate the harmonic currents and sum them to get the primary, secondary and magnetising-current waveforms. Each rotor (secondary) harmonic current will react with its own harmonic magnetising-current to produce a steady-component of torque. It will react with all the other harmonic magnetising currents to produce a component of the pulsating torque. In fact, only the interaction of the 5th and 7th harmonic rotor currents, I'_5 and I'_7 say, with the fundamental magnetising current I_{m1} will be of great significance. The phase sequences of the 5th and 7th harmonics are opposite in sense, so their steady-torque components tend to cancel. These approximations will be justified later. At this stage we will make one further approximation, namely: that the fundamental harmonic magnetising current I_{m1} is given by $(V_1/X_m)\ \underline{/-90^\circ}$, from the approximate circuit, though the exact calculation of I_{m1} is not very much more difficult.

For the rotor currents, we require the slip, and if this is s for the fundamental, it will be $6 - s$ for the 5th harmonic producing a backwards rotating field at $5\omega_s$, and $6 + s$ for the 7th harmonic producing a forwards rotating field at speed $7\omega_s$ (see Figure 7.14a). Hence, taking V_1 as reference: and using the approximate circuits:

$$I'_5 = \frac{V_1}{5}\,\frac{1}{R_1 + R'_2/(6-s) + \mathrm{j}5(x_1 + x'_2)} \quad \text{and } i'_5 = \sqrt{2}I'_5\sin(5\omega t - \varphi_5),$$

$$I'_7 = \frac{V_1}{7}\,\frac{1}{R_1 + R'_2/(6+s) + \mathrm{j}7(x_1 + x'_2)} \quad \text{and } i'_7 = \sqrt{2}I'_7\sin(7\omega t - \varphi_7).$$

Consider the instant of time, $\omega t = 0$, when the fundamental and harmonic voltages are momentarily zero and initially are all increasing positively as t increases. Note this means that the 5th harmonic, of reverse phase-sequence, must be shown in antiphase with $\mathbf{V}_1$ and $\mathbf{V}_7$ as on Figure 7.14b, which is the phasor diagram for this instant. Now when the fundamental rotor-current $\mathbf{I}'_2$ reacts with its own magnetising current $\mathbf{I}_{m1}$, its phasor leads the $\mathbf{I}_{m1}$ phasor if producing forwards motoring-torque. Hence, $\mathbf{I}'_7$ reacting with $\mathbf{I}_{m1}$ at the instant shown is producing a positive torque and the angle between these phasors, taking $\omega t = 0$ as reference is:

$$(-\varphi_7 + 7\omega t) - (-90 + \omega t) = 6\omega t + (90 - \varphi_7).$$

The angle of I'_5 (lagging V_5 of reverse phase-sequence) leads I_{m1} by:

$$(180 + \varphi_5 - 5\omega t) - (-90 + \omega t) = 270 + \varphi_5 - 6\omega t$$
$$= -[6\omega t + (90 - \varphi_5)].$$

Thus, the total torque due to 5th and 7th harmonic rotor-currents reacting with the fundamental magnetising current is, from eqn (7.9):

$$3M' \frac{V_1}{X_{m1}} [-I'_5 \sin(6\omega t + 90 - \varphi_5) + I'_7 \sin(6\omega t + 90 - \varphi_7)] \times \text{pole pairs.} \quad (7.8)$$

The expression shows that 5th and 7th harmonics combine to give a pulsating torque at 6 times fundamental frequency. In Section 8.2, as an

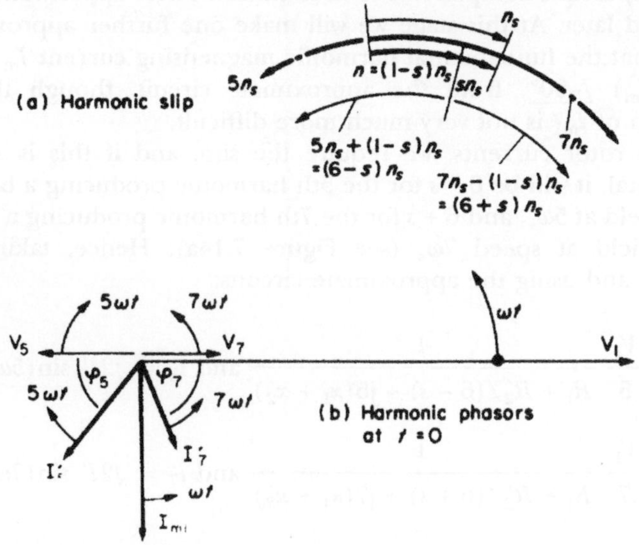

Figure 7.14 *Pulsating-torque calculations: (a) Harmonic slip; (b) Harmonic phasors at t = 0.*

illustration of a.c. machine simulation, the computed results from an 'exact' solution, including all the harmonics, show that this 6th harmonic pulsation is almost completely dominant (Figure 8.4, see also T7.5). If the motor had been current fed, the division of 5th and 7th harmonic currents into magnetising and rotor currents would have had to invoke the standard parallel-circuit relationships, as in Section 4.3, Examples 4.12–4.16.

Current-source inverter

Current-source inverters, in which a large choke in the d.c. input forces an almost constant d.c. input current and hence square wave a.c. output currents, find use in very high power drives, for which the ratings of available 'turn-off' devices, such as bipolar transistors and GTOs, would be inadequate. Rugged rectifier-grade SCRs of very high rating can be used, provided that 'forced commutation' is used to remove the current from each device in turn to allow it to turn off. Figure 7.15 shows a current

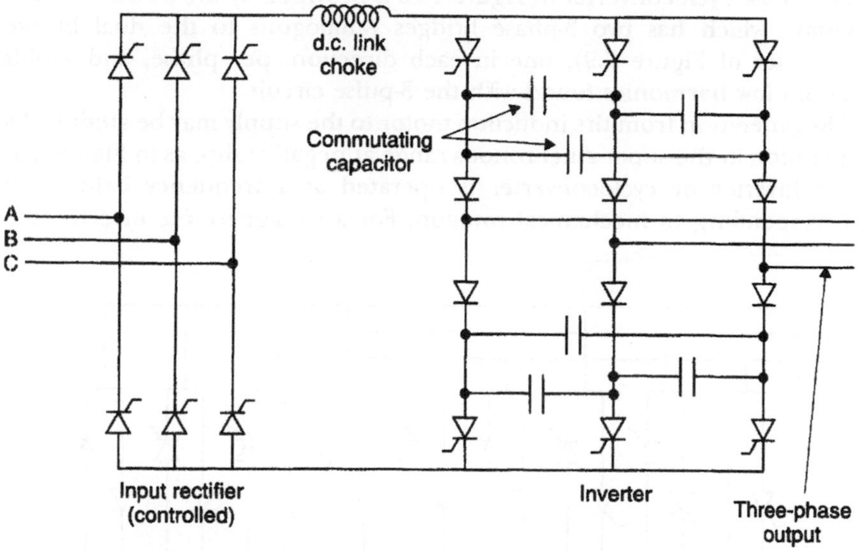

Figure 7.15 *Current-source inverter.*

source inverter. Commutation of current from one SCR to the next occurs when the second one turns on, causing reverse bias to the previous device and diversion of its current through the capacitor network. As one device causes its predecessor to turn off, the conduction interval is approximately 120° per SCR and PWM cannot be applied. Voltage control is carried out in the controlled rectifier. The presence of the large d.c. line inductance and a fully controlled rectifier bridge allow straightforward regeneration of energy back to the supply.

Cycloconverter

A cycloconverter provides direct a.c./a.c. conversion, at the price of a larger number of power devices and poorer output waveforms. The frequency range is restricted to half of mains frequency to maintain tolerable waveshapes. However, the cycloconverter has certain features which make it superior to inverters for large, slow-speed drives. It allows energy to be transferred in either direction, and the large number of devices is no handicap when load currents are so large that parallel devices would have to be used in an inverter. The a.c. input waveform is used as the commutating medium by arranging that a device is turned on when it is the most forward-biased. It therefore takes over the current in that phase from the preceding device. Cycloconverters may have varying numbers of devices, from 18 upwards. An improvement on the waveforms shown for the 3-pulse cycloconverter in Figure 7.16 is provided by the 36-SCR 6-pulse design, which has two 3-phase bridges (analogous to the dual bridge converter of Figure 7.9), one in each direction, per phase, and avoids certain low harmonics found with the 3-pulse circuit.

Regeneration from the induction motor to the supply may be attained by operation in the super-synchronous range at negative slip, as in Figure 4.1. The inverter or cycloconverter is operated at a frequency below that corresponding to mechanical rotation. For a voltage-source inverter, the

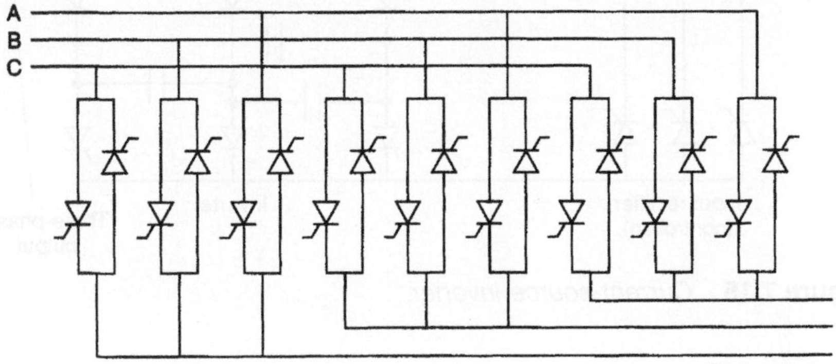

(a)

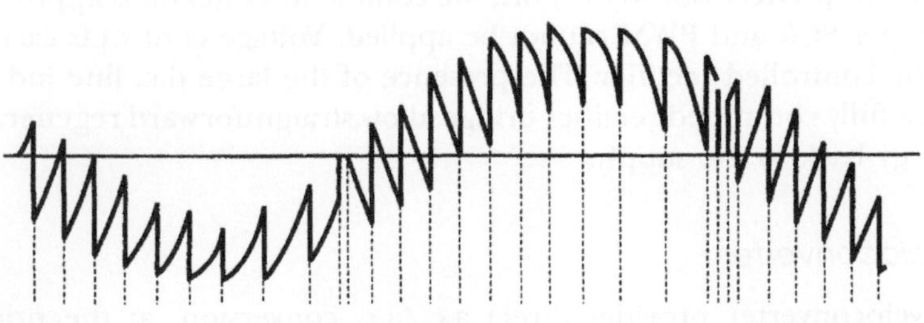

(b)

Figure 7.16 *Cycloconverter: (a) Circuit; (b) Output waveform.*

average current through the freewheel diodes will be greater than that through the power switches. The regenerated energy must either be absorbed by a resistor across the d.c. link capacitor, or returned to the supply via a rectifier operating in the inversion mode.

Induction motor control

If the stator resistance in eqn (4.5) is considered to be small enough to be neglected in comparison with other impedances,

$$T_e = \frac{3}{2\pi \times f/p} \times \frac{V_1^2}{(R_2'/s)^2 + (x_1 + x_2')^2} \times \frac{R_2'}{s}$$

$$2\pi f/p = \omega_s,\ s = \frac{n_s - n}{n_s} = \frac{\omega_s - \omega_m}{\omega_s} = \frac{\omega_{sl}}{\omega_s}, \text{ say,}$$

then $$T_e = \frac{3V_1^2}{\omega_s} \times \frac{R_2'\omega_s/\omega_{sl}}{(R_2'\omega_s/\omega_{sl})^2 + \{\omega_s(L_1 + L_2')p\}^2} \times \frac{(\omega_{sl})^2}{(\omega_{sl})^2}$$

$$= \frac{3V_1^2}{\omega_s^2} \times \frac{R_2'\omega_{sl}}{(R_2')^2 + \{\omega_{sl}(L_1 + L_2')p\}^2} \qquad (7.9)$$

In eqn (7.9), the relationship between slip frequency ω_{sl} and T_e is approximately constant if the ratio of supply voltage V_1 to supply frequency ω_s is constant. It approaches the condition shown by the speed/torque curves of Figure E.4.11, where the flux level ($\propto E/\omega_s$) is actually maintained constant over the constant-torque region. Figure 7.17 shows a family of

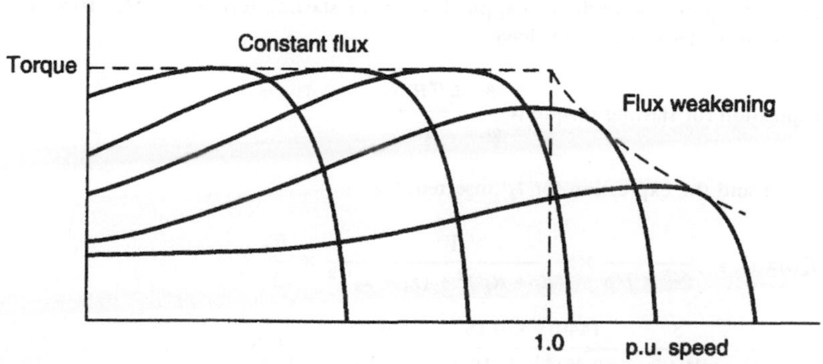

Figure 7.17 *Induction motor torque/speed curves at variable frequency.*

torque/speed curves for a range of frequencies from a few Hz up to the point at which maximum inverter output voltage is reached. At low voltage and frequency, additional voltage is necessary to allow for the voltage drop in the stator resistance, which was neglected above. This voltage drop is significant at low frequency when reactance voltage drops are small. Note that control of harmonics is only feasible with a voltage source inverter.

Beyond the point of maximum voltage, only ω_s can be increased.

The input impedance neglecting stator resistance is

$$\frac{R_2'}{s} + j\omega_s(L_1 + L_2')p = \frac{R_2\omega_s}{\omega_{sl}} + j\omega_s(L_1 + L_2')p$$

$$= \omega_s\{R_2'/\omega_{sl} + j(L_1 + L_2')p\}$$

If, in this region, with voltage fixed at maximum, the frequency is increased under current control in such a manner that I_1 is maintained constant, then the above impedance must also be constant. There will therefore be a constant-power range (VI_1 constant) but with reducing flux (approximately $\propto V/f$) and torque in a similar manner to the d.c. machine. I_1 can be maintained constant so long as the slip frequency can increase, keeping the impedance constant, and operation is stable. Beyond this limit, I_1 will have to decrease with consequent additional reductions in flux and torque. This is summarised later in Figure 7.24.

Example 7.6

Example 4.11 showed an induction motor being driven at 60% of supply voltage and frequency, as is typically the case for constant volts/Hz control using a PWM voltage source inverter.

For the same machine as in Example 4.11, find the starting torque and current if normal voltage and frequency were directly applied. Find the starting torque at 5 Hz (10% of rated frequency) and 14% of normal voltage.

The expression for starting torque is $\frac{3}{\omega_s}\frac{I_2'^2R_2'}{s}$

With $s = 1$ and the expression for I_2' inserted:

$$T_e(\text{start}) = \frac{3}{2\pi \times f/p} \times \frac{V^2}{(R_1 + R_2')^2 + (x_1 + x_2')^2} \times \frac{R_2'}{1}$$

$$= \frac{3}{104.7} \times \frac{(440)^2 \times 0.18}{(0.38)^2 + (1.16)^2} = \frac{104544}{156} = \underline{670\,\text{Nm}}$$

and $I_2' = \underline{152.9\,\text{A}}$

At reduced voltage and frequency,

$$T_e(\text{start}) = \frac{3}{10.47} \times \frac{(440 \times 0.14)^2 \times 0.18}{(0.38)^2 + (1.16 \times 0.1)^2} = \frac{2049}{1.65}$$

$$= \underline{1240\,\text{Nm}}$$

and $I_2' = \underline{155\,\text{A}}$

Note that the starting torque is almost doubled at approximately the same current, the reason being the improved power factor following the reduction of reactance. The voltage has been decreased by a smaller percentage than the frequency to allow for the proportionately greater effect of resistance at the lower frequency. If the voltage had been reduced to 10% of the rated value, a current of only 110 A would be present.

Electromagnetic interference

The frequencies encountered in pulse-width-modulated switching of power semiconductors range up to 50 kHz in quite ordinary applications. Each square pulse at this frequency has high dV/dt, representing higher harmonics, and very high frequency oscillations are common in driver circuits and power devices due to parasitic effects. These lead to electromagnetic interference (EMI), which can pose problems with modern low-power integrated devices. EMI may be conducted or radiated, and also occurs from natural sources. Modern electronic devices are available with rise times which both create EMI and reduce EMC (electromagnetic compatibility), which is defined as the ability of a circuit to operate satisfactorily in the presence of EMI, without itself producing excessive EMI. Improvement of EMC in the presence of radiated or conducted EMI may be achieved by screening, filtering, decoupling and path layout. However, internal sources of EMI can be reduced by the use of devices which operate at the speed required by the circuit specification, rather than at the highest speed available. Note that the problem may be less significant with analogue components, which have limited bandwidth, and with large scale integration, such as microcomputers.

7.5 Vector (field-oriented) control

While many induction motor drives employ current or speed control, fast torque response is improved if, under transient conditions, flux is kept constant and torque is developed by a separate component of excitation. In separately excited d.c. and synchronous machines, the field is largely 'decoupled' from the controlled winding, but the squirrel-cage induction motor is singly excited and the stator input is the only source of excitation.

We must therefore find a means of expressing the induction machine currents in such a way that they may be manipulated by a control system to produce a situation similar to the d.c. machine, whereby the air-gap flux is 'decoupled' from (i.e. at right angles to) the m.m.f. axis of the rotor torque-producing current. In this section, the transformation of induction-motor input currents into equivalent d.c. components will be explained. A simple mathematical model will be developed which is adequate for estimating torque and flux and is thus suitable for use in closed-loop control. Finally, it will be explained how rapid changes of current are obtained, as necessary to achieve a fast torque response. A clear, non-mathematical description of field-oriented control is provided in Reference 9.

Resolving the three stator-phase m.m.f.s, represented by the currents, along the two orthogonal (at right-angles) axes of two 'replacement' stator-phases α and β, with phase α coincident with phase A, as in Figure 7.18a, the equations are:

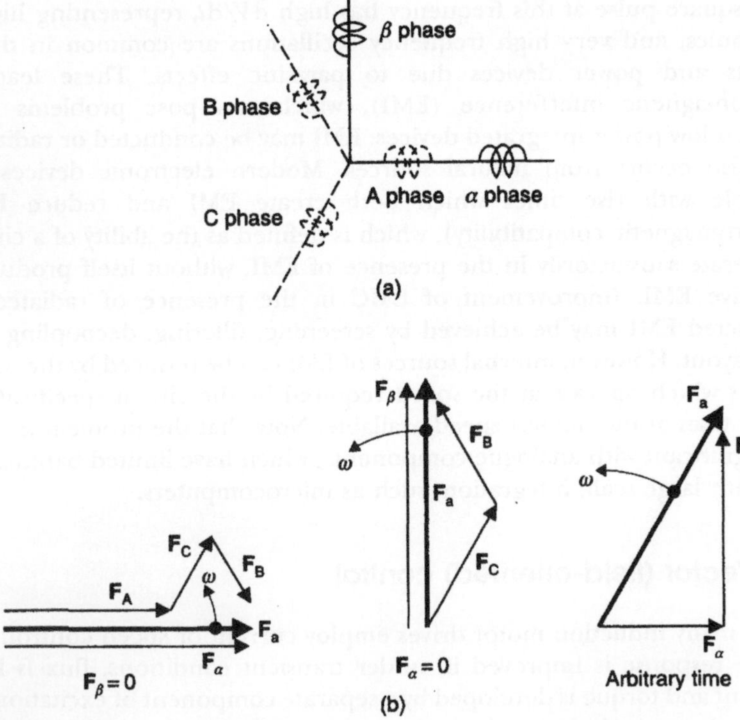

Figure 7.18 *3-phase/2-phase transformation: (a) Phase axes; (b) M.M.F. components F_A and F_α at maximum when $t = 0$.*

$$\mathbf{i}_\alpha = \mathbf{i}_A + \mathbf{i}_B \cos 120° + \mathbf{i}_c \cos 240°$$

$$= \mathbf{i}_A - 0.5\mathbf{i}_B - 0.5\mathbf{i}_c \qquad (7.10a)$$

and $\mathbf{i}_\beta = \mathbf{i}_B \cos 30° + \mathbf{i}_c \cos 150°$

$$= \frac{\sqrt{3}}{2}\mathbf{i}_B - \frac{\sqrt{3}}{2}\mathbf{i}_c \qquad (7.10b)$$

The resultant m.m.f. wave is now due to the two orthogonal component m.m.f.s of $\mathbf{i}_\alpha$ and $\mathbf{i}_\beta$ instead of the three components due to $\mathbf{i}_A$, $\mathbf{i}_B$ and $\mathbf{i}_C$, but with the same magnitude and the same angular shift from the same stator reference point. Figure 7.18b shows the individual stator-phase m.m.f.s at various different instants, varying sinusoidally in time and assumed to be sinusoidally distributed in space, so that they can be combined in the manner of vectors.

In the simpler two-phase equivalent, the resultant m.m.f. can be expressed as:

$$\mathbf{F}_a = \mathbf{F}_\alpha + j\mathbf{F}_\beta$$

Since the m.m.f.s are proportional to currents, the stator m.m.f. $\mathbf{F}_a$ can therefore be defined by, say:

$$\mathbf{i}_s = \mathbf{i}_\alpha + j\mathbf{i}_\beta$$

We now make a further transformation, which again gives the same resultant m.m.f. wave, but as viewed from an observer moving at synchronous speed. The α and β components are resolved along two orthogonal axes, d and q, which are in a reference frame moving in synchronism with the m.m.f. wave. Figure 7.19 shows the instant at which the d-axis is at an angle ρ to the α-axis. ρ is changing at the same rate as ωt but is not necessarily equal to ωt. For the stator currents, therefore:

along the d-axis: $\mathbf{i}_{ds} = \mathbf{i}_\alpha \cos \rho + \mathbf{i}_\beta \sin \rho$ (7.11a)

along the q-axis: $\mathbf{i}_{qs} = -\mathbf{i}_\alpha \sin \rho + \mathbf{i}_\beta \cos \rho$ (7.11b)

In a steady-state operating condition, these transformed (d,q) currents are steady d.c. values (as demonstrated in the table on Figure 7.19), which provide magnetisation on an axis moving at synchronous speed and combine to produce the same m.m.f as the actual currents in the original three-phase system.

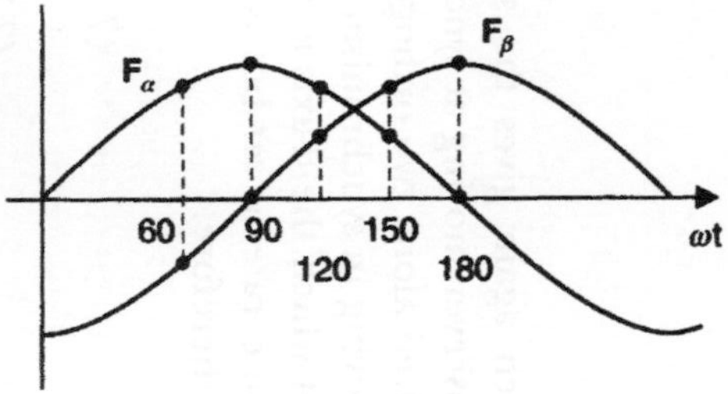

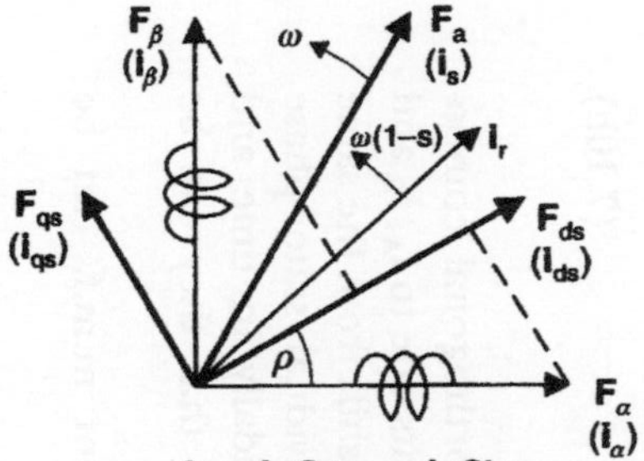

$$i_{ds} = i_\alpha \cos\rho + i_\beta \sin\rho$$

$$i_{qs} = -i_\alpha \sin\rho + i_\beta \cos\rho$$

Space angle ρ measured from α^-(and A^-)phase axis	$-60°$	$-30°$	$0°$	$30°$	$60°$
Time angle ωt	$60°$	$90°$	$120°$	$150°$	$180°$
M.M.F. F_α (prop. to sin ωt)	$\sqrt{3}/2$	1	$\sqrt{3}/2$	1/2	0
M.M.F. F_β (prop. to cos ωt)	$-1/2$	0	1/2	$\sqrt{3}/2$	1
Sin ρ (space angle)	$-\sqrt{3}/2$	$-1/2$	0	1/2	$\sqrt{3}/2$
Cos ρ (space angle)	1/2	$\sqrt{3}/2$	1	$\sqrt{3}/2$	1/2
$i_\alpha \cos\rho$ } i_{ds}	$\sqrt{3}/2 \times 1/2$	$1 \times \sqrt{3}/2$	$\sqrt{3}/2 \times 1$	$1/2 \times \sqrt{3}/2$	0
$i_\beta \sin\rho$ } i_{ds}	$-1/2 \times -\sqrt{3}/2$	0	0	$\sqrt{3}/2 \times 1/2$	$1 \times \sqrt{3}/2$
$-i_\alpha \sin\rho$ } i_{qs}	$-\sqrt{3}/2 \times -\sqrt{3}/2$	$-1 \times -1/2$	0	$-1/2 \times 1/2$	0
$i_\beta \cos\rho$ } i_{qs}	$-1/2 \times 1/2$	0	$1/2 \times 1$	$\sqrt{3}/2 \times \sqrt{3}/2$	$1 \times 1/2$
$i_{ds} =$	$\sqrt{3}/2$	$\sqrt{3}/2$	$\sqrt{3}/2$	$\sqrt{3}/2$	$\sqrt{3}/2$
$i_{qs} =$	1/2	1/2	1/2	1/2	1/2

Figure 7.19 *Transformation to synchronous d–q axes.*

The resultant stator m.m.f. will be obtained in the same way as $\mathbf{F_a}$ was represented by $\mathbf{i_s} = \mathbf{i_\alpha} + j\mathbf{i_\beta}$, but now it is made up of the two orthogonal components $\mathbf{i_{ds}}$ and $j\mathbf{i_{qs}}$. Substituting the values of these in terms of $\mathbf{i_\alpha}$, $\mathbf{i_\beta}$:

$$\mathbf{i_{ds}} + j\mathbf{i_{qs}} = \mathbf{i_\alpha} \cos\rho + \mathbf{i_\beta} \sin\rho - j\mathbf{i_\alpha} \sin\rho + j\mathbf{i_\beta} \cos\rho$$

$$= (\mathbf{i_\alpha} + j\mathbf{i_\beta})(\cos\rho - j\sin\rho)$$

$$= \mathbf{i_s} e^{-j\rho}$$

The 'vector' $\mathbf{i_s} e^{-j\rho}$ represents the resultant stator m.m.f. transformed through angle ρ to be stationary with respect to the chosen synchronous axes. The transformations may be computed practically by using a commercial vector-rotation integrated circuit.

The term $e^{-j\rho}$ is a vector operator which has transformed the currents expressed in axes at rest with respect to the stator, to synchronous axes, behind which the stator α-axis (or A-axis) lags by angle ρ. A similar transformation will be used later to transform between other axes displaced in angle, e.g. from rotor axes to synchronous axes or vice versa, to have stator and rotor currents in the same reference frame so that they can be combined. The rotor axes will lag by some angle γ behind the synchronous axes when motoring, with positive slip, so the vector operator will be $e^{-j\gamma}$ to move from the rotor axes to the synchronous axes. The inverse operator $e^{j\gamma}$ will be needed to rotate, backwards, the synchronous-axis stator currents $\mathbf{i_{ds}}$ and $\mathbf{i_{qs}}$ to the rotor axes to give $\mathbf{i_{dsr}}$ and $\mathbf{i_{qsr}}$, for use in eqns (7.12).

The table in Figure 7.19 shows the progression of the various m.m.f.s as the space angle ρ rotates synchronously with the time angle ωt. Arbitrarily, the alignment of the d-axis with the A-axis is chosen to occur at a time angle $\omega t = 120°$. The combination of the α- and β-m.m.f.s and that of the $\mathbf{i_{ds}}$ and $\mathbf{i_{qs}}$ component m.m.f.s calculated from eqn (7.11) are both shown in the table. It can be seen that the resultant is the same for the two methods, since the directions of $\mathbf{i_s}$ (calculated from $\mathbf{i_{ds}}$, $\mathbf{i_{qs}}$) and $\mathbf{F_a}$ are the same in each column. The values of $\mathbf{i_{ds}}$ and $\mathbf{i_{qs}}$ are constant (d.c.) throughout, but move with the synchronous axes: the α and β components vary in magnitude, but are constant in directional sense along their axes.

The usefulness of the double transformation from measurable three-phase quantities to two-phase equivalents and then to two-phase quantities but viewed from a rotating reference frame appears when we remember that, if we move round with the flux wave (a synchronously rotating reference frame at constant speed), the vectors appear stationary. This corresponds to the situation in a d.c. or brushless d.c. machine.

In a similar way, the resultant rotor m.m.f. may be expressed in equivalent fashion to give a rotor 'vector' $\mathbf{i}_r$, relative to the rotor reference frame, i.e.

$$\mathbf{i}_r = \mathbf{i}_{drr} + j\mathbf{i}_{qrr}$$

We can now consider how, by manipulating the equations between axes, we can derive control signals sensitive to torque and to flux so that these can feed back information through their two control loops, by means of which orthogonality of the machine's flux and armature m.m.f. axes (as in a d.c. machine) can be achieved during transient conditions, giving the required fast response.

The voltage in a circuit carrying current i which is magnetically coupled with a circuit carrying current i' can be described by:

$$v = M\,\mathrm{d}i'/\mathrm{d}t + iR + L\,\mathrm{d}i/\mathrm{d}t$$

where M is the mutual inductance between the circuits.

As the rotor of a squirrel-cage induction machine is a short circuit, there is no voltage source other than that induced by mutual transformer action from the stator. Thus, $v = 0$.

In Laplace form,

$$0 = M\mathrm{p}i' + (R + \mathrm{p}L)i$$

In a rotor reference frame, there will be corresponding equations for each axis:

$$\text{For the rotor d-axis: } 0 = M\mathrm{p}\mathbf{i}_{dsr} + (R_2 + \mathrm{p}L_2)\mathbf{i}_{drr} \tag{7.12a}$$

$$\text{For the rotor q-axis: } 0 = M\mathrm{p}\mathbf{i}_{qsr} + (R_2 + \mathrm{p}L_2)\mathbf{i}_{qrr} \tag{7.12b}$$

where $\mathbf{i}_{dsr}$ and $\mathbf{i}_{qsr}$ are $\mathbf{i}_{ds}$ and $\mathbf{i}_{qs}$ referred to the rotor reference frame. $\mathbf{i}_{drr}$ and $\mathbf{i}_{qrr}$ are the corresponding rotor currents with the same reference and p is the Laplace operator ($\equiv \mathrm{d}/\mathrm{d}t$).

Similar equations can be written down for each stator-axis, referred here to the synchronously rotating reference frame and including the voltages applied to the stator terminals:

$$\text{For the stator d-axis: } \mathbf{v}_{ds} = M\mathrm{p}\mathbf{i}_{drs} + (R_1 + \mathrm{p}L_1)\mathbf{i}_{ds} \tag{7.12c}$$

$$\text{For the stator q-axis: } \mathbf{v}_{qs} = M\mathrm{p}\mathbf{i}_{qrs} + (R_1 + \mathrm{p}L_1)\mathbf{i}_{qs} \tag{7.12d}$$

where $\mathbf{i}_{drs}$ and $\mathbf{i}_{qrs}$ are $\mathbf{i}_{dr}$ and $\mathbf{i}_{qr}$ referred to the synchronously rotating reference frame. $\mathbf{v}_{ds}$ and $\mathbf{v}_{qs}$ are the corresponding stator applied voltages with the same reference. Eqns (7.12c) and (7.12d) will be employed later.

Combining eqn (7.12a) and j × eqn (7.12b):

$$\therefore \quad 0 = Mp(\mathbf{i}_{dsr} + j\mathbf{i}_{qsr}) + (R_2 + pL_2)(\mathbf{i}_{drr} + j\mathbf{i}_{qrr})$$

Also (except at synchronous speed, when no torque is developed) the rotor is moving relatively to the rotating flux, displaced by instantaneous angle γ

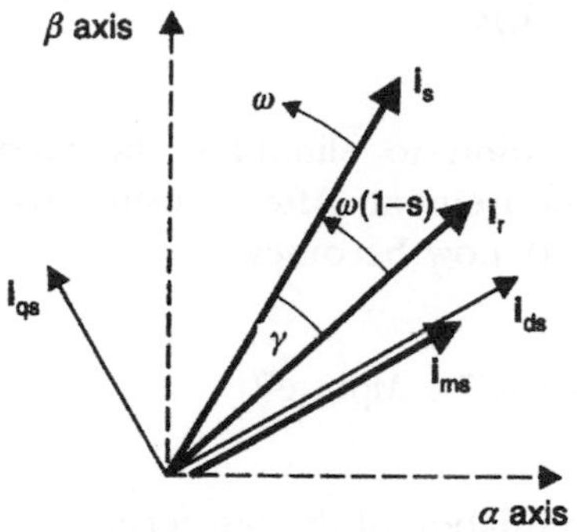

Figure 7.20 *Vector relationships.*

(see Figure 7.20, which has been supplemented and extracted from Figure 7.19). To refer the stator currents from synchronous- to rotor-axes, we apply the vector operator $e^{j\gamma}$ as mentioned earlier:

$$\mathbf{i}_{dsr} + j\mathbf{i}_{qsr} = \mathbf{i}_s e^{j\gamma}$$

$$\therefore \quad 0 = R_2\mathbf{i}_r + p(L_2\mathbf{i}_r + M\mathbf{i}_s e^{j\gamma}) \tag{7.13}$$

which, after rearranging, becomes

$$0 = R_2\mathbf{i}_r + p(\mathbf{i}_s + \frac{L_2}{M}\mathbf{i}_r e^{-j\gamma})Me^{j\gamma} \tag{7.14}$$

The term in brackets represents the combination of the stator m.m.f. ($\mathbf{i}_s$) and the rotor m.m.f., since $\mathbf{i}_r e^{-j\gamma}$ is $\mathbf{i}_r$ transformed to the same synchronous axes as $\mathbf{i}_s$. The term will be given the symbol $\mathbf{i}_{ms}$ since it represents the resultant m.m.f. referred to synchronous axes; neglecting the iron-loss component, it will be in phase with the flux, i.e. along the d-axis.

$$\mathbf{i}_{ms} = \mathbf{i}_s + \frac{L_2}{M}\mathbf{i}_r e^{-j\gamma}$$

Rearranging:

$$\mathbf{i}_{ms}e^{j\gamma} = \mathbf{i}_s e^{j\gamma} + \frac{L_2}{M}\mathbf{i}_r$$

which is back to the rotor reference frame and therefore

$$\mathbf{i}_r = \frac{M}{L_2}(\mathbf{i}_{ms} - \mathbf{i}_s)e^{j\gamma}$$

We are now in a position to eliminate the rotor current, which is not accessible for measurement. After substituting this expression and rearranging, eqn (7.14) now becomes:

$$0 = \frac{MR_2}{L_2}(\mathbf{i}_{ms} - \mathbf{i}_s)e^{j\gamma} + Mp\mathbf{i}_{ms}e^{j\gamma}$$

Performing the differentiation of the last term:

$$0 = \frac{MR_2}{L_2}(\mathbf{i}_{ms} - \mathbf{i}_s)e^{j\gamma} + Me^{j\gamma}p\mathbf{i}_{ms} + Me^{j\gamma}\mathbf{i}_{ms}jp\gamma$$

$p\gamma = d\gamma/dt$ is ω_{sl}, the slip speed, which is proportional to f_2, the slip frequency.

Dividing throughout by $Me^{j\gamma}$ and expressing $\mathbf{i}_s$ as $\mathbf{i}_{ds} + j\mathbf{i}_{qs}$,

$$0 = \frac{R_2}{L_2}(\mathbf{i}_{ms} - \mathbf{i}_{ds} - j\mathbf{i}_{qs}) + p\mathbf{i}_{ms} + j\omega_{sl}\mathbf{i}_{ms}$$

Separating into real and imaginary parts,

$$\underline{\text{Re}}: \quad 0 = \frac{R_2}{L_2}(\mathbf{i}_{ms} - \mathbf{i}_{ds}) + p\mathbf{i}_{ms}$$

$$\therefore \quad \mathbf{i}_{ds} = \mathbf{i}_{ms}(1 + pL_2/R_2)$$

$$\therefore \quad \mathbf{i}_{ms} = \frac{\mathbf{i}_{ds}}{1 + pL_2/R_2} \tag{7.15}$$

The term including p is a transient component; once this has died away, the steady-state value of $\mathbf{i}_{ms}$ is equal to $\mathbf{i}_{ds}$.

Im: $0 = \frac{R_2}{L_2}(-j\mathbf{i}_{qs}) + j\omega_{sl}\mathbf{i}_{ms}$

$$\therefore \quad \omega_{sl} = \frac{R_2\mathbf{i}_{qs}}{L_2\mathbf{i}_{ms}} \tag{7.16}$$

The quantity $\mathbf{i}_{ms}$ found from eqn (7.15) is used to represent flux and can be controlled by a closed-loop system, as described later. L_2/R_2 must be known; $\mathbf{i}_{ds}$ may be calculated from the three-phase currents via the two transformations in eqns (7.10) and (7.11), provided that the angle ωt required for the transformations is known instantaneously. Referring back to Figure 7.19, the time in the cycle when the synchronously rotating d-axis was in line with the α-phase axis was chosen arbitrarily for illustration purposes. But $d\rho/dt$ is nevertheless the same as the synchronous speed $\omega = d\theta/dt$. So, to detect the angle for control purposes, we can integrate ω, i.e. $\int\omega\, dt = \omega t +$ constant and, since the synchronous axes are not attached to any physical member,(2) we can choose ρ to be zero when $t = 0$, in which case $\rho = \omega t$. To measure this angle, we can combine $\omega_m + \omega_{sl} = \omega$ and integrate this (see Figure 7.22a). Since the equation for ω_{sl} eqn (7.16) includes $\mathbf{i}_{ms}$, the process of control proceeds by updating each in turn.

The e.m.f. produced by a field system is proportional to (flux × speed) and the flux in this case is taken to be proportional to the magnetising current $\mathbf{i}_{ms}$.

i.e. $E = K_\phi \times \text{speed} \times \mathbf{i}_{ms}$

The torque produced is the vector product of e.m.f. and current intersecting the flux, divided by speed, i.e.

$$T_e = K_\phi \times \mathbf{i}_{ms} \times \mathbf{i}_s$$

The vector ('cross') product of $\mathbf{i}_{ms}$ and $\mathbf{i}_s$ involves the vector products $\mathbf{i}_{ms} \times \mathbf{i}_{qs}$ and $\mathbf{i}_{ms} \times \mathbf{i}_{ds}$. The latter term is zero, as, referring again to Figure 7.20, $\mathbf{i}_{ds}$ is in line with $\mathbf{i}_{ms}$.

$$\therefore \quad T_e = K_\phi \mathbf{i}_{ms}\mathbf{i}_{qs} \tag{7.17}$$

Hence torque may be controlled by $\mathbf{i}_{qs}$ via eqn (7.16) using the latest values of $\mathbf{i}_{ms}$ and ω_{sl}. If $\mathbf{i}_{ms}$ and ω_{sl} are constant, $T_e \propto \mathbf{i}_{qs}$. A change in $\mathbf{i}_{qs}$ is reflected in the rotor, due to the coupling between stator and rotor windings.

Equations (7.15) to (7.17) represent a simple control model which may be used to *estimate* flux and torque for providing feedback signals. Note that the equations do not represent an analysis of normal machine operation. Numerical scaling factors may be necessary in implementation of the model in a digital control system.

Implementation

The achievement of fast torque response in a field-oriented, or vector, control scheme thus depends on obtaining almost instantaneous current response (see eqn (7.17)). To see how this occurs, consider the equations for the two coupled coil systems from eqns (7.12). The rotor is short-circuited and a step voltage is applied to the stator.

Combining eqn (7.12c) and j × eqn (7.12d):

$$\therefore \quad \mathbf{v}_{ds} + j\mathbf{v}_{qs} = Mp(\mathbf{i}_{drs} + j\mathbf{i}_{qrs}) + (R_1 + pL_1)(\mathbf{i}_{ds} + j\mathbf{i}_{qs})$$

Using the same process as for eqns (7.12a) and (7.12b), the referred rotor current components are transformed to the synchronous coordinates by the vector operator $e^{-j\gamma}$.

$$\mathbf{i}_{drs} + j\mathbf{i}_{qrs} = \mathbf{i}_r e^{-j\gamma}$$

$$\mathbf{i}_{ds} + j\mathbf{i}_{qs} = \mathbf{i}_s .$$

Similarly, let $\mathbf{v}_{ds} + j\mathbf{v}_{qs} = \mathbf{v}_s$.

$$\therefore \quad \mathbf{v}_s = R_1\mathbf{i}_s + L_1 p\mathbf{i}_s + Mp\mathbf{i}_r e^{-j\gamma}$$

Divide eqn (7.13) by $e^{j\gamma}$:

$$0 = R_2\mathbf{i}_r e^{-j\gamma} + p(L_2\mathbf{i}_r e^{-j\gamma} + M\mathbf{i}_s)$$

For simplicity, define the referred rotor current $\mathbf{i}_r e^{-j\gamma}$ as $\mathbf{i}_r'$

$$\therefore \quad \mathbf{v}_s = (R_1 + L_1 p)\mathbf{i}_s + Mp\mathbf{i}_r' \qquad (7.18a)$$

$$\text{and} \quad 0 = (R_2 + pL_2)\mathbf{i}_r' + Mp\mathbf{i}_s \qquad (7.18b)$$

From eqn (7.18b), $\mathbf{i}_r'(R_2 + pL_2) = -Mp\mathbf{i}_s$

$$\therefore \quad \mathbf{i}_r' = \frac{-Mp\mathbf{i}_s}{R_2 + pL_2}$$

Substituting for $\mathbf{i}_r'$ in eqn (7.18a):

$$\mathbf{v}_s = \mathbf{i}_s\left(R_1 + pL_1 - \frac{M^2p^2}{R_2 + pL_2}\right)$$

$$= \mathbf{i}_s\left[\frac{R_1R_2 + p(L_1R_2 + L_2R_1) + p^2(L_1L_2 - M^2)}{R_2 + pL^2}\right]$$

This is a second-order function. However, for coils with perfect coupling, $M^2 = L_1L_2$, which would mean that the leakage flux was zero.

Assuming perfect coupling:

$$\frac{\mathbf{i}_s}{\mathbf{v}_s} = \frac{R_2 + pL_2}{R_1R_2 + p(L_1R_2 + L_2R_1)}$$

If $\tau_1 = L_1/R_1$ and $\tau_2 = L_2/R_2$,

$$\frac{\mathbf{i}_s}{\mathbf{v}_s} = \frac{1 + p\tau_2}{R_1[1 + p(\tau_1 + \tau_2)]} \tag{7.19a}$$

This is first-order transfer function. If $\mathbf{v}_s$ is a step input, the response of $\mathbf{i}_s$ will be of the form of curve (a) in Figure 7.21. A boosted voltage would be necessary to offset the initial drop below the steady-state value of $\mathbf{i}_s$.

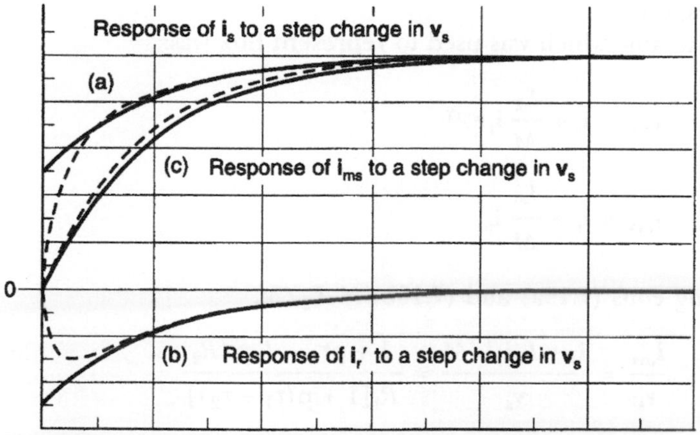

Note:
Solid lines represent perfect coil coupling (a coupling factor of 1.0)
Dotted lines represent a coupling factor of 0.8

Figure 7.21 *Transient response for coupled coils.*

Similarly, from eqn (7.18b),

$$\mathbf{i}_s = \frac{-(R_2 + pL_2)\mathbf{i}_r'}{Mp}$$

Eliminating $\mathbf{i}_s$ in eqn (7.18a):

$$\mathbf{v}_s = \left[\frac{-(R_1 + pL_1)(R_2 + pL_2)}{Mp} + Mp\right]\mathbf{i}_r'$$

$$= \left[\frac{p^2(M^2 - L_1L_2) - p(L_1R_2 + L_2R_1) - R_1R_2}{Mp}\right]\mathbf{i}_r'$$

Again assuming perfect coupling, $M^2 = L_1L_2$

$$\therefore \quad \frac{\mathbf{i}_r'}{\mathbf{v}_s} = \frac{-Mp}{R_1R_2[1 + p(L_1R_2 + L_2R_1)/R_1R_2]}$$

$$= \frac{-Mp}{R_1R_2[1 + p(\tau_1 + \tau_2)]} \qquad (7.19b)$$

The 'p' term in the numerator of this transfer function indicates that there is no steady-state value once the transient has died away. The transfer function is also negative, indicating that the second coil opposes the effect produced by $\mathbf{v}_s$. The response of $\mathbf{i}_r'$ to a step change in $\mathbf{v}_s$ is shown as curve (b) in Figure 7.21.

The variable which was used to represent flux was:

$$\mathbf{i}_{ms} = \mathbf{i}_s + \frac{L_2}{M}\mathbf{i}_r e^{-j\gamma}$$

i.e.

$$\mathbf{i}_{ms} = \mathbf{i}_s + \frac{L_2}{M}\mathbf{i}_r'$$

Summing eqns (7.19a) and (7.19b) × L_2/M,

$$\frac{\mathbf{i}_{ms}}{\mathbf{v}_s} = \frac{\mathbf{i}_s + L_2\mathbf{i}_r'/M}{\mathbf{v}_s} = \frac{1 + p\tau_2 - L_2p/R_2}{R_1[1 + p(\tau_1 + \tau_2)]}$$

$$= \frac{1}{R_1[1 + p(\tau_1 + \tau_2)]} \qquad (7.20)$$

The response of $\mathbf{i}_{ms}$, which represents the shape of the response of the combined m.m.f., is shown as curve (c) in Figure 7.21. Note the presence of the sum of the time-constants $(\tau_1 + \tau_2)$ when the response to a step change in voltage is shown. In the motor model (see eqn (7.15)), we considered only the response of magnetising current $\mathbf{i}_{ms}$ to stator current $\mathbf{i}_{ds}$, assuming that $\mathbf{i}_{ds}$ would respond instantaneously to a change in voltage $\mathbf{v}_s$. Curve (a) in Figure 7.21 shows this to be roughly the case if the windings are closely coupled. If $\mathbf{i}_{ms}/\mathbf{v}_s$ in eqn (7.20) is divided by $\mathbf{i}_s/\mathbf{v}_s$ from eqn (7.19a), the result is:

$$\frac{\mathbf{i}_{ms}}{\mathbf{i}_s} = \frac{1}{1 + p\tau_2}$$

which is the same as eqn (7.15), except that $\mathbf{i}_s$ includes $j\mathbf{i}_{qs}$. However, $\mathbf{i}_{qs}$ has no effect on $\mathbf{i}_{ms}$, as the 'vectors' are at right-angles to each other. Hence, if a large voltage change is available to drive the change in $\mathbf{i}_s$, the flux will respond approximately with the rotor time-constant.

If the coil-coupling is ideal, a step change in $\mathbf{v}_s$ will give an instantaneous change in $\mathbf{i}_s$ (Figure 7.21, curve (a)), but not to the final value, and hence a step change in torque response, according to eqn (7.17).

Thus a desired current can be injected rapidly into the short-circuited winding of a squirrel-cage rotor, while the air-gap flux, which has a relatively long time-constant, will be virtually unchanged until the next stator current impulse is applied; if no further impulse is applied, then the rotor current will decay and revert to normal operation.

If the coupling between stator and rotor coils is not perfect, the rise of $\mathbf{i}_s$ is not instantaneous. In this case,

$$\frac{\mathbf{i}_s}{\mathbf{v}_s} = \frac{1 + p\tau_2}{R_1[1 + p(\tau_1 + \tau_2) + p^2(\tau_1\tau_2 - M^2/R_1R_2)]}$$

$$\text{and } \frac{\mathbf{i}_r'}{\mathbf{v}_s} = \frac{-Mp}{R_1R_2[1 + p(\tau_1 + \tau_2) + p^2(\tau_1\tau_2 - M^2/R_1R_2)]}$$

For the case of a coupling factor $M/\sqrt{(L_1L_2)}$ having a typical value of 0.8, the step responses are shown dotted in Figure 7.21. Again, presence of a high driving voltage will enhance the response of $\mathbf{i}_s$, and hence the torque.

Figure 7.22a shows how the model based on eqns (7.15) to (7.17) may be incorporated into a typical vector-control system using a flux control loop. The model is employed as an estimator for flux and torque control loops in the system shown in Figure 7.22b. Note that, neglecting the calculation time of the processor and response time of the inverter and transducers, all loops appear as instantaneous with the exception of the 'flux' signal output from the signal $\mathbf{i}_{ds}$. This has a time-constant L_2/R_2. There is an analogy to

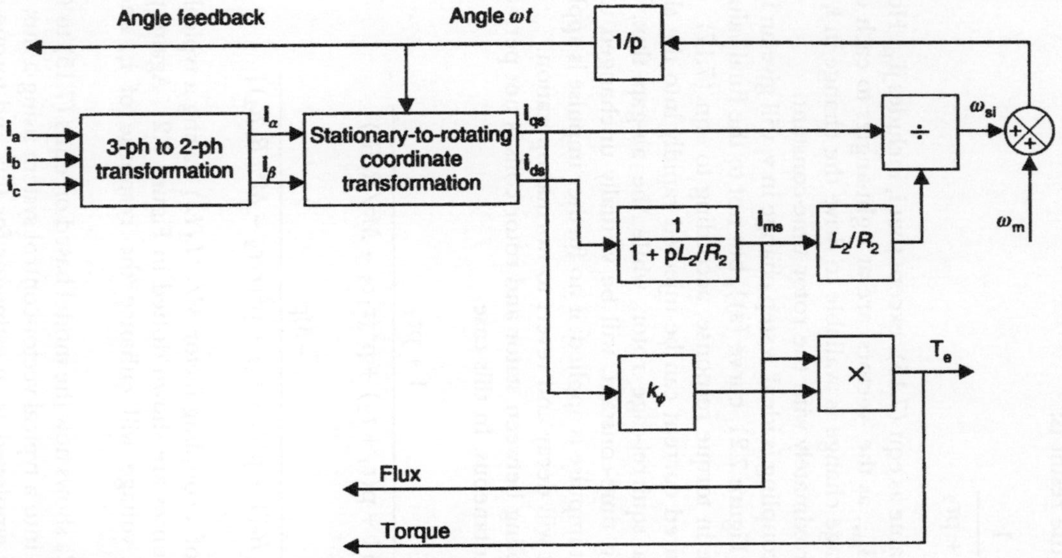

Figure 7.22 *Vector-control: (a) Flux model*

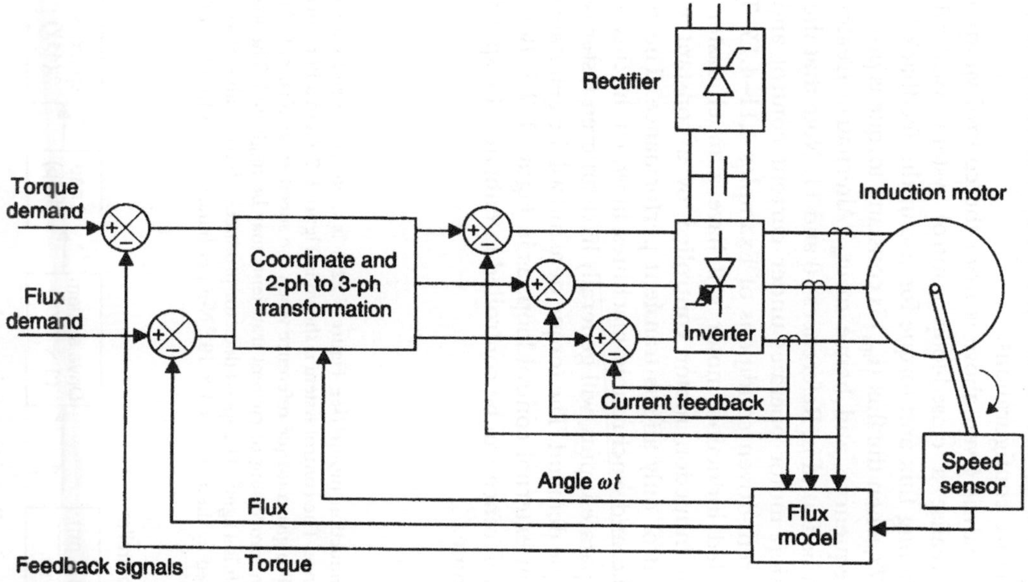

Figure 7.22 *(b) Control scheme.*

the transfer function for a d.c. drive, as mentioned in Chapter 6. The vector-controlled induction motor has separate torque (as in armature current) and flux (as in field current) control inputs. The flux control loop has a substantial electrical time-constant L_2/R_2 which must be known accurately or estimated by the control processor. Reference 10 provides coverage of alternative configurations.

The simplified analysis given above is from the control viewpoint, the object being to realise a closed-loop control system with calculable quantities representing flux and torque for use in the feedback loops. As explained on p. 277, when the flux ($\mathbf{i}_{ms}$) is constant, torque is proportional to f_2 (the slip frequency), and hence to ω_{sl}. Alternative treatments of vector-control can be found in References 10 and 11. Note that the vector-controlled induction motor operates under current control and corresponds to the current-driven conditions of Examples 4.11–4.15. Though the vector-controlled induction motor will have a much faster torque response than an induction motor controlled by a constant volts/Hz system, vector control only affects transient performance. The transient performance of d.c. and synchronous machines, in which the flux is largely determined by separate inputs, will generally have an even faster response to changes in torque demand. The torque demand will be generated by the 'error' signal of an external control loop, as in Figure E.7.7. Beyond full voltage, the flux reference can be controlled to enhance the speed range, as for a d.c. machine.

Example 7.7

A vector-controlled induction motor (see Figure 7.22b) is used in a speed control system requiring rapid response. The control system is shown in Figure E.7.7 and a PI controller with gain $K = 8$ is included to give a torque reference when the speed error is zero. Delays or lags due to vector rotation, processing or current transients may be neglected. The load inertia seen at the motor shaft is $4\,\text{kgm}^2$. The speed detector produces a binary value of 417 per rad/s. The torque produced by the motor is 3×10^{-3} Nm per binary unit output from the digital

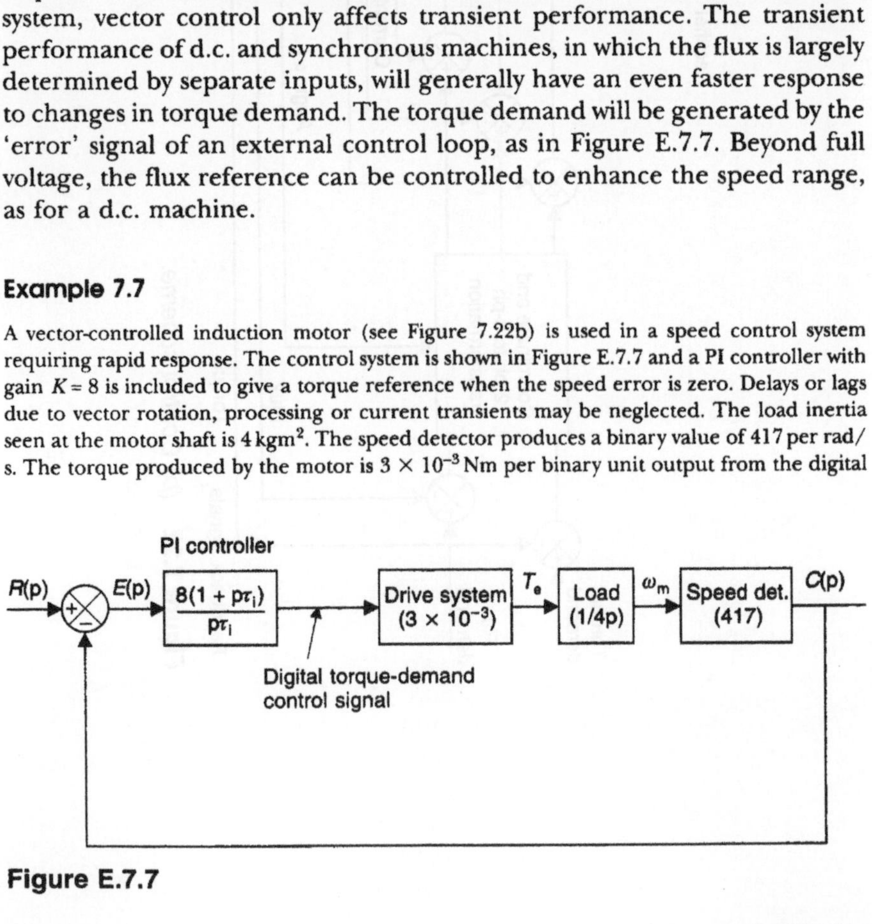

Figure E.7.7

PI controller. Considering only the change in speed produced by a change in input reference (i.e. neglecting steady-state load torque), find the value of the integral constant τ_i which will give the system a damping factor of 0.4. Find the settling time to within 5% of the final value and the damped frequency of oscillation.

Figure E.7.7 shows the system's block diagram, on which the transfer functions of the various blocks have been identified from the data. From the diagram,

$$\frac{C(p)}{T_e(p)} = \frac{417}{4p}$$

and $$\frac{T_e(p)}{E(p)} = \frac{8(1 + p\tau_i) \times 3 \times 10^{-3}}{p\tau_i}$$

The open-loop transfer function is obtained by multiplying the above expressions:

$$\frac{C(p)}{E(p)} = \frac{10(1 + p\tau_i)}{4\tau_i p^2} = G(p)$$

$$\therefore \quad \frac{C(p)}{R(p)} = \frac{G(p)}{1 + G(p)} = \frac{10(1 + p\tau_i)}{4\tau_i p^2 + 10p\tau_i + 10}$$

$$= \frac{2.5/\tau i + p/4}{p^2 + 2.5p + 2.5/\tau_i}$$

Comparing the denominator with that of eqn (6.5),

$$2\zeta\omega_n = 2.5 \text{ and } \omega_n^2 = 2.5/\tau_i.$$

If $\zeta = 0.4,\ \omega_n = 2.5/0.8 = 3.125.$

$\therefore \quad \omega_n^2 = (3.125)^2 = 9.77 = 2.5/\tau_i$

$\therefore \quad \tau_i = 2.5/9.77 = \underline{0.255}$

The decay rate is $\zeta\omega_n = 1.25$ seconds^{-1}

and the settling time to 5% is $3/1.25 = \underline{2.4 \text{ seconds}}$

The damped frequency of oscillation is:

$$\sqrt{(1 - \zeta^2)}\omega_n = \sqrt{(1 - 0.16)} \times 3.125 = \underline{2.86 \text{ rad/s}}$$

Slip-power recovery

In Section 4.5, this method of induction motor speed control was illustrated by examples which assumed that auxiliary machines were available which operated at slip frequency. Example 4.23 showed that, for particular types of load where torque falls appreciably with speed, the auxiliary-machine rating is relatively small, even when a wide range of speed control is required. Access to the rotor circuit means that the more expensive slip-ring motor design is needed but nevertheless this type of system is economical for certain large-power drives. It is much more conveniently implemented by using power-electronic circuits.

Two main methods are used. Figure 7.23 shows the Static-Kramer system in which the rotor currents are rectified and then inverted to the supply, using a fully controlled rectifier working in the inversion mode. Slip power is extracted from the rotor to give sub-synchronous speeds. A large line

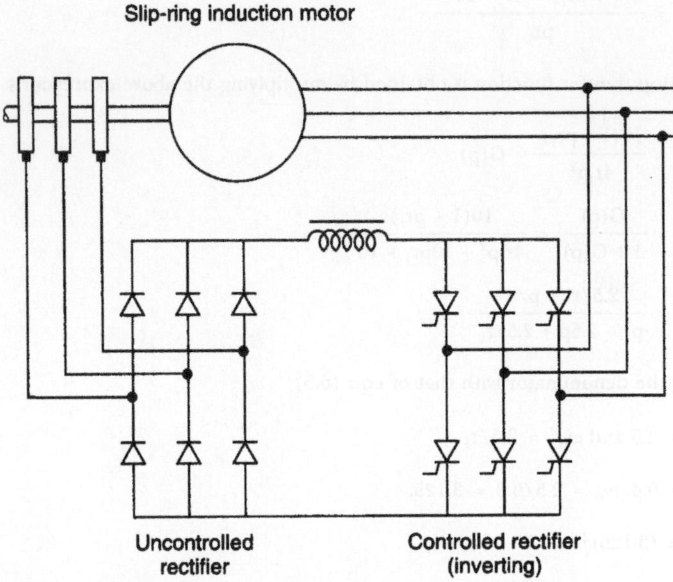

Figure 7.23 *Static-Kramer slip energy recovery system.*

inductor is required. In the Static-Scherbius system, the twin rectifier arrangement is replaced by a cycloconverter which performs the frequency conversion and the power flow can be reversed to give both sub- and super-synchronous speeds. As long as the frequency ratio is large, the cycloconverter waveforms are acceptable.

Example 7.8

The 3-phase, 440-V, star-connected, slip-ring induction motor of Example 4.22, with equivalent circuit parameters per phase of:

$$R_1 = 0.2\,\Omega;\ x_1 = 0.8\,\Omega;\ R_2 = 0.06\,\Omega;\ x_2 = 0.25\,\Omega$$

and stator/rotor turns ratio of 2:1, is connected in a Static-Kramer system. Using the answers of Example 4.22 where necessary, determine the firing delay-angle required to achieve the same speed reduction as in part (a) of that example, namely 25% slip, while still developing full-load torque.

From part (a) of Example 4.22, the injected voltage V_3 required for 25% slip at full-load torque is

$$16.2\,\text{V per phase} = 28.06\ \text{line V.}$$

The mean output voltage of the 3-phase rectifier bridge, neglecting internal voltage drops is:

$$\frac{3 \times \sqrt{2} \times 28.06}{\pi} = 37.89\,\text{V}$$

Hence the controlled rectifier bridge must invert from a d.c. voltage of -37.89 V to an a.c. line voltage of 440 V.

The output voltage of the fully controlled bridge must therefore be:

$$\frac{3 \times \sqrt{2} \times 440 \times \cos\alpha}{\pi} = -37.89$$

$$\therefore \quad \cos\alpha = -0.0638 \quad \therefore \alpha = \underline{93°66}$$

7.6 Synchronous, brushless and reluctance drives

Synchronous machine drives

A synchronous machine under steady-state operation will maintain the constant speed set by the stator supply frequency. Control schemes using synchronous motors and power-electronic converters are most common in large drives in which the excitation of the motor may be controlled, often by rotor angle measurement, to produce a leading power factor. This allows rectifier-grade SCRs to be used as switching devices, as the current will commutate naturally from one device to another, due to the favourable current/voltage phase relationships at such power factors. At low speeds, when the commutating e.m.f. is inadequate, inverter device turn-off may be achieved, albeit slowly, by turning off the devices in the controlled rectifier which supplies d.c. to the inverter until the d.c. current falls to zero, before restoring rectifier conduction. Examples 5.14 and 5.15 illustrate calculations for variable voltage and frequency control of synchronous motors and earlier Examples 5.9–5.13 cover the effects on performance, e.g. maximum torque, of varying different parameters.

Under transient conditions, the rate of frequency increase must be limited to prevent the load angle falling beyond 90° as the rotor attempts to keep up with the rotating field. As a margin should be allowed, maximum torque would not be attainable. Closed-loop control, using detection or estimation of rotor position to determine torque angle, will allow the margins to be reduced and transient performance to be improved.

Section 5.5 dealt with constant-current operation of the synchronous machine. If the angle φ_{ma} between I_m and I_a in Figure 5.2b is controlled to be a constant 90°, then I_a would be at unity power factor. If I_m is at its rated value, 1 *per unit*, and I_a is controlled to be 2 *per unit*, the torque will be 2 *per unit* until rated voltage is reached. To complete the right-angled triangle, the field m.m.f. would have to be $\sqrt{5}$ *per unit.* Beyond the point where maximum voltage is reached, increasing frequency will bring about a reduction of I_m and flux ($\phi \propto E/f \approx V/f$). If I_a can be maintained, there is constant power until the increase of leakage reactance voltage drop due to increasing frequency becomes significant. It is also likely that field current will eventually have to be reduced if φ_{ma} is to be maintained at 90°.

In the synchronous machine, an alternative control strategy would be to take the flux into saturation, as described in Section 5.5 and would give improved performance over a limited speed range where the excess flux would not lead to unacceptable iron losses. If, for example, I_m was 1.5 *per unit*, giving a flux of about 1.25 *per unit*, allowing for saturation, as in Example 5.15(j), equal currents $I_a = I_f'$ of about 1.75 *per unit* would give a torque of 2 *per unit* with a better torque per ampere than described previously, i.e. 2/3.5 as against $2/(2 + \sqrt{5}) = 2/4.2$, though not quite as good as the d.c. machine at $2/(2 + 1) = 2/3$.

The synchronous machine will motor or generate, depending on whether mechanical power at the coupling is being extracted or injected, so it can regenerate easily to the supply, provided that the power-electronic converter can handle the reverse power effectively.

Comparison of main machine types

The above remarks about synchronous-machine torque capacity are of special importance for variable-speed drives where, in addition, rapid speed response is often required. With such features in mind, comparisons of different machine types are informative and are summarised in Figure 7.24. Electromagnetic torque-overload-capacity governs the maximum acceleration (and deceleration) rates. A unique feature of the d.c. machine is its overload capacity; e.g. a doubling of armature current would virtually double the torque for any particular value of field current. This does not follow for a.c. machines because the torque angle between stator and rotor m.m.f.s is not fixed but depends on the load, and the machine could pull out of step. Thus, if a short-term overload of 2 *per unit*, or even more, is required, as in some steel-mill and traction drives, an a.c. machine might have to be derated to meet it, i.e. made larger, so that, at full load, it is under-used in terms of its continuous rating. The d.c. machine does not usually have to be derated but, if supplied from an SCR converter, the mains power factor falls as d.c. voltage is reduced, because the firing-delay angle has to be increased for this purpose. This problem is often overcome by using multiple bridge rectifiers in series.

In Figure 7.24a, an overload torque of 2 *per unit* is chosen up to 1 *per unit* (base) speed. This means an armature current of 2 *per unit* in this constant-torque region. After reaching full voltage, further increase beyond base speed requires field weakening, which, at constant armature current, would cause torque capacity to fall inversely with flux reduction. The torque × speed product would be constant in this constant-power region. Beyond 2 *per unit* speed, the armature current might have to be reduced because of commutation and stability limitations, but field weakening ranges of up to 4/1 or more have been used in certain industrial drives. Speed control by field weakening, in its simplicity of application, has always been an attractive feature. Nevertheless, as d.c. machines carry a heavy burden of maintenance, because of the commutator and brushes, large-power drives have virtually been taken over by a.c. machines, for which many present-day control schemes are of relatively recent origin, following rapid developments in power-electronic and microelectronic technology.

Figure 7.24b for the induction machine is based on the work done in Section 4.3 and Examples 4.11–4.16 and assumes an overload capacity the same as for the d.c. machine at 2 *per unit*, though it would require about 3 *per unit* current, based on full-load current (see Example 4.13). The slip frequency is assumed to be adjusted to bring about a constant flux per pole which in turn occurs with a constant E/f ratio. The current would have to be maintained at the overload value needed to give 2 *per unit* torque at starting. As for the d.c. machine, further increase of speed when maximum voltage is reached requires flux weakening which occurs as frequency is reduced at the same, maintained current-overload. This is the constant-power region. As frequency increases, the torque for a particular slip is less (eqn (4.5)), and larger slip is required to get a large enough rotor current, so the speed regulation curve becomes steeper, as indicated. With vector control, a better management of the torque angle can be achieved during transient changes and, since this can be obtained with the simpler and cheaper squirrel-cage motor, the d.c. machine has to yield another advantage in terms of its rapid response to torque demand. On medium- to low-power drives however, it may still be able to compete on price.

Synchronous machine capabilities have already been discussed and the availability of field control makes it possible to operate at higher power-factors and lower currents than induction motors. Figure 7.24c shows a close comparison with the d.c. machine. Nevertheless, for these short-term overloads, the synchronous machine must be designed and rated for greater increases in field and/or armature current than for the d.c. machine, because the torque per ampere is lower, as explained earlier.

For power-electronic drives generally, although the waveforms are far from pure d.c. or sinusoidal a.c., the performance may be calculated with

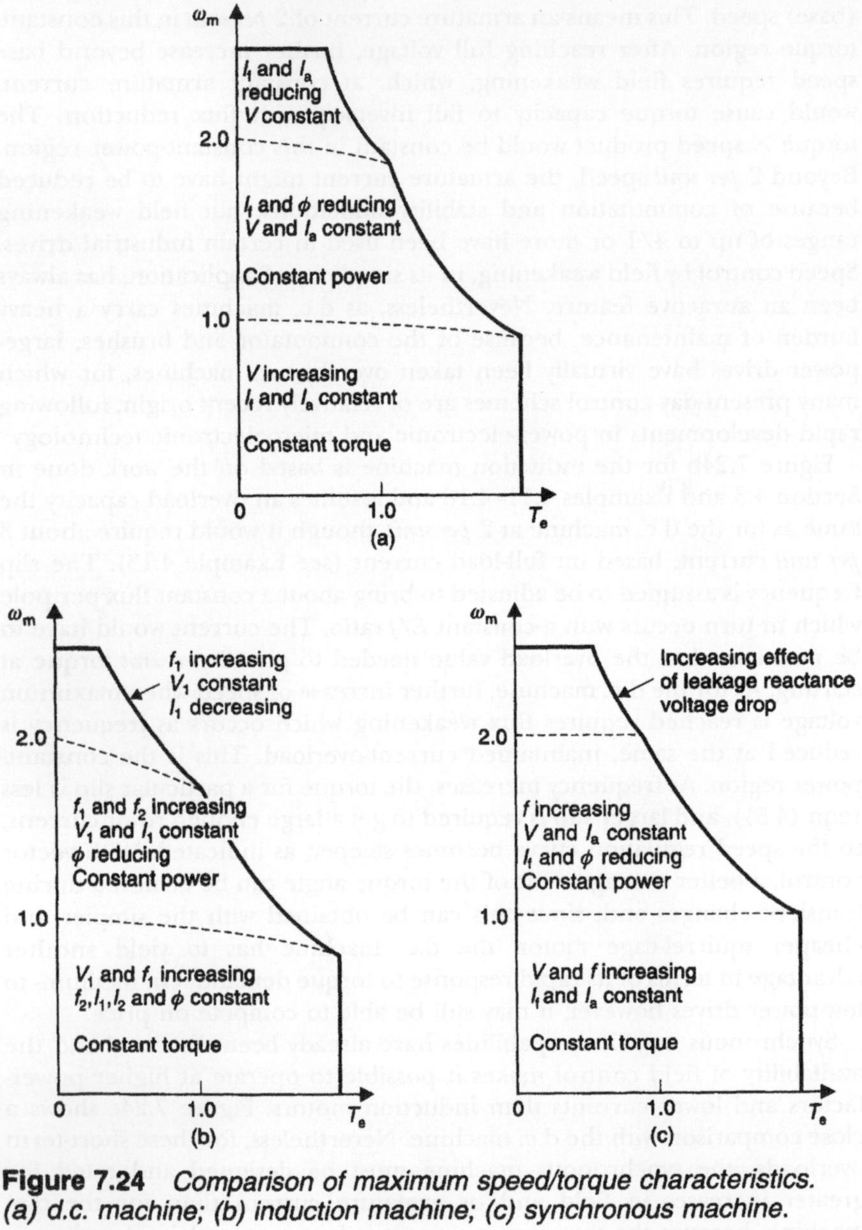

Figure 7.24 *Comparison of maximum speed/torque characteristics. (a) d.c. machine; (b) induction machine; (c) synchronous machine.*

reasonable accuracy by averaging out the harmonics and assuming that the change of mean (r.m.s.) voltage is the only consideration. The methods used in Chapters 3, 4 and 5, when variation of voltage and/or frequency were involved, did not specify the power source, which today would commonly be a power-electronic circuit. Although neglecting the harmonics means neglecting the extra machine losses, commutation problems and the presence of torque pulsations, this would not usually lead to significant errors in speed/mean-torque calculations. The worked examples in this present chapter follow this procedure, though, for the chopper circuit, the current waveforms were calculated and, subsequently, mean-torque values were worked out.

It is perhaps worth pointing out that, even with sinusoidal supplies, certain assumptions have been made in the performance calculations. For example, during the starting of an induction motor, the peak currents and torques can far exceed those calculated from voltage divided by equivalent-circuit impedance. In Chapter 8, this is illustrated by computer simulations of starting and synchronising transients for which the a.c. machine equations are developed from first principles and the organisation of the computer program explained.

Brushless motor drives

These motors attempt to copy electronically the action of brushes and commutator on a d.c. machine. This arrangement ensures that the armature-coil currents are reversed (commutated) when the coils rotate from under the influence of one field polarity to the opposite polarity. Thus the total force and torque produced maintain the same directional sense. The commutator and brushes in a d.c. machine act as a shaft-position sensor. The armature and field m.m.f.s are at a fixed angular displacement δ, sometimes called the torque angle (φ_{fa}), which is shown schematically in Figure 7.25a, where the armature is assumed to be so wound that its total m.m.f. is in the same direction as the current flow into the brush.

For a totally brushless machine, for which the field would have to be a permanent magnet, the armature coils are wound on the stationary (outside) member (Figure 7.25b) and are connected through semi-conductor switches which are activated from shaft position (Figure 7.25c), so that their currents are similarly reversed to match the polarity of the rotating field pole. The frequency of switching is thus automatically in synchronism with the shaft speed, as for a conventional d.c. motor. At $\delta = 90°$, the torque is proportional to $F_a \times F_f$ and, at any other angle, assuming sinusoidal m.m.f. distributions, the torque is proportional to $F_aF_f \sin \delta$. As the rotor moves, δ goes from 0° to 180°; the supply is then switched to return δ to zero again and the cycle repeats. The torque will thus pulsate like a single-phase rectified sine wave (Figure 7.25d). This arrangement is equivalent to a d.c. machine with only two commutator segments, and has a minimum torque value of zero. Commonly there are at least three tappings from a three-phase type of winding, in turn supplied from a three-phase bridge inverter. This is triggered under the control of the position detector so that its output frequency is automatically governed by the shaft speed. The torque pulsations would now be similar to the

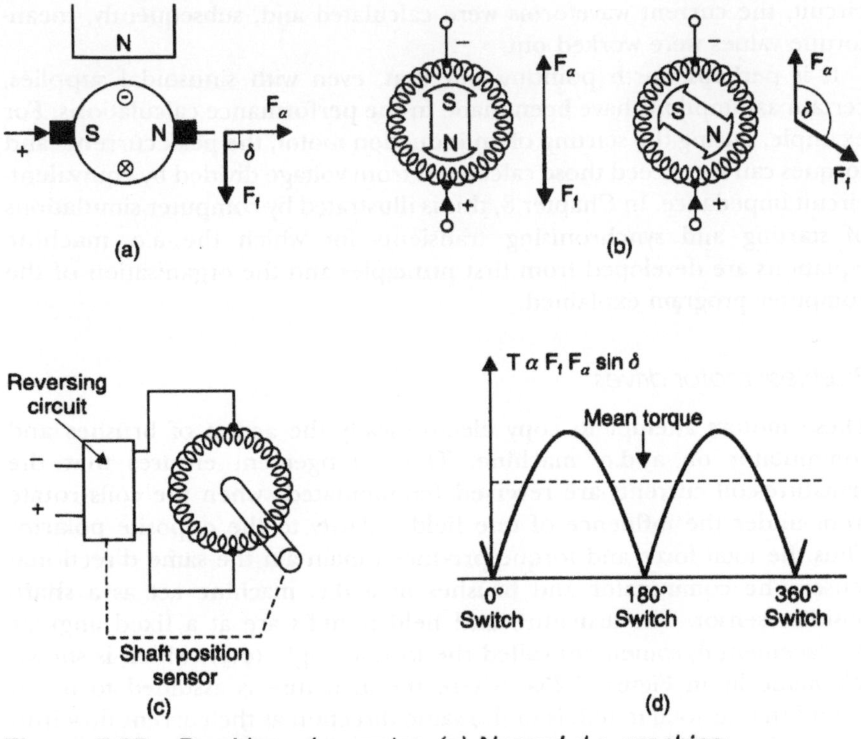

Figure 7.25 *Brushless d.c. motor. (a) Normal d.c. machine; (b) armature on stator; (c) main-circuit control scheme; (d) torque.*

output waveform of a three-phase bridge rectifier; as there is no torque zero, starting torque is always available. Profiling of the magnet pole-face further improves the smoothness of the torque over a complete cycle. The instants of switching can easily be changed to give effects similar to brush-axis shifting, which is sometimes used in moderate measure on conventional d.c. machines. See Example 3.1. The speed/load characteristic of the brushless machine is similar to a d.c. machine with fixed excitation, i.e. the speed falls a small amount as the torque increases.

Brushless d.c. drives are in common use for position-controlled applications in the industrial control area. As the length of the cycle depends on rotor movement, PWM is not normally applied to these drives. The rotor flux is established by permanent magnets on the rotor, providing a trapezoidal m.m.f. A variant with shaped magnets to produce a sinusoidal m.m.f. is known as 'brushless a.c.'. A brushless machine is normally supplied by a three-phase inverter and regeneration is again straightforward if a suitable power-electronic-converter arrangement is provided. Although considerable research effort has been expended on improving the response speed of, or removing the need for expensive sensors on brushless d.c. motors, most industrial controllers use simple Hall-effect shaft sensors and fixed conduction angles, with variable d.c. link voltage. Commercial units frequently include PI or PID controllers (p. 197).

Switched reluctance motor drives

A further variant in the family of synchronous machines is the reluctance motor, as outlined in Section 5.8. Switched reluctance motors step the stator supply voltages in response to rotor position in the same way as brushless machines. The characteristics are similar to those of a series d.c. motor or stepper motor (Figure 5.5) if a constant angle criterion is used for triggering the stator power electronic switches. In some cases, fewer switches can be used than in an inverter. The switched reluctance drive is most commonly used in medium-power variable speed applications. It is also a competitor, along with other brushless machines, for the coming lucrative market in drives for electric and hybrid road vehicles. This was formerly the province of the d.c. machine, which is currently being challenged by induction motor drives.(13)

Conclusion

In summary, the basic d.c. machine provides the best acceleration performance and simplest control characteristics and the basic induction machine the poorest. This reflects the physical complexity of one relative to the other; the cage-rotor induction machine is cheaper, more rugged and virtually maintenance-free. The d.c. machine possesses commutation limits and, with the slip-ring synchronous and induction motors, requires brushgear maintenance. With the addition of power-electronic converter(s) and microelectronic controllers, any machine can be controlled to provide, at certain cost, similar characteristics. Advances in high-power, fast-switching semiconductors such as the IGBT have offered improved PWM and other waveshaping techniques to reduce harmonic losses to low levels. Although d.c. machines remain popular for smaller precision drives, some manufacturers have ceased production of d.c. drives. The vector-controlled induction motor has substantially increased its market share and traction drives, for long a traditional market for large d.c. series motors, are now largely supplied with three-phase induction motors; the induction motor, starting with low stator frequency, avoids the commutator-burning or excessive rating of a single semiconductor associated with a stalled d.c. or brushless d.c. motor respectively. Although the presence of a sophisticated microelectronic controller adds cost, it is possible to standardise the converter and customise the drive to a particular machine or set of characteristics by user input to the software and incorporate additional condition-monitoring or control loops without the expense of a custom-designed system.

8 Mathematical and computer simulation of machine drives

Machine-drive systems have become more complex in recent years, especially with the proliferation of power-electronic circuits and their consequential introduction of substantial harmonic content in the waveforms. Most of the text has been devoted to illustrations of behaviour with pure d.c. or pure sinusoidal supplies and, even with harmonics present, these methods still give a good idea of the overall electromechanical performance. However, for the keen student, the specialist and for post-graduate studies, it is important to indicate how a more accurate solution of a system may be obtained. This inevitably involves the use of computational aids and commercial software packages which can be applied to simulations of various power-electronic and other circuits are available. It may be useful, however, in this final chapter to provide enough information, in terms of methods and equations, for individual project students to develop their own solutions, assuming a knowledge of computer programming. Particular emphasis is given to the equations for polyphase machines expressed in terms of phase parameters (real-axis simulation), which need make no assumptions about sinusoidal m.m.f. distributions as in d–q simulations. However, for comparison, for reference purposes and as an alternative method, the d–q matrices are given in Appendix C.

A simple example of a single-phase, thyristor bridge/d.c.-machine system is included as an illustration of the allowance for the effects of power-electronic switching in such circuits. If machine equations are organised in such a manner as to allow for this, or other non-linearities such as saturation, the possibility of a general solution is usually lost. However, iterative[4] and step-by-step numerical solutions can be used and as they proceed, any change in the system condition or configuration can be included.

8.1 Organisation of step-by-step solutions

Figure 8.1 shows a flow diagram for a section of a computer program calculating the electromechanical performance following some system change, e.g. acceleration from zero speed; or a transient following a load change or a voltage/frequency change. Following the specification of the

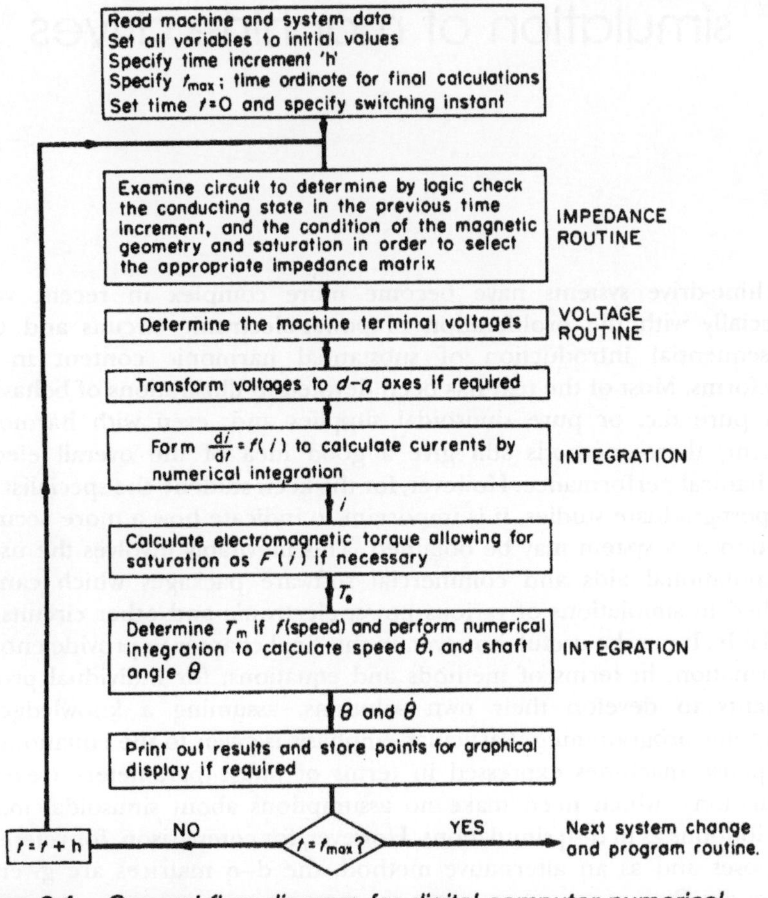

Figure 8.1 *General flow diagram for digital-computer numerical solutions of electromechanical performance.*

data, the time limits of the calculation and the system change being investigated, the machine and circuits are scanned to determine the appropriate equations. If power semiconductors are involved, then this will require a logic check, at each time increment, to find the conducting state of every circuit branch including such switching devices. For a fairly simple representation:

A diode will conduct if the forward voltage drop across it is equal to or greater than 1 V (say).

For a GTO, IGBT or power transistor, a firing pulse must be present for conduction, AND the forward voltage drop must be greater than 1 V.

For a thyristor in a naturally commutated system, conduction will continue if it was conducting in the previous time increment, with a current greater than the holding value – say 100 mA; OR it will conduct if the forward voltage drop is greater than 1 V, AND a firing pulse is present.

For a thyristor in a forced-commutated system, a GTO or a power transistor, the switch-off time can be specified.

If a no-conduction condition is detected, it may be necessary to calculate the voltage across such an element arising from the rest of the circuit and apply this to the element in the simulation, so that its current will calculate zero. It may be possible, with certain types of simulation, to set this semiconductor impedance to several thousand ohms say, and solve the circuit without mathematical instabilities.

With the conducting paths known, the equations can now be set up, making allowance for the semiconductor voltage drops if these are of relative significance. The machine terminal-voltages and the impedance matrices may be affected by the nature of the conducting paths so these must now be determined. If a d–q simulation is being used, transformation of these voltages to d–q axes must be performed. For a current-fed system, the determination of any unknown currents is usually fairly simple. In any case an expression for $\mathrm{d}i/\mathrm{d}t$ is required in terms of the circuit parameters. From the circuit equation:

$$v = Ri_1 + L_{11}\,\mathrm{d}i_1/\mathrm{d}t + L_{12}\,\mathrm{d}i_2/\mathrm{d}t \ldots + f(\text{speed, currents, inductances}),$$

the general form:

$$\frac{\mathrm{d}i_1}{\mathrm{d}t} = \frac{v - Ri_1 - f(\text{speed, currents, inductances}) - \sum_{2}^{n} L_{1_n}\mathrm{d}i_n/\mathrm{d}t}{L_{11}}$$

is obtained. Numerical integration methods predict the change of i during the time increment, from the initial value and the rate of change. This is done for all the conducting paths, using matrix techniques if there are several of these. The electromagnetic torque can then be calculated.

The next step is to calculate the speed $\dot{\theta}$ and the shaft angle θ if this is required, e.g. to allow for change of inductances with shaft orientation. θ will normally be an electrical angle. Referring to the earlier equations for the mechanical system:

$$\frac{d\omega_m}{dt} = \frac{T_e - T_m}{J}$$

and hence:

$$\frac{d\dot{\theta}}{dt} = \frac{T_e - T_m}{J} \times \text{pole pairs} \quad \text{(in electrical rad/s}^2\text{)} \tag{8.1}$$

and:

$$\frac{d\theta}{dt} = \dot{\theta} \tag{8.2}$$

are the two relevant first-order equations required for numerical solution.

If the mechanical load is a function of speed, it can be continually updated throughout the solution at each incremental step. In a similar manner, inductances can be changed if there is a variation with position or with current value.

The time increment is now added, the new circuit conduction condition checked and the new driving function (voltage or current) at this new time, is inserted to repeat the calculation procedure, until t_{max} is reached.

8.2 Simulations for polyphase a.c. machines

These simulations are usually more complex than for d.c. machines. For steady-state (constant-speed) calculations, equivalent-circuit methods can be used of course, as already demonstrated. The equivalent-circuit model is based on a viewpoint which notes that for a uniform air-gap machine, or along an axis fixed with respect to a salient pole, e.g along the field-coil (d-) axis or the interpole (q-) axis, the magnetic permeance does not change with flux-wave position, so inductances are not time dependent. The induced voltages shown on the circuit take account of the flux-wave movement and are usually expressed in terms of the winding carrying the line-frequency current. It must be remembered that inductance variation is the source of the motional e.m.f.s ($\Sigma i \cdot dL/dt$), see p. 5, but these can be replaced by terms which are a function of a fixed inductance, a current and the rotational speed ($\dot{\theta}$).

A formal transformation to d–q axes, which must be stationary with respect to the flux wave for the salient-pole machine, but need not necessarily be so for the uniform air-gap machine, permits much greater flexibility in the analytical solution of both steady and transient states. Time dependence of inductances can be removed by the transformation. This can be achieved providing certain conditions are fulfilled[(2)]; e.g. sinusoidal distribution of d-axis and q-axis flux and any winding imbalance confined to one side of the air gap. Analytical transient solutions at a particular speed for current and torque say, are then possible. For an electro-mechanical transient, including speed as a variable, a numerical solution would be necessary. Even then, the d–q simulation permits the greatest economy in computation time because the equations are so concise and the inductances are constant. However, if the necessary conditions are not fulfilled, the transformation will not lead to time-independent inductances and it may then be convenient to use a real-axis simulation (A-B-C, a-b-c, for 3-phase), sometimes known as the method of phase coordinates. No extensive additional knowledge is required to follow the development and setting out of the equations. They are preferably in matrix form since this is the way they are manipulated in the computer program. The method is explained in sufficient detail to enable such a program to be written using standard inversion, multiplication and integration procedures.

Development of equations

The circuit equations for all the phase windings of a machine can be expressed very simply and concisely in matrix form as:

$$\mathbf{v} = \mathbf{Ri} + \mathrm{p}(\mathbf{Li}) \tag{8.3}$$

It will be noticed that time changes in both the currents and the inductances must be allowed for. For a 3-phase, wound-rotor machine which is going to be used to illustrate the method, the inductance matrix will be 6 × 6 since there are 6 coupled windings, stator A-B-C and rotor a-b-c. Each winding has a self inductance and 5 mutual inductance terms, e.g. for the A-phase winding:

$$\begin{aligned} v_A = {} & R_A i_A + \mathrm{p}(L_{AA} i_A) + \mathrm{p}(L_{AB} i_B) + \mathrm{p}(L_{AC} i_C) \\ & + \mathrm{p}(L_{Aa} i_a) + \mathrm{p}(L_{Ab} i_b) + \mathrm{p}(L_{Ac} i_c) \end{aligned} \tag{8.3a}$$

The other winding equations can be written down in the same straight-forward fashion.

The A-phase axis will be taken as the reference position and the rotor a-phase will be at some electrical angle θ to the stator A phase, Figure 8.2. θ varies with rotation and time so that the stator/rotor mutual inductances vary through a complete cycle for every pole-pair pitch of movement. With a uniform air-gap machine, all the self inductances will be unaffected by rotor position. This applies also to the mutuals between stator phases and between the rotor phases.

When parameter tests are conducted, they yield an equivalent-circuit stator inductance $L_s = (x_1 + X_m)/\omega$ which is 50% higher than the stator-phase

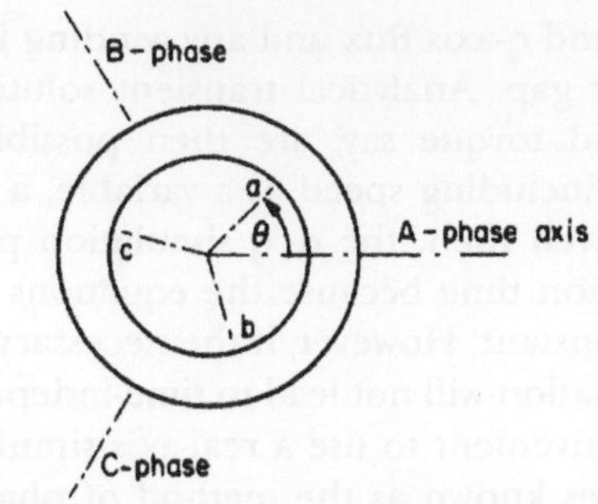

Figure 8.2 *Phase axes for uniform air-gap wound rotor machine.*

self-inductance L_{AA},[1] because the other two phases are excited and contribute this extra flux. So:

$$L_{AA} = L_{BB} = L_{CC} = L_s \times 2/3.$$

The mutual inductance between stator phases is half of this because they are displaced by 120° and cos 120° = $-\frac{1}{2}$. A similar situation exists for the rotor phases which in referred terms are often taken to be the same as the stator inductance, i.e. $L'_{aa} = L_{AA}$. The actual rotor parameters can be used in which case the voltages and currents for the rotor will be those for the winding itself. The mutual inductance derived from equivalent circuit referred to the stator, will be $X_m/(\omega \cdot N_{stator}/N_{rotor})$. If this is designated M, then, for the same reason as in the case of self inductance, the maximum value between the A-phase and the a-phase when these are in alignment physically and additive magnetically will be $M \times 2/3$. For any angle θ:

$$L_{Aa} = (2/3)M \cos\theta.$$

The other stator/rotor mutual inductances vary similarly but with a phase displacement in accordance with the $\mathbf{L}_{sr}$ matrix to be given. This can be checked from Figure 8.2 by considering when a particular pair of stator and rotor phases are in alignment and give maximum mutual inductance $(2/3)M$. It can also be verified on studying Figure 8.2 at different angular positions, that $\mathbf{L}'_{rs}$ is the same as $\mathbf{L}_{sr}$. This is a consequence of the choice of axes as shown, and recommended in Reference 2 where it is explained that several advantages arise in the manipulation of the equations and the application to different machine types. In displaying the matrices, it is convenient[12] to delineate the stator and rotor sections, eqn (8.3b), so that the four quarters of the inductance matrix can be shown separately. If all parameters are referred to the stator, only two quarters need be given in detail if $\mathbf{L}'_{ss} = \mathbf{L}'_{rr}$ (referred). In this case also, the value of M in $\mathbf{L}'_{rs} = \mathbf{L}_{sr}$

must now be the referred value, i.e. X_m/ω. The equations described can be summarised as:

$$\mathbf{v} = \begin{bmatrix} \mathbf{v}_{stator} \\ \mathbf{v}_{rotor} \end{bmatrix} = \begin{bmatrix} \mathbf{R}_s + p\mathbf{L}_{ss} & p\mathbf{L}_{sr} \\ p\mathbf{L}'_{rs} & \mathbf{R}_r + p\mathbf{L}'_{rr} \end{bmatrix} \times \begin{bmatrix} \mathbf{i}_{stator} \\ \mathbf{i}_{rotor} \end{bmatrix} \qquad (8.3b)$$

where:

$$\mathbf{L}_{ss} = \frac{2}{3} \times \begin{array}{c|ccc} & A & B & C \\ \hline A & L_s & -L_s/2 & -L_s/2 \\ B & -L_s/2 & L_s & -L_s/2 \\ C & -L_s/2 & -L_s/2 & L_s \end{array} = \mathbf{L}_{ir} \text{ (say)}$$

and

$$\mathbf{L}_{sr} = \mathbf{L}'_{rs} = \frac{2}{3} \times \begin{array}{c|ccc} & a & b & c \\ \hline A & M\cos\theta & M\cos(\theta - 2\pi/3) & M\cos(\theta - 4\pi/3) \\ B & M\cos(\theta - 2\pi/3) & M\cos(\theta - 4\pi/3) & M\cos\theta \\ C & M\cos(\theta - 4\pi/3) & M\cos\theta & M\cos(\theta - 2\pi/3) \end{array}$$

The assumption in the above equations is that the flux distribution is sinusoidal but this is not an essential condition. Any known flux distribution can be allowed for providing this is reflected in the expressions for inductance variation. Harmonic terms could be added, or, since the equations are to be solved numerically, a non-analytical relationship of inductance variation with angle could be referred to at every incremental step, for the appropriate inductance and its angular rate of change. If one member is salient for example, the other member will have self- as well as mutual-inductance variation.

Organisation of computed solution

In solving the equations for current, given the voltages, the impedance matrix has to be inverted; $\mathbf{i} = \mathbf{Z}^{-1} \cdot \mathbf{v}$, and it would appear that for the system described, a 6 × 6 inversion is necessary. However, this would ignore the fact that for a 3-line, 3-phase circuit, the sum of the three currents is zero and hence one current can be expressed in terms of the other two. Indeed, an attempt to invert the 6 × 6 impedance matrix would result in failure because there is redundant information. Substituting the conditions that:

$$i_C = -(i_A + i_B) \text{ and } i_c = -(i_a + i_b)$$

and using line voltage across two loops, will reduce the equations from 6 to 4, giving much shortening of computer time. Actually, it is much more convenient to reduce the equations to 4, by employing routine matrix techniques as will be shown in main outline. The stator phase voltages:

$$v_A\ \hat{V} \sin \omega t,\ v_B = \hat{V} \sin(\omega t - 120°) \text{ and } v_C = \hat{V} \sin(\omega t + 120°),$$

will come out in the transformation as two line voltages. A similar treatment can be accorded to the rotor voltages though this winding will here be assumed short circuited so $\mathbf{v}_r = 0$.

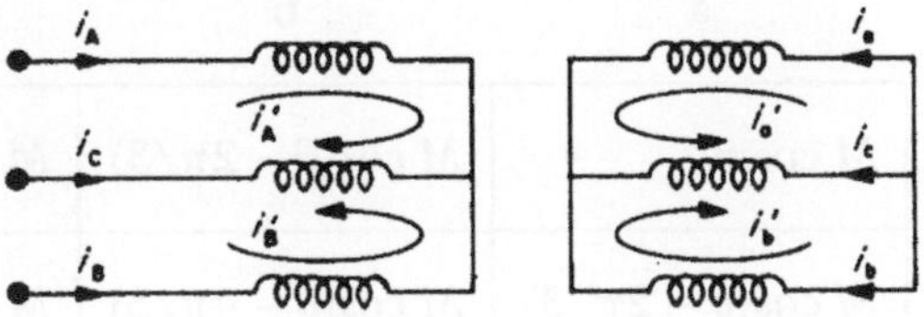

Figure 8.3

We require a connection matrix **C**, relating the six old currents i_A, i_B, i_C, i_a, i_b and i_c, to the four new currents i_A', i_B', i_a' and i_b' say. With this we proceed to find the corresponding transformations for voltage and impedance using standard techniques. Figure 8.3 shows the physical relationship of the currents as chosen, for star-connected windings.

From the figure:

$$i_A = i_A' \qquad i_a = i_a'$$
$$i_B = i_B' \qquad i_b = i_b'$$
$$i_C = -(i_A' + i_B') \qquad i_c = -(i_a' + i_b')$$

In matrix form:

$$
\begin{bmatrix} i_A \\ i_B \\ i_C \\ i_a \\ i_b \\ i_c \end{bmatrix}
=
\begin{array}{c} \\ A \\ B \\ C \\ a \\ b \\ c \end{array}
\begin{array}{c} \begin{array}{cccc} A' & B' & a' & b' \end{array} \\
\begin{bmatrix} 1 & & & \\ & 1 & & \\ -1 & -1 & & \\ & & 1 & \\ & & & 1 \\ & & -1 & -1 \end{bmatrix} \end{array}
\begin{bmatrix} i'_A \\ i'_B \\ i'_a \\ i'_b \end{bmatrix}
\quad \text{or; } \mathbf{i} = \mathbf{C} \cdot \mathbf{i}' \qquad (8.4)
$$

This is the transformation for currents. For voltages and impedance, maintaining constant power through the transformation, ($\mathbf{i}_t \cdot \mathbf{v} = \mathbf{i}'_t \cdot \mathbf{v}'$) the corresponding equations are[2]:

$$\mathbf{v}' = \mathbf{C}_t \cdot \mathbf{v} \text{ and } \mathbf{Z}' = \mathbf{C}_t \cdot \mathbf{Z} \cdot \mathbf{C}$$

where $\mathbf{C}_t$ is the transpose of $\mathbf{C}$. Applying these and assuming a short-circuited rotor so that $\mathbf{v}_r = \mathbf{v}'_r = 0$:

$$
\mathbf{v}' =
\begin{bmatrix} 1 & & -1 & & & \\ & 1 & -1 & & & \\ & & & 1 & & -1 \\ & & & & 1 & -1 \end{bmatrix}
\begin{bmatrix} v_A \\ v_B \\ v_C \\ 0 \\ 0 \\ 0 \end{bmatrix}
=
\begin{bmatrix} v_A - v_C \\ v_B - v_C \\ 0 \\ 0 \end{bmatrix}
=
\begin{bmatrix} v'_A \\ v'_B \\ v'_a \\ v'_b \end{bmatrix}
\qquad (8.5)
$$

These are the line voltages to the common line C, for which the current has been eliminated from the equations. The phase (and line) voltages could have any time variation specified in the problem.

For the impedance, the 6 × 6 matrix will have to be assembled in detail. It will be convenient to make the substitutions:

$$\alpha_1 = \theta,\ \alpha_2 = (\theta - 2\pi/3) \text{ and } \alpha_3 = (\theta + 2\pi/3) = (\theta - 4\pi/3).$$

In the matrix multiplications, certain expressions will arise for which the following further substitutions will simplify the intermediate equations:

$$\mathrm{X} = \cos\alpha_1 - \cos\alpha_2;\ \mathrm{Y} = \cos\alpha_1 - \cos\alpha_3 \text{ and } \mathrm{Z} = \cos\alpha_2 - \cos\alpha_3$$

and because of the 120° relationship between the angles, the combinations of X, Y and Z which arise can, by trigonometric manipulation, be simplified as follows:

$$(\mathrm{X}+\mathrm{Y}) = 3\cos\alpha_1;\ (\mathrm{Z}-\mathrm{X}) = 3\cos\alpha_2 \text{ and } (\mathrm{Y}+\mathrm{Z}) = -3\cos\alpha_3.$$

Hence, performing first the operation **Z · C** on the whole matrix expanded from eqn (8.3b), including the *R* terms:

$$\mathbf{Z}\cdot\mathbf{C} = \frac{1}{3}\cdot\begin{bmatrix} 3R_s+2pL_s & -pL_s & -pL_s & 2pM\cos\alpha_1 & 2pM\cos\alpha_2 & 2pM\cos\alpha_3 \\ -pL_s & 3R_s+2pL_s & -pL_s & 2pM\cos\alpha_2 & 2pM\cos\alpha_3 & 2pM\cos\alpha_1 \\ -pL_s & -pL_s & 3R_s+2pL_s & 2pM\cos\alpha_3 & 2pM\cos\alpha_1 & 2pM\cos\alpha_2 \\ 2pM\cos\alpha_1 & 2pM\cos\alpha_2 & 2pM\cos\alpha_3 & 3R_r+2pL_r & -pL_r & -pL_r \\ 2pM\cos\alpha_2 & 2pM\cos\alpha_3 & 2pM\cos\alpha_1 & -pL_r & 3R_r+2pL_r & -pL_r \\ 2pM\cos\alpha_3 & 2pM\cos\alpha_1 & 2pM\cos\alpha_2 & -pL_r & -pL_r & 3R_r+2pL_r \end{bmatrix}$$

$$\times\cdot\begin{bmatrix} 1 & & & \\ & 1 & & \\ -1 & -1 & & \\ & & 1 & \\ & & & 1 \\ & & -1 & -1 \end{bmatrix}$$

To complete the transformation $\mathbf{Z}' = \mathbf{C}_t \cdot \mathbf{Z} \cdot \mathbf{C}$:

$\mathbf{Z}' =$

1		–1			
	1	–1			
			1		–1
				1	–1

$\cdot \times$

$\frac{1}{3}$

$3(R_s + pL_s)$	0	$2pM(Y)$	$2pM(Z)$
0	$3(R_s + pL_s)$	$2pM(-X)$	$2pM(-Y)$
$-3(R_s + pL_s)$	$-3(R_s + pL_s)$	$2pM(-Z)$	$2pM(X)$
$2pM(Y)$	$2pM(Z)$	$3(R_r + pL_r)$	0
$2pM(-X)$	$2pM(-Y)$	0	$3(R_r + pL_r)$
$2pM(-Z)$	$2pM(X)$	$-3(R_r + pL_r)$	$-3(R_r + pL_r)$

giving finally:

$\mathbf{Z}' =$ (8.6)

	A′	B′	a′	b′
A′	$2(R_s + pL_s)$	$R_s + pL_s$	$-2pM\cos\alpha_3$	$2pM\cos\alpha_2$
B′	$R_s + pL_s$	$2(R_s + pL_s)$	$2pM\cos\alpha_2$	$-2pM\cos\alpha_1$
a′	$-2pM\cos\alpha_3$	$2pM\cos\alpha_2$	$2(R_r + pL_r)$	$R_r + pL_r$
b′	$2pM\cos\alpha_2$	$-2pM\cos\alpha_1$	$R_r + pL_r$	$2(R_r + pL_r)$

For numerical integration, an expression for $\mathrm{d}i/\mathrm{d}t$ is required. Rearranging the matrix voltage equation (8.3) in terms of the motional and transformer components:

$$\mathbf{v} = \mathbf{R} \cdot \mathbf{i} + \mathbf{i} \cdot \mathrm{d}\mathbf{L}/\mathrm{d}t + \mathbf{L} \cdot \mathrm{d}\mathbf{i}/\mathrm{d}t$$

$$= \mathbf{R} \cdot \mathbf{i} + \mathbf{i} \cdot \frac{\mathrm{d}\mathbf{L}}{\mathrm{d}\theta} \cdot \frac{\mathrm{d}\theta}{\mathrm{d}t} + \mathbf{L} \cdot \mathrm{d}\mathbf{i}/\mathrm{d}t$$

from which:

$$\frac{\mathrm{d}\mathbf{i}}{\mathrm{d}t} = \mathbf{L}^{-1} \times \left[\mathbf{v} - \left(\mathbf{R} + \frac{\mathrm{d}\mathbf{L}}{\mathrm{d}\theta} \cdot \dot{\theta}\right)\mathbf{i}\right] \tag{8.7}$$

It is seen that the speed $\dot{\theta}$, in electrical radians per second, comes into the motional voltage term and also the variation of inductance with angle must be known in order to obtain $\mathrm{d}\mathbf{L}/\mathrm{d}\theta$. In the present case, differentiating the inductive terms of $\mathbf{Z}'$ is very simple because L_s and L_r are constant and the mutual terms involving θ vary sinusoidally with angular position. Performing this differentiation and combining the R terms gives:

$\mathbf{R} + \frac{\mathrm{d}\mathbf{L}}{\mathrm{d}\theta}\dot{\theta} =$

$2R_s$	R_s	$2M\dot{\theta}\sin\alpha_3$	$-2M\dot{\theta}\sin\alpha_2$
R_s	$2R_s$	$-2M\dot{\theta}\sin\alpha_2$	$2M\dot{\theta}\sin\alpha_1$
$2M\dot{\theta}\sin\alpha_3$	$-2M\dot{\theta}\sin\alpha_2$	$2R_r$	R_r
$-2M\dot{\theta}\sin\alpha_2$	$2M\dot{\theta}\sin\alpha_1$	R_r	$2R_r$

(8.7a)

This expression is now multiplied by the transformed $\mathbf{i}'$ matrix, then subtracted from the transformed $\mathbf{v}'$ matrix corresponding to the particular instant, the whole being multiplied by the inverse of the inductance matrix, i.e. $\mathbf{Z}'^{-1}$ omitting the p's and the R terms. This gives the $\mathrm{d}\mathbf{i}/\mathrm{d}t$ matrix from which numerical integration by any appropriate method, e.g. Runge Kutta or even Modified Euler, will yield the new currents $\mathbf{i}'$ for the end of the integration interval. These are now ready for the next incremental step to repeat the above procedure. The actual winding currents are readily obtained since for example: $i_A = i'_A$ and $i_C = -(i'_A + i'_B)$, eqn (8.4).

The electromagnetic torque is obtained at each integration step from the calculated currents using the matrix expression[2]:

$$T_e = \frac{1}{2} \mathbf{i}_t \frac{d\mathbf{L}}{d\theta} \mathbf{i} \times \text{pole pairs or, with a less concise formulation}$$

of the equations:

$$= \frac{1}{2} \cdot \begin{bmatrix} \mathbf{i}'_{stator} & \mathbf{i}'_{rotor} \end{bmatrix} \cdot \frac{d}{d\theta} \begin{bmatrix} & \mathbf{L}_{sr} \\ \mathbf{L}'_{rs} & \end{bmatrix} \cdot \begin{bmatrix} \mathbf{i}'_{stator} \\ \mathbf{i}'_{rotor} \end{bmatrix} \times \text{pole pairs} \quad (8.8)$$

The differential of the inductance matrix is the same as eqn (8.7a) except that the R terms and $\dot{\theta}$ are omitted. The actual machine winding currents and the original 6 × 6 inductance matrix differentiated could be used to calculate T_e without much additional work since no inversion is needed.

For the mechanical system, the two 1st-order equations required to obtain numerical solutions for speed and angle have already been given in terms of electrical radians as:

$$\frac{d\dot{\theta}}{dt} = \frac{T_e - T_m}{J} \times \text{pole pairs} \quad \text{and} \quad \frac{d\theta}{dt} = \dot{\theta} \qquad (8.1) \quad (8.2)$$

If the machine is supplied through semiconductor switching circuits, these may occasion phase connections to be broken during part of the cycle so that during certain periods, not all phases are in circuit. The equations would require reformulation to cover all connection modes, determined by a logic check of conduction conditions during the course of the program, as indicated on p. 297 and illustrated for a simple case in Example 8.1. The program could be modified to allow for saturation in a similar manner to the provision made on Figure 8.1 to read in a functional relationship of T_m with speed. Such refinements are part of the usefulness of numerical solutions, where constant monitoring of the system condition can be made. Figure 8.1 shows the steps in the computed solution. The machine terminal voltages are the line voltages of eqn (8.5), no assumption about the potential of the neutral connection having to be made. Eqn (8.7) requires the multiplication of eqn (8.7a) by the previous calculation of $\mathbf{i}'$ – or the initial value. This is deducted from eqn (8.5) and the whole multiplied by the inverse of eqn (8.6) omitting the p's and the R terms, to get $d\mathbf{i}/dt$. Solutions for $\mathbf{i}$, T_e, $\dot{\theta}$ and θ follow as indicated on the figure.

Equations in per-unit terms

Sections 3.3, 5.3 and Example 4.2 have dealt with the steady-state *per-unit* equations. To cover the dynamic state, more information is required to deal with inertia, inductance and terms involving shaft angle. For the first quantity, the inertia torque is divided by T_{base} to produce the *per-unit* value:

$$\frac{J \,.\, \mathrm{d}\omega_{\text{m}}/\mathrm{d}t}{\dfrac{\text{Power}_{\text{base}}}{\omega_{\text{m(base)}}} \dfrac{\omega_{\text{m(base)}}}{\omega_{\text{m(base)}}}} = \frac{(\omega_{\text{m(base)}})^2 J \cdot \dfrac{\mathrm{d}}{\mathrm{d}t} \dfrac{\omega_{\text{m}}}{\omega_{\text{m(base)}}}}{\text{Power}_{\text{base}}}$$

$$= \frac{J \cdot (\omega_{\text{m(base)}})^2}{\text{Power}_{\text{base}}} \times \frac{\mathrm{d}\omega_{\text{m}}\ (\textit{per unit})}{\mathrm{d}t} \qquad (8.9)$$

So if ω_{m} is in *per-unit*, the corresponding expression for *per-unit* inertia is the first term, which is twice the stored energy at base speed, divided by rated VA, i.e. twice the stored-energy constant.

For the second quantity, the inductive voltage is divided by V_{base} to produce the *per-unit* value:

$$\frac{L \cdot \mathrm{d}i/\mathrm{d}t}{V_{\text{base}}} \frac{I_{\text{base}}}{I_{\text{base}}} = \frac{L}{V_{\text{base}}/I_{\text{base}}} \frac{\mathrm{d}i/I_{\text{base}}}{\mathrm{d}t} = \frac{L}{Z_{\text{base}}} \frac{\mathrm{d}i\ (\textit{per unit})}{\mathrm{d}t} \qquad (8.10)$$

Both *per-unit* J and L are not dimensionless but the associated torque and voltage terms become so, on multiplication by *per-unit* rate of change of speed and current respectively.

Solution of eqn (8.1) using *per-unit* inertia and *per-unit* torque will give $\dot{\theta}$ as a fraction of base speed, i.e. in *per unit.* Base speed is usually taken as the synchronous value, namely $\omega_{\text{m(base)}} = \omega_{\text{s}} = 2\pi f/\text{pole pairs}$. Consequently to find the angle θ in electrical radians, $\dot{\theta}$ in eqn (8.2) must be multiplied by $\omega = 2\pi f$ when integrating, if $\dot{\theta}$ is in *per unit.*

For a 3-phase machine, rated voltage per phase V_{R} is usually taken as V_{base} and either rated VA *or* rated current may be chosen as the third base value from the expression: $P_{\text{base}} = 3 \cdot V_{\text{base}} \cdot I_{\text{base}}$. For the induction motor, the rated output power P_{R} is usually taken as P_{base} and this gives $I_{\text{base}} = P_{\text{R}}/3V_{\text{R}}$ which is not equal to I_{R} because of power factor and efficiency. The torque base would then follow from the above choices as:

$$\frac{P_{\text{base}}}{\omega_{\text{m(base)}}} = \frac{P_{\text{R}}}{\omega_{\text{s}}} = \frac{3 \cdot V_{\text{base}} \cdot I_{\text{base}}}{\omega_{\text{s}}}.$$

To express electromagnetic torque in *per-unit*, referring back to eqn (8.8):

$$\frac{T_e}{T_{base}} = \frac{\frac{1}{2} \cdot \mathbf{i}_t \cdot \dfrac{d\mathbf{L}}{d\theta} \cdot \mathbf{i} \cdot \text{pole pairs}}{\dfrac{3 \cdot V_{base} \cdot I_{base}}{\omega_s} \cdot \dfrac{I_{base}}{I_{base}}}$$

$$= \frac{1}{2}\,\mathbf{i}_{t(p.u.)} \cdot \frac{d\mathbf{L}_{(p.u.)}}{d\theta} \cdot \mathbf{i}_{(p.u.)} \cdot \frac{\omega}{3} \qquad (8.11)$$

$\omega = \omega_s \times$ pole pairs and could be regarded as converting θ, which is in radians, to a *per-unit* measure on division by ω. The number 3 arises because phase voltage and currents have been taken as base values. $\mathbf{L}_{(p.u.)}$ is the inductance divided by base impedance V_{base}/I_{base}. Note that although $\dot{\theta}$ is in *per unit* in the relevant equations, $d\mathbf{L}_{(p.u.)}/d\theta$ must be the variation of *per-unit* inductance with electrical angle in radians, as obtained by integrating $d\theta/dt = \dot{\theta}_{(p.u.)} \times \omega$.

All the equations are now available to incorporate in a computer program, to be expressed either in actual or *per-unit* terms. They refer specifically to the 3-phase, uniform air-gap machine but can be applied to any polyphase machine with appropriate modifications. The only significant difficulty is in defining the inductance/angle function from which $d\mathbf{L}/d\theta$ is deduced. This is a matter for either design calculations or actual measurements. The usual assumptions made when obtaining the d–q equations yield analytical expressions of the inductance variation,[(2)] but if further refinement is sought, some extra complication is only to be expected. The method is in a sense generalised, not being restricted by winding unbalance or asymmetry, nor by the magnetic geometry such as the presence of saliency on both sides of the air gap or even allowance for rotor and stator slots. For a simple, uniform air-gap machine one can compare the computer time required to achieve comparable accuracy. Based on the time for the more economical d--q simulation, where the constant inductance matrix does not have to be inverted at every integration interval, the real-axis simulation is two to ten times longer, depending on the voltage waveform. Such a difference, though becoming less important as computers continue to improve, would not be justified. But if the situation is such that the d–q transformation does not yield constant inductances, then a lengthy computation time is endemic in the problem and a real-axis simulation may be the most convenient. For single-phase machines, unlike the model for analytical solutions, the real-axis equations are simpler than for 3-phase machines because there are fewer of them.

The next two figures use the simulation just described to illustrate some instructive electromechanical transients. Figure 8.4 is for an inverter-fed

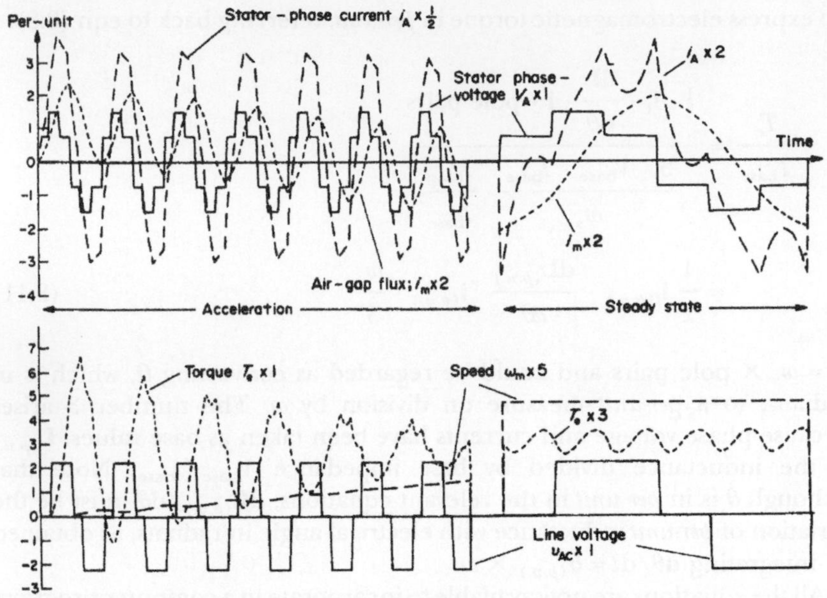

Figure 8.4 *Acceleration and steady-state waveforms of inverter-fed induction motor. (D.C. link voltage set to give fundamental V_1 = rated value)*

induction motor, accelerated from zero speed up to the steady-state speed where the slip is about 6%, full-load mechanical torque being coupled throughout. The bottom figures show the quasi-square-wave line-voltage which is applied at full value and frequency. In practice, such a source would be applied at reduced voltage and frequency, but is computed as shown to illustrate the nature of current and torque transients for direct-on-line start. A sine-wave supply would have given similar peak values but without the 6-pulse ripple. The torque peaks reach nearly 7 *per unit* and the first six torque cycles are shown. The computer graph plot is then stopped until the steady-state is reached and then the expanded waveforms for one cycle are plotted. The speed is then constant but the torque pulsations are considerable at about ±0.33 *per unit.* The dominant frequency is 6th harmonic as deduced from Figure 7.14a and eqn (7.8). But the waveform shows a slight even-harmonic asymmetry, and this will be due to the small effect of 11th and 13th rotor harmonic currents giving a 12th-harmonic component of torque. The top diagrams show the phase-voltage waveform, deduced from the calculated currents and the sum of the impedance drops in accordance with eqn (8.3a). It is the same shape as already found in Figure 7.13. The stator-phase current reaches starting peaks of over 7 *per unit* and on steady state is rich in 5th and 7th harmonic components. The peak currents reached on steady state are somewhat higher than those occurring with a sinusoidal-voltage supply which would give a current sine wave running through the distorted wave shown here. The remaining waveform is for the flux, neglecting saturation and deduced by combining the stator A-phase current with the referred and transformed a-phase rotor current; namely the waveform of I_m. Because of the parameters of this particular machine, 1 *per-unit* I_m corresponds to about normal *peak* flux and the diagram shows that the *peak* flux on steady state is a few per cent higher than the normal value for a sine-wave supply. More interesting is that in spite of the violent i_A fluctuations, the induced rotor-current acts to damp these out as shown by the nearly sinusoidal flux-waveform. The rotor current in fact has a pronounced 6th harmonic component superimposed on its slip-frequency sine-wave variation. This ripple, which is only to be expected since there are six switching changes per cycle, is similar to the torque pulsation. This is also to be expected since it is the reaction between the nearly sinusoidal flux wave and the rotor current which produces the torque. On observing the flux transient, it is seen that this dies out from a high peak in a few cycles, the value then being about half the normal rated value for a short time, due to the high voltage drop across the primary impedance during starting. It is also seen that the flux wave is approximately 90° lagging on the phase-voltage wave.

Figure 8.5 is for the same machine, this time running at rated torque as an induction motor from a 50-Hz sinusoidal-voltage supply. The top diagram shows that the phase current lags the voltage by about 45° initially

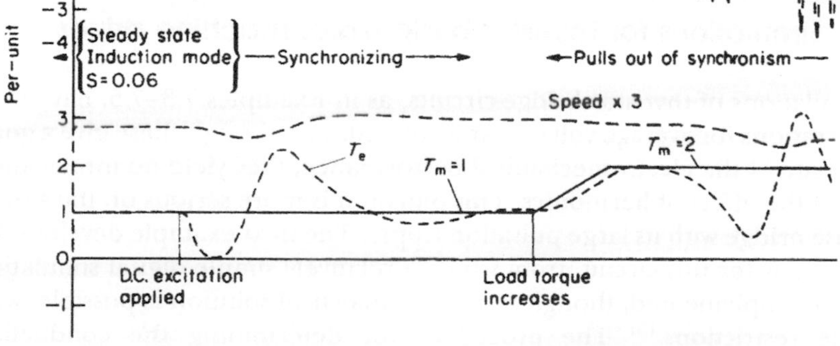

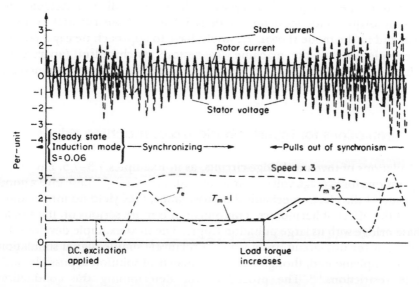

Figure 8.5 *3-phase wound-rotor, a.c. machine. First loaded as an induction motor, then synchronised, then overloaded to pull-out.*

and the rotor current can be seen oscillating at the very much lower slip frequency of about 3 Hz. At time 0.14 sec from the beginning of the plot, the rotor is supplied with a d.c. voltage, sufficient to cause it to pull into step as a synchronous motor. This takes about 0.5 sec, during which time the stator and rotor currents exceed their rated values quite considerably. When synchronisation is successfully achieved, at time 0.6 sec from the beginning of the plot, the mechanical torque T_m is increased linearly up to 2 *per unit*. It is appropriate now to look at the bottom diagram and see the electromagnetic torque T_e initially in balance with T_m, the speed being about 0.94 times the synchronous value. The electromagnetic-torque-transient is quite violent during synchronising and the speed dips, then overshoots and at 0.6 second, it is virtually synchronous, with T_e again almost in balance with T_m. The mechanical oscillations are somewhat exaggerated because the inertia was lowered artificially below its true value, for the computation of Figures 8.4 and 8.5, to demonstrate the torque and speed changes in a short computation time. The increase of T_m is then slowly followed by T_e as the load angle increases and speed falls. However, the load angle must have increased well beyond 90° and the torque commences another violent oscillation which is sufficient to pull the machine out of synchronism. The armature current and rotor current again rise to excessive values, but the combination of induction torque produced by the induced rotor-currents reacting with the stator currents, and the pulsating torque produced by the reaction of the d.c. component of rotor current reacting with the stator currents, is still sufficient to hold the load torque though with an appreciable speed oscillation. Actually, the pull-out torque in the induction mode is 3.2 *per-unit*, see Tutorial Example T7.5, and in the mode shown, T_m would have to approach nearly 3 *per unit* to completely stall the machine. To restore synchronism (apart from reducing the load to about 1.5 *per unit*), it would be necessary to increase the excitation by about 50%.

8.3 Simulations for Thyristor bridge/d.c. machine drives

Calculations of thyristor bridge-circuits, as in Examples 7.3–7.5, have used expressions for average voltages and currents and though these give a good estimate of the electromechanical performance, they yield no information about the effect of harmonics. This omission is more serious on the single-phase bridge with its large pulsation ripple. The next example develops the equations for this circuit, from which a relatively simple digital simulation can be implemented, though in fact an analytical solution is possible, with some restrictions.[3] The procedure for determining the conduction condition is illustrated by a logic-check diagram on Figure E.8.1 and another such exercise is provided by Tutorial Example T8.2.

Example 8.1

Set up the equations for the single-phase thyristor bridge/d.c. motor circuit and explain how the conducting conditions can be determined and a computer simulation implemented. Allow for all circuit resistances and inductances.

The first figure shows the circuit; currents and voltages being indicated. The transformer is represented by its secondary e.m.f. e_{ac} behind the leakage and supply impedances referred to the secondary, r + lp, where p represents d/dt. This is a naturally commutated circuit since the a.c. voltage developed across the thyristors v_f reverses every half cycle and reduces the thyristor current to zero naturally. There is a period of 'overlap', however, when the incoming thyristors share the total load current until the outgoing thyristors have their currents eventually reduced to zero. For the single-phase bridge, this means that the d.c. terminals are short circuited through the thyristors so that the terminal voltage is zero during overlap. If the equations are set up for this case when all thyristors are conducting, it is merely a question of setting the appropriate terms to zero, if a current falls to zero.

Consider a path through thyristors 1 and 3, when all thyristors are conducting:

$$e_{ac} = r(i_1 - i_2) + l\text{p}(i_1 - i_2) + 1\,\text{V} + L_L\text{p}i_a + R_L i_a + e_m + 1\,\text{V}.$$

from which:

$$\text{p}i_1 = \frac{e_{ac} - Ri_1 - R'i_2 - e_m - 2\,\text{V}}{L} - \frac{L'}{L}\text{p}i_2$$

where $R = R_L + r$; $R' = R_L - r$; $L = L_L + l$; $L' = L_L - l$ and $i_a = i_1 + i_2$.

For a path through thyristors 2 and 4 for the same condition:

$$-e_{ac} = 1\,\text{V} + L_L\text{p}i_a + R_L i_a + e_m + 1\,\text{V} - l\text{p}(i_1 - i_2) - r(i_1 - i_2)$$

from which:

$$\text{p}i_2 = \frac{-e_{ac} - Ri_2 - R'i_1 - e_m - 2\text{V}}{L} - \frac{L'}{L}\text{p}i_1.$$

pi is thus available for each current to perform the numerical integration as required.

To determine the conducting condition, the voltages developed by the circuit across each thyristor must be monitored whilst it is switched off, since when the gate firing pulse is applied, conduction will not start unless v_f is greater than the normal forward volt drop, taken to be 1 V. It is assumed that v_{f1} is the same for thyristors 1 and 3 and v_{f2} is the same for thyristors 2 and 4.

For a path through thyristors 1 and 3 when they are not conducting:

$$v_{f1} = -e_m - R_L i_2 - L_L\text{p}i_2 - v_{f1} + l\text{p}i_2 + ri_2 + e_{ac}$$

from which:

$$2v_{f1} = -e_m + e_{ac} - R'i_2 - L'\text{p}i_2.$$

Similarly:

$$2v_{f2} = -e_m - e_{ac} - R'i_1 - L'\text{p}i_1.$$

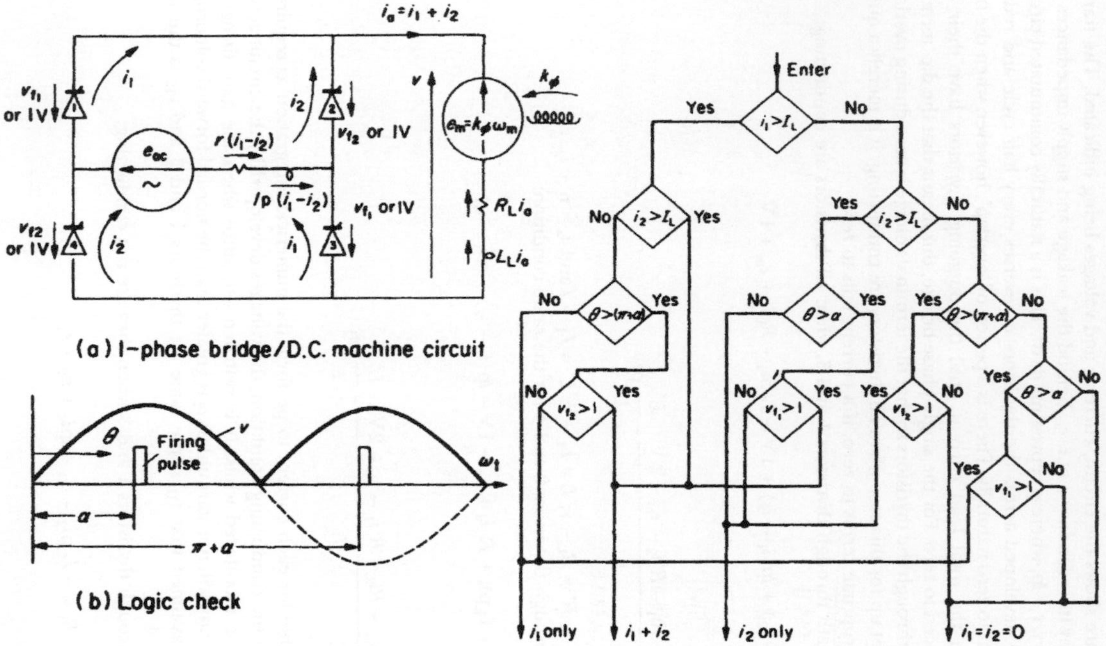

(a) 1-phase bridge/D.C. machine circuit

(b) Logic check

Figure E.8.1

The phase-delay angle α (or $\pi + \alpha$ for thyristors 2 and 4) and the value of flux k_φ must be specified in the data. i_1 and i_2, ω_m and e_m will be known from the previous time increment and solution and e_{ac} will be known from its time-function expression. The pattern of logical checks is shown, by means of which, working through the various questions, each thyristor path is checked for previous conduction ($i > I_L$, the *latching* or *holding* value) and the possibility of starting conduction.

The normal condition is for i_1 or i_2 to be conducting so these questions are asked first since if the answer is YES the remaining questions are unnecessary. If the answer is NO, for either or both currents, then the phase delay of the firing pulses is checked against the progression of angle $\theta = \omega t$, measured from the beginning of the cycle, to see if a firing pulse is present. If not, then that thyristor will be non-conducting in the next increment. If the answer is YES, the forward voltage across that thyristor must be checked to see if it is greater than 1 V (say). There are four possible answers to the overall question pattern. Either circuit can be conducting, both can be conducting (overlap), or both currents can be zero (a discontinuous-current condition). With these conditions known, the appropriate equations can be formed and i_1 and/or i_2 can be determined by numerical integration. The solution thereafter follows the pattern of the general flow diagram of Figure 8.1.

Again the time-constants of the electrical system are usually very much shorter than for the mechanical system and it may be sufficient to study the circuit for a particular speed, flux and constant e.m.f. Natural commutation may take more than 1 millisecond and full-scale simulations may be necessary however to check that this aspect of circuit operation is working correctly; it is liable to break down under certain conditions.

3-phase converter bridge

This is one of the most versatile power-electronic circuits, employed up to high power levels and frequently using a dual converter, as described in Section 7.3. It forms a link between a.c. and d.c. systems with power flow in either direction. It is used in H.V.D.C. transmission systems with a converter at each end of the d.c. transmission line. For drives, it provides a controlled d.c. voltage supply for current-source inverters and also for large synchronous machines, excited for leading power factor and with their inverter frequencies controlled by shaft-position switching so that speed varies with voltage, as in a d.c. machine. The cycloconverter is really another example, since it has effectively one dual converter per phase, this time forming a direct a.c./a.c. link between systems of different frequency. All these applications share the common features of providing forward and reverse power flow and employing natural commutation which permits the use of SCRs, as pointed out in Section 7.6.

The simulation of such systems is rather more complex than for the circuit of Figure E.8.1, though, for the d.c. drive, illustrated on Figure 7.10, this simulation was achieved through solving several simple equations of the form: $v_x = f(R_x, L_x, i_x)$, following a logic check through nearly 100 possible circuit configurations arising from the action of the switching elements. A more sophisticated simulation is given in Reference 14. However, this figure covers many practical aspects of converter operation and is meant to illustrate the various machine/converter modes as an aid to understanding since, because of the waveforms, the operation is not very easy to explain satisfactorily to the keen student without such a simulation. The one shown is for a non-circulating current scheme with all the thyristors controlled, logic monitoring to ensure that only one bridge can be fired at any one time and requiring a short 'dead' time between bridge changeover. There are other schemes where, to avoid the 'dead' time, the bridges are fired together but with the firing delay angles α_1 and α_2 summing to 180° so that the average bridge voltages are the same, to prevent any d.c. current circulating between them. Due to the waveforms, the instantaneous voltages are not equal, so there is an a.c. circulating current which must be limited by additional inductors. There are simpler schemes using fewer controlled rectifiers and more complicated schemes using twice as many thyristors and giving a 12-pulse output voltage to reduce the harmonics on H.V.D.C. systems.

The choice of schemes and of the machine type for a particular drive is largely governed by economic considerations but, technically, with the availability of so many high-performance electronic switches and micro-electronic control elements, any desirable drive characteristic can be designed.

Appendix A
Table of Laplace transforms

F(p)	$f(t)$, $t \geq 0$
1	$\delta(t_0)$, unit impulse at $t = t_0$
$1/p$	1, unit step, or constant for $t > 0$
$\dfrac{n!}{p^{n+1}}$	t^n
$\dfrac{1}{(p + a)}$	e^{-at}
$\dfrac{1}{(p + a)^n}$	$\dfrac{t^{n-1}e^{-at}}{(n - 1)!}$
$\dfrac{a}{p(p + a)}$	$1 - e^{-at}$
$\dfrac{a^2}{p(p + a)^2}$	$1 - e^{-at} - ate^{-at}$
$\dfrac{1}{(p + a)(p + b)}$	$\dfrac{e^{-at} - e^{-bt}}{(b - a)}$
$\dfrac{p + \alpha}{(p + a)(p + b)}$	$\dfrac{(\alpha - a)e^{-at} - (\alpha - b)e^{-bt}}{(b - a)}$
$\dfrac{ab}{p(p + a)(p + b)}$	$1 + \dfrac{ae^{-bt} - be^{-at}}{(b - a)}$

$\dfrac{1}{(p+a)(p+b)(p+c)}$	$\dfrac{e^{-at}}{(b-a)(c-a)} + \dfrac{e^{-bt}}{(c-a)(a-b)} + \dfrac{e^{-ct}}{(a-c)(b-c)}$
$\dfrac{p+\alpha}{(p+a)(p+b)(p+c)}$	$\dfrac{(\alpha-a)e^{-at}}{(b-a)(c-a)} + \dfrac{(\alpha-b)e^{-bt}}{(c-b)(a-b)} + \dfrac{(\alpha-c)e^{-ct}}{(a-c)(b-c)}$
$\dfrac{ab(p+\alpha)}{p(p+a)(p+b)}$	$\alpha - \dfrac{b(\alpha-a)e^{-at} + a(\alpha-b)e^{-bt}}{(b-a)}$
$\dfrac{\omega}{p^2+\omega^2}$	$\sin \omega t$
$\dfrac{p}{p^2+\omega^2}$	$\cos \omega t$
$\dfrac{\omega}{(p+a)^2+\omega^2}$	$e^{-at} \sin \omega t$
$\dfrac{p+a}{(p+a)^2+\omega^2}$	$e^{-at} \cos \omega t$
$\dfrac{\omega_n^{\,2}}{p(p^2+2\zeta\omega_n p+\omega_n^{\,2})}$	$1 - \dfrac{\exp(-\zeta\omega_n t)\,\sin\{\omega_n \surd(1-\zeta^2)\,t + \cos^{-1}\zeta\}}{\surd(1-\zeta^2)}$ (for $\zeta<1$)

Appendix B Voltage/current/power relationships for bridge-rectifier circuits

The 3-phase bridge circuit is shown on Figure B.1 together with the voltage and current waveforms for the 'ideal' assumptions of high d.c. circuit-inductance (so that I_d is of constant magnitude) and instantaneous transfer (commutation) of currents from semiconductor element to element in sequence, without voltage loss. Elements A_+ and C_- conduct in period 1 and it can be checked that across all other elements, the voltage is negative, e.g. V_{BA} across B_+ and V_{CB} across B_-. At the end of this period – instant 2 – V_{BA} becomes positive and B_+ takes over (commutating naturally) from A_+ across which appears V_{AB} which is negative. Similar voltage changes in sequence give rise to the d.c. voltage waveform pulsating 6 times per fundamental cycle, at values following the maximum forward line-voltage.

The secondary-current waveforms are square sections of 120° duration and referring to the development of eqn (6.16) give an r.m.s. value of:

$$\sqrt{\frac{I_d^2(1+1+1+1)}{1+1+1+1+1+1}} = \sqrt{\frac{2}{3}}\,I_d = 0.816I_d$$

The primary phase current will be of the same shape – neglecting magnetising current – and with a delta connection, the line current will be nearer to a sine wave as shown.

The 'ideal' no-load voltage deduced from the waveform will have an average value designated as E_{do}, being the 'Thevenin' e.m.f. behind the d.c. terminals and internal impedance. Its value is deduced on Figure B.2a for m = 6 pulses per cycle, the general expression being:

$$\frac{m}{\pi}\sin\frac{\pi}{m} \times \text{Peak line voltage,} \qquad \text{(B.1)}$$

e.g. for a single-phase bridge with m = 2,

$$E_{do} = \frac{2}{\pi}\sin\frac{\pi}{2}E_p = 0.637E_p.$$

With controlled rectifiers, 'firing' can be delayed, by $\alpha°$ say, from the natural point of current transfer. Figure B.2b shows how this reduces E_{do} to $E_{do}\cos a$, eqn (B.2), if I_d is a smooth current.

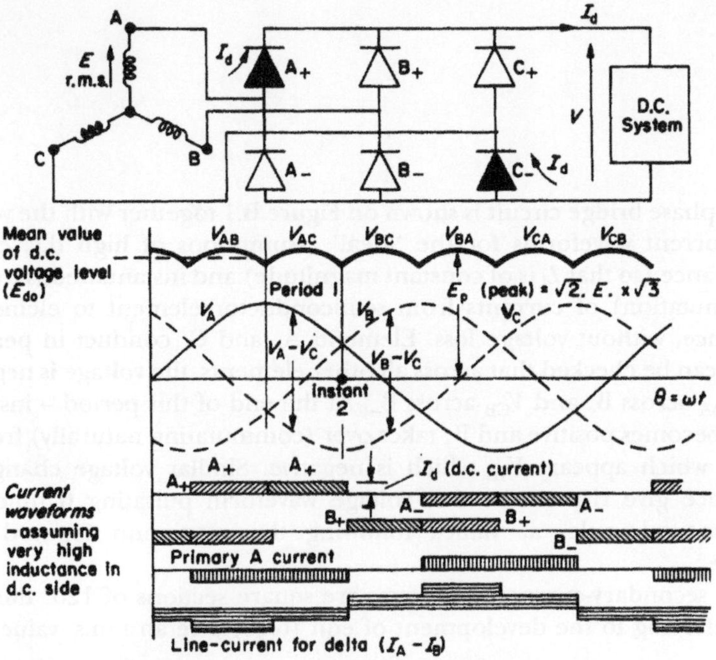

Figure B.1 *3-phase bridge circuit and waveforms.*

A further reduction in voltage occurs due to the time taken to commutate the current – the overlap period, for an angle μ – the delay being due to the inductance on the a.c. side, mostly due to transformer leakage. If this is l per phase for a star connection, between terminals the commutating inductance will be $2l$ and the corresponding reactance is X_c. From the waveforms and circuit of Figure B.2c, the voltage loss is proportional to the shaded area averaged out over 1/6 of a cycle, i.e.

(a) *Mean* d.c. *voltage* E_{do} (with zero current)

$$E_{do} = \frac{1}{2\pi/6} \int_{\frac{-\pi}{6}}^{\frac{+\pi}{6}} E_p \cdot \cos\theta$$

$$= E_p \cdot \frac{6}{\pi} \sin\frac{\pi}{6}$$

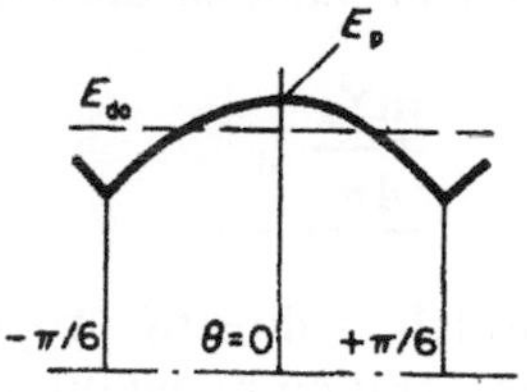

since $E_p = \sqrt{2} \cdot E \times \sqrt{3}$; $E_{do} = \underline{2.34E}$.

(b) *Effect of firing-delay-angle* (α) (with zero current)
Average d.c. voltage

$$= \frac{1}{2\pi/6} \int_{\alpha}^{\frac{2\pi}{6}} E_p \cdot \cos\left(\theta - \frac{\pi}{6}\right)$$

$$= \frac{6}{2\pi} \cdot E_p \left[\sin\left(\frac{2\pi}{6} + \alpha - \frac{\pi}{6}\right) - \sin\left(\alpha - \frac{\pi}{6}\right)\right]$$

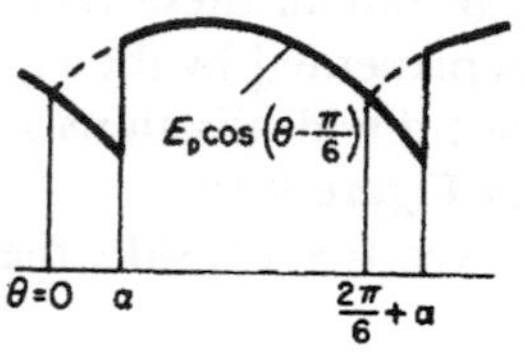

$$= E_p \cdot \frac{6}{\pi} \cdot \sin\frac{\pi}{6} \cdot \cos\alpha - \text{in general} = \underline{E_{do}\cos\alpha} \qquad \text{(B.2)}$$

(c) *Effect of overlap angle* μ *during commutation*

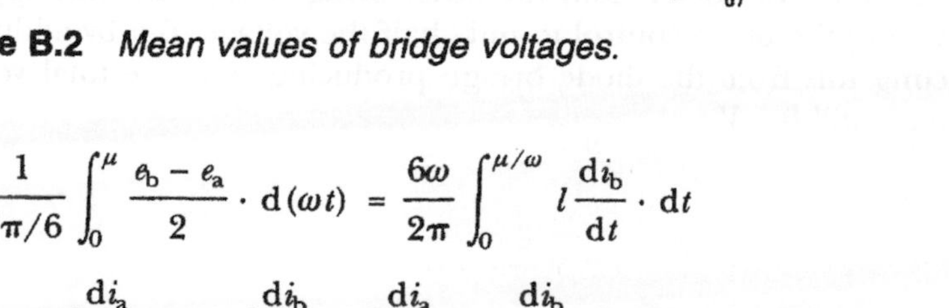

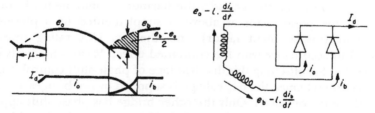

Figure B.2 *Mean values of bridge voltages.*

$$\frac{1}{2\pi/6} \int_0^{\mu} \frac{e_b - e_a}{2} \cdot d(\omega t) = \frac{6\omega}{2\pi} \int_0^{\mu/\omega} l\frac{di_b}{dt} \cdot dt$$

(since $e_a - l\frac{di_a}{dt} = e_b - l\frac{di_b}{dt}$ and $\frac{di_a}{dt} = -\frac{di_b}{dt}$ when $i_a + i_b = I_d$ is constant)

$$= \frac{6\omega l}{2\pi} \int_0^{I_d} di_b = \frac{6X_c}{4\pi} I_d$$

For any bridge circuit of m pulses per cycle, the general expression for the 'mean' value of the voltage loss due to overlap is:

$$\frac{mX_c}{4\pi} I_d \tag{B.3}$$

which is directly proportional to the d.c. current value. An alternative integration based on the instantaneous expressions for e_a and e_b gives the loss as $E_{do} \sin^2 \mu/2$, see Tutorial Example T.7.3. There is a further voltage loss due to the transformer resistance which can be referred to the d.c. side as: Transformer copper loss/I_d. A voltage loss of 1 V for each side of the bridge is a typical allowance for the approximately constant forward voltage drop of a semiconductor.

From all these relationships, the output from a bridge rectifier can be represented by the equation for average voltage as $V = E_{do} \cos \alpha - kI_d - 2$, but the idealisations above should be compared with the computed graph of Figure 8.10.

On the a.c. side, the secondary VA is:

$$3 \times E \times I_{rms} = 3 \times \frac{E_{do}}{\sqrt{2}\sqrt{3} \times 3/\pi} \sqrt{\frac{2}{3}} I_d = \frac{\pi}{3} \times \text{d.c. power.}$$

This means that even without a phase shift of current due to firing delay or overlap, the power factor is $3/\pi = 0.955$ as a maximum. Actually, the control of voltage by α variation is brought about by delaying the phase of the current so that on the a.c. side, the harmonic component (distortion factor) of the power-factor as above is supplemented by a phase-shift component (displacement factor) and the a.c. kVA rating must be increased in accordance with the combined effect. Various methods have been devised for offsetting this disadvantage of phase-shift control, e.g. on some traction schemes, the d.c. voltage is supplied by two bridges in series, one of them uncontrolled. Only the other bridge has phase shift applied, thus limiting the phase control to only half the voltage. But by adding or subtracting this from the diode bridge producing $V/2$, the total voltage variation is still $0 - V$.

Appendix C
Summary of equations for d–q simulations

The effect of the d–q transformation is to replace the angular rate of inductance variation $\mathbf{dL}/\mathrm{d}\theta$ in eqns (8.3) and (8.7) by fixed inductances $\mathbf{G}$, similarly associated with the angular speed $\dot{\theta}$ and defining the rotational voltages. Eqn (8.3) becomes:

$$\mathbf{v} = \mathbf{R} \cdot \mathbf{i} + \mathbf{G} \cdot \dot{\theta} \cdot \mathbf{i} + \mathbf{L} \cdot \mathbf{di}/\mathrm{d}t$$

and from this, the equation corresponding to (8.7) is:

$$\frac{\mathbf{di}}{\mathrm{d}t} = \mathbf{L}^{-1} \times [\mathbf{v} - (\mathbf{R} + \mathbf{G} \cdot \dot{\theta}) \cdot \mathbf{i}]$$

The electromagnetic torque $T_e = \mathbf{i}_t \cdot \mathbf{G} \cdot \mathbf{i} \times$ pole pairs which compares with eqn (8.8).

The transformations for voltages and currents do not have such an obvious physical relationship as those of eqns (8.4) and (8.5) and they depend on the reference frame to which the d–q axes are fixed and along which the phase axes are resolved to produce the transformation; see Figure C.1. For the uniform-air-gap machine, this is usually, though not necessarily, the stator, on which the line-connected winding is placed. For the salient-pole machine, the d–q axes must be fixed to the salient-pole member (for the transformations to result in simpler equations), and this is usually the rotor for the synchronous machine. The transformed, transient impedance-matrix can be filled in by a standard technique,[2] the main diagonal taking the self impedance of the corresponding axis, using the same resistances and inductances as eqn (8.6). Windings on a common

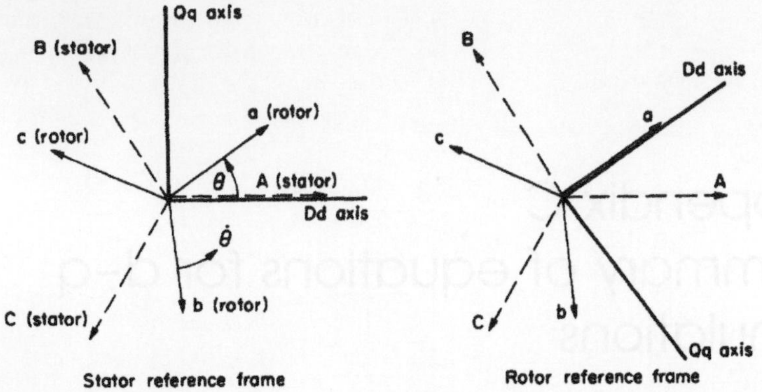

Figure C.1 *Machine axes.*

axis have mutual inductance which must be entered. The rotational voltage terms appear only in the windings which are rotating with respect to the reference frame, the inductance being taken from the d to q axis with a negative sign and from q to d axis with positive sign. Consider the *stator reference frame* and let the symbol 'dss' mean the d axis of the stator referred to the stator; 'qrs' mean the q axis of the rotor referred to the stator; the remaining symbols 'qss' and 'drs' following the same code. The voltage equation is:

V = **Z** **i**

in detail:

V		D	Q	d	q	i
vdss	D	$R_s + L_s p$		Mp		*idss*
vqss	Q		$R_s + L_s p$		Mp	*iqss*
vdrs	d	Mp	$M\dot{\theta}$	$R_r + L_r p$	$L_r\dot{\theta}$	*idrs*
vqrs	q	$-M\dot{\theta}$	Mp	$-L_r\dot{\theta}$	$R_r + L_r p$	*iqrs*

for a uniform air-gap machine referred to the stator reference frame. The **G** matrix consists only of the terms associated with $\dot{\theta}$ on the d and q rows. The **L** matrix has all the remaining inductive terms, as deduced directly from the normal, steady-state equivalent circuit.

In this equation, the transformation of voltages from the three, phase-to-neutral voltages of stator and rotor is given by resolving ABC abc axes along direct and quadrature axes fixed to the stator, Dd and Qq:

$$
\begin{bmatrix} v_{dss} \\ v_{qss} \\ v_{drs} \\ v_{qrs} \end{bmatrix}
= \sqrt{\frac{2}{3}}
\begin{array}{c} \\ D \\ Q \\ d \\ q \end{array}
\begin{array}{cccccc}
A & B & C & a & b & c \\
1 & -\frac{1}{2} & -\frac{1}{2} & & & \\
0 & \sqrt{3}/2 & -\sqrt{3}/2 & & & \\
 & & & \cos\theta & \cos(\theta - 2\pi/3) & \cos(\theta - 4\pi/3) \\
 & & & \sin\theta & \sin(\theta - 2\pi/3) & \sin(\theta - 4\pi/3)
\end{array}
\begin{bmatrix} v_A \\ v_B \\ v_C \\ v_a \\ v_b \\ v_c \end{bmatrix}
$$

The backward transformation, which would be needed to get the phase currents from the solution for d–q currents, is:

$$
\begin{bmatrix} i_A \\ i_B \\ i_C \\ i_a \\ i_b \\ i_c \end{bmatrix}
= \sqrt{\frac{2}{3}}
\begin{array}{c} \\ A \\ B \\ C \\ a \\ b \\ c \end{array}
\begin{array}{cccc}
D & Q & d & q \\
1 & 0 & & \\
-\frac{1}{2} & \sqrt{3}/2 & & \\
-\frac{1}{2} & -\sqrt{3}/2 & & \\
 & & \cos\theta & \sin\theta \\
 & & \cos(\theta - 2\pi/3) & \sin(\theta - 2\pi/3) \\
 & & \cos(\theta - 4\pi/3) & \sin(\theta - 4\pi/3)
\end{array}
\begin{bmatrix} i_{dss} \\ i_{qss} \\ i_{drs} \\ i_{qrs} \end{bmatrix}
$$

It will be noticed that the inverse transformation is the same as the transpose, one of the advantages of this particular arrangement.[2] The $\sqrt{(2/3)}$ factor is associated with maintaining constant power throughout the transformation. The solution is organised in a similar manner as that for real axes and in accordance with Figure 8.1. The mechanical equations remain as in eqns (8.1) and (8.2) and the *per-unit* system may be used if desired.

Extracting **G** from the transient impedance matrix and abbreviating stator current symbols to i_D, i_Q; rotor current symbols to i_d, i_q, the electromagnetic torque is:

$$
\begin{bmatrix} i_D & i_Q & i_d & i_q \end{bmatrix}
\begin{array}{c} \\ D \\ Q \\ d \\ q \end{array}
\!\!\begin{array}{c} \begin{array}{cccc} D & Q & d & q \end{array} \\ \begin{bmatrix} & & & \\ & & & \\ & M & & L_r \\ -M & & -L_r & \end{bmatrix} \end{array}
\begin{array}{c} \\ \begin{bmatrix} i_D \\ i_Q \\ i_d \\ i_q \end{bmatrix} \end{array} \times \text{pole pairs}
$$

which when multiplied out gives:

$$(i_d M i_Q + i_d L_r i_q - i_q M i_D - i_q L_r i_d) \times \text{pole pairs}$$

$$T_e = M(i_d i_Q - i_q i_D) \times \text{pole pairs.}$$

If *per-unit* quantities are used, the term $\omega/3$ replaces pole pairs in the above expression, as for eqn (8.11).

The rotor reference frame

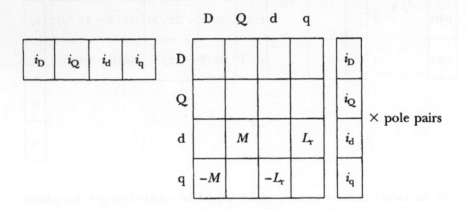

It is possible of course to solve the uniform air-gap machine using d–q axes fixed to the rotor instead of the stator, but this reference frame will be used now, primarily to introduce saliency and a rotor d-axis field winding, so that the equations will have an immediate connection with the salient-pole synchronous machine. The rotor d and q windings will represent the dampers. The difference between the permeances on d and q axes is reflected in a difference between the mutual inductances M_D and M_Q say, which may be in the ratio 'r' as high as 4:1 on reluctance machines. Referring all inductances to one winding, the direct and quadrature axis values will be.

stator, D axis	leakage + mutual	$= L_D - M_D + M_D$	$= L_D$
stator, Q axis	$= L_D - M_D + \dfrac{M_D}{r}$	$= L_D + \left(\dfrac{1}{r} - 1\right) M_D$	$= L_Q$
rotor, d axis	$=$	$= L_d - M_D + M_D$	$= L_d$
rotor, q axis	$= L_d - M_D + \dfrac{M_D}{r}$	$= L_d + \left(\dfrac{1}{r} - 1\right) M_D$	$= L_q$
rotor field winding	$L_f - M_{td} + M_{fd}$		$= L_f$

Note: the mutual between field and stator D winding $\neq M_{fd}$ but $= M_{fD}$.

The voltage equation can be written down as before, with an extra row and column to accommodate the field winding and noting that rotational voltages now appear in the stator winding which is rotating with respect to the rotor reference frame.

		D	Q	d	q	f	
$vdsr$	D	$R_D + L_D p$	$L_Q \dot{\theta}$	$M_D p$	$M_Q \dot{\theta}$	$M_{fD} p$	$idsr$
$vqsr$	Q	$-L_D \dot{\theta}$	$R_Q + L_Q p$	$-M_D \dot{\theta}$	$M_Q p$	$-M_{fD} \dot{\theta}$	$iqsr$
$vdrr$	$=$ d	$M_D p$		$R_d + L_d p$		$M_{fd} p$	$idrr$
$vqrr$	q		$M_Q P$		$R_q + L_q P$		$iqrr$
vfr	f	$M_{fD} p$		$M_{fd} p$		$R_f + L_f p$	ifr

Resolving the axes along rotor d and q gives:

$$\begin{bmatrix} vdsr \\ vqsr \\ vdrr \\ vqrr \\ vfr \end{bmatrix} = \sqrt{\frac{2}{3}} \begin{array}{c} \\ D \\ Q \\ d \\ q \\ f \end{array} \begin{array}{ccccccc} A & B & C & a & b & c & f \\ \hline \cos\theta & \cos(\theta - 2\pi/3) & \cos(\theta - 4\pi/3) & & & & \\ \sin\theta & \sin(\theta - 4\pi/3) & \sin(\theta - 4\pi/3) & & & & \\ & & & 1 & -\frac{1}{2} & -\frac{1}{2} & \\ & & & 0 & \sqrt{3}/2 & -\sqrt{3}/2 & \\ & & & & & & 1 \end{array} \begin{bmatrix} v_A \\ v_B \\ v_C \\ v_a \\ v_b \\ v_c \\ v_f \end{bmatrix}$$

The backwards transformation is the transpose of the above, as for the stator reference frame.

It will be noted that the parameters can be defined in terms of the D and d axes together with the ratio r, i.e. $M_Q = M_D/r$ so the effect of different r ratios can be conveniently tried out. Usually, $R_D = R_Q = R_s$, the stator resistance per phase but R_d would not normally be the same as R_q (the direct and quadrature damper-winding resistances). However, if for example the rotor reference frame was used for solution of an induction motor, then these two would be equal to the rotor resistance per phase R_r. The field winding row and column would be omitted. The parameters are usually referred to one particular winding though this is not essential providing there is consistency in the use of voltages and currents. A 4 × 4 matrix would also apply to the reluctance machine, with d and q representing the starting/damping winding in the salient poles.

Extracting the **G** matrix from the top two rows and multiplying out to get $\mathbf{i}_t\mathbf{G}\mathbf{i}$, will give the following expressions for instantaneous electromagnetic torque:

Salient-pole machine:

$$T_e = [-(L_D - L_Q)(i_Q i_D) + M_Q(i_q i_D) - M_D(i_d i_Q) - M_{fD}(i_f i_Q)] \times \text{pole pairs}$$

Reluctance machine:

$$T_e = [-(L_D - L_Q)(i_Q i_D) + M_Q(i_q i_D) - M_D(i_d i_Q)] \times \text{pole pairs}$$

Induction machine with

$$L_D = L_Q \text{ and } M_D = M_Q = M$$

$$T_e = M(i_q i_D - i_d i_Q) \times \text{pole pairs}$$

It is sometimes useful to express the equations in terms of the flux linkage ($\lambda = N\phi = Li$) on direct and quadrature axes, e.g. if saturation of parameters had to be allowed for. The voltage equations for the excited salient-pole machine would become, using the abbreviated notation:

$$v_D = R_D i_D + d(\lambda_D)/dt + \lambda_Q \dot{\theta}$$

$$v_Q = R_Q i_Q + d(\lambda_Q)/dt - \lambda_D \dot{\theta}$$

$$v_d = R_d i_d + d(\lambda_d)/dt$$

$$v_q = R_q i_q + d(\lambda_q)/dt$$

$$v_f = R_f i_f + d(\lambda_f)/dt$$

where:

$$\lambda_D = L_D i_D + M_D i_d + M_{fD} i_f$$

$$\lambda_Q = L_Q i_Q + M_Q i_q$$

$$\lambda_d = M_D i_D + L_d i_d + M_{fd} i_f$$

$$\lambda_q = M_Q i_Q + L_q i_q$$

$$\lambda_f = M_{fD} i_D + M_{fd} i_d + L_f i_f$$

The electromagnetic torque is given by:

$$T_e = (\lambda_Q i_D - \lambda_D i_Q) \times \text{pole pairs}$$

It can be seen that all the torque equations reduce to one simple form, since the common-axis terms have been grouped together, expressed as the total flux linkage on each axis.

Stator and rotor reference frames are the most commonly used though for certain purposes, e.g. study of oscillatory conditions, some simplification follows if the reference frame is fixed to the rotating field (synchronous axes), even though either stator or rotor may not be stationary with respect to this reference.[2] This reference frame has also been used in Section 7.5 when developing the equations for vector (field orientation) control. Where such control is achieved by relying entirely on the mathematical model of the machine, the d–q equations lend themselves to an economical solution.

Appendix D
Tutorial examples with answers

The following examples are provided so that there is opportunity to check understanding of the various problems encountered in the main text. Most of them are at least slightly different from these, but the necessary background has been covered for all. Chapter 1 on basic theory should be thoroughly understood and the recommendation to incorporate the question data on a simple diagram should be followed. This is especially important when dealing with 3-phase circuits so that mistakes of interpretation are avoided when extracting phase voltages, currents and impedances from the data given. Note that the 'rating' refers to the total power or VA, the line voltage and the line current. Chapter 2 on Transformers revises all these points and the practice of declaring the no-load voltage ratio is followed, so that this is also the turns ratio when reduced to phase values. Answers, and some part answers, are given to all numerical problems, together with an indication in some cases of theoretical points which are raised. Accuracy is generally to three figures, though sometimes more, where appropriate. Although much care has been taken, there may be a few errors and the authors would be very grateful to be informed about any which are discovered.

Chapter 1. Introduction

No examples are set for this revisory chapter but two convention problems were raised therein. On p. 6 the motor convention equation always has a positive sign and the generator convention equation always has a negative sign.

Referring to the question on p. 9 concerning Figure 1.5 for the generator phasor diagrams; with a motor convention, the current phasor in Figures 1.5a and 1.5b would be in the lower half of the diagrams. **E** would still be $\mathbf{V} - jX\mathbf{I} - R\mathbf{I}$, but because of the lower position of the current phasor, **E** would now lead **V** in both cases. For a generator convention, the **I** phasor would be reversed, together with the sign in the above equation but the phasor diagrams would be otherwise unchanged in shape, the new current phasor being in the upper half of the diagram, with the power-factor angle less than 90°.

Chapter 2. Transformers

In the following examples, the approximate equivalent-circuit can be used and in some cases, the magnetising impedance may be omitted altogether. For determination of turns ratio and for 3-winding transformers, all impedances of the transformer are neglected.

T2.1. A 3-phase, 50-Hz, 6600/400-V, star/zigzag transformer has a core cross-sectional area of 0.04 m^2. Calculate the number of turns required for each coil of primary- and secondary-winding sections, if the maximum flux density is not to exceed 1.1 T. Note, each secondary phase is made up of two equal sections, taken from neighbouring phases and connected in opposition so that the phase voltage is the phasor difference of the two sections; i.e. phase voltage = $\sqrt{3} \times$ secondary-section voltage.

T2.2. A 3-phase, 3-winding, 1000-kVA, 50-Hz transformer is required to meet the following specification.

Primary 66 kV (line) star connected.
Secondary 6.6 kV (line) delta connected.
Tertiary 440 V (line) star connected.

The maximum flux is not to exceed 0.1 Wb.

(a) Determine the required number of turns per phase on each winding.
(b) Taking the magnetising current as 5% of the rated current and the iron loss as 10 kW total, determine the input current, power factor, kVA and kW when the secondary is loaded at 600 kVA, 0.9 pf lagging and the tertiary is loaded at 400 kVA, 0.8 pf lagging.

T2.3. In a certain transformer, the winding leakage reactances are six times the winding resistances. Estimate the power factor at which the voltage regulation is zero. If the leakage reactance is 10%, what is the maximum regulation in *per unit* for a current of 1 *per unit*?

T2.4. A single-phase, 50-Hz, 500-kVA, 33 000/3300-V transformer has the following parameters: $R_1 = 8.6\,\Omega$, $R_2 = 0.08\,\Omega$, $x_1 = 52\,\Omega$, $x_2 = 0.46\,\Omega$.

On no load, the h.v. current at 33 kV is 0.3 A and the power input is 3.5 kW. Calculate:

(a) The regulation at rated current and power factors of: 0.866 lag, u.p.f. and 0.866 lead.
(b) The efficiencies at the same load current and power factors (neglect regulation).
(c) The maximum efficiency.
(d) The in-phase and quadrature components of the no-load current.
(e) The primary current and power factor and the secondary current when the secondary power-factor is 0.8 lagging and the efficiency is at its maximum for this power factor.

T2.5. A 3-phase, 50-Hz, 300-kVA, 11 000/660-V, star/delta transformer gave the following line-input readings during light-load tests:

Open circuit	660 V,	8 A,	2.4 kW
Short circuit	500 V,	15 A,	4.1 kW

Calculate and show on a per-phase equivalent circuit the parameters deduced from these tests, referred to the h.v. side. Also show all currents, input power-factor and secondary terminal voltage when the secondary is supplying rated current at 0.8 p.f. lagging. Allow for the voltage regulation.

T2.6. An 800-kVA transformer at normal voltage and frequency requires an input of 8 kW on open circuit. The input on short circuit at rated current is 15 kW. The transformer has to operate on the following duty cycle:

6 h at 450 kW, 0.8 p.f.
4 h at 650 kW, 0.9 p.f.
5 h at 250 kW, 0.95 p.f.
Remainder of 24-h day on no load.

Calculate the all-day efficiency defined as: output in kWh/(output in kWh + losses in kWh). What is the efficiency at a constant load of 800 kVA at 0.8 p.f.?

T2.7. Two single-phase transformers A and B operate in parallel. $\mathbf{E}_A = 200$ V and $\mathbf{E}_B = 203$ V, $\mathbf{Z}_A = \mathbf{Z}_B = 0.01 + j0.1\ \Omega$, all data referred to the secondary side. Calculate $\mathbf{I}_A$ in terms of $\mathbf{I}_B$ (magnitude and phase), when the load impedance has the following values:

(a) $\mathbf{Z} = 10 + j1\ \Omega$.
(b) $\mathbf{Z} = 0.1 + j1\ \Omega$.

In what mode is transformer A operating in case (a)? Phasor diagrams would be informative.

T2.8. The following data refer to two 3-phase, delta/star connected transformers:

Transformer	kVA rating	Line voltage	Short-circuit test at rated current
A	600	2300/398	160 V, 4.2 kW
B	900	2300/390	100 V, 5.1 kW

Calculate the total current supplied by the two transformers connected in parallel to a star-connected load of 0.132 Ω per phase and of power factor 0.8 lagging.

If the load is disconnected, what will be the secondary circulating current and the change in secondary terminal-voltage from the loaded condition?

T2.9. A single-phase transformer supplies a full-load secondary current of 300 A at 0.8 p.f. lagging when the load voltage is 600 V. The transformer is rated at 6600/600 V on no load and is provided with tappings on the h.v. side to reduce this ratio. It is supplied from a feeder for which the sending-end voltage is maintained at 6.6 kV. The impedances are as follows:

Feeder (total)	= 1 + j4 Ω
Transformer primary	= 1.4 + j5.2 Ω
Transformer secondary	= 0.012 + j0.047 Ω

To what value must the turns ratio be adjusted if the secondary terminal voltage is to be 600 V under the full-load condition? Neglect the magnetising current and the effect of the changed turns-ratio on the referred impedance.

Chapter 3. D.C. machines

In the following questions, the brush voltage-drop, the magnetising effects of armature reaction and the mechanical losses may be neglected unless specifically stated otherwise.

T3.1. A 25-hp, 500-rev/min d.c. shunt-wound motor operates from a constant supply voltage of 500 V. The full-load armature current is 42 A. The field resistance is 500 Ω and the armature resistance is 0.6 Ω. The magnetisation characteristic was taken at 400 rev/min as follows:

Field current	0.4	0.6	0.8	1.0	1.2 A
Generated e.m.f.	236	300	356	400	432 A

Calculate the mechanical loss torque using the full-load data (1 hp = 746 W), and assume it is constant in the following calculations.

(a) Calculate the field current and the external field resistance for operation at rated load and speed.
(b) What is the speed when the load is removed, leaving just the loss torque?
(c) With the excitation of part (a), at what speed must the machine be driven to regenerate at rated current?
(d) What extra field-circuit resistance is necessary to cause the machine to run on no load at 600 rev/min?
(e) What extra armature-circuit resistance is needed to cause the machine to develop half its rated electromagnetic torque at a speed of 300 rev/min with field current as in (a)? What would be the output power at the coupling?

T3.2. A 500-V, d.c. separately excited generator at its normal rating gives an output of 50 kW at 1000 rev/min. The armature-circuit resistance is 0.4 Ω.

The machine is to be run as a motor, but with the voltage reduced to 200 V. At what speed will it run if set at rated flux and taking rated armature current?

What reduction of flux will be necessary to run at 1000 rev/min with the same armature current and how will the electromagnetic torque differ from that developed as a generator at normal rating?

T3.3. A 500-V d.c. shunt motor has a rated output at the coupling of 40 kW when running at 500 rev/min, the efficiency then being 90%. The armature- and field-circuit resistances are 0.23 Ω and 400 Ω respectively. The following magnetisation curve was taken when running as an open-circuited generator at 600 rev/min:

Field current	0.25	0.5	0.75	1.0	1.25	1.5 A
Generated e.m.f.	170	330	460	531	572	595 V

Determine, at the rated condition:

(a) the armature current;
(b) the electromagnetic torque;
(c) the loss torque.

Find also the additional field-circuit resistance necessary to give a speed of 1200 rev/min, the armature current and loss torque being assumed constant and rated voltage being applied. What will be the output power at this condition?

T3.4. Derive the speed/torque expression for a d.c. machine. Hence explain the various methods of controlling the speed of such a motor, commenting on their advantageous and disadvantageous features.

A d.c. generator at its normal rating gives an output of 100 kW at 500 V when driven at 500 rev/min. It is to be run as a motor from a 500-V supply at a speed of 750 rev/min, the additional cooling at this speed permitting an armature-current increase of 10% above the rated value. Calculate the electromagnetic torque the motor will develop under this condition. What value of flux is necessary and what total mechanical power will the motor produce? Express this flux, torque and power as fractions of rated flux, torque and power. The armature resistance is 0.15 Ω.

T3.5. A d.c. motor has a full-load armature rating of 7.5 kW, 200 V, 45 A at 800 rev/min. The armature resistance is 0.5 Ω. Determine the mechanical loss torque and assume this is constant, independent of speed. The mechanical load torque has a characteristic such that beyond 800 rev/min, it falls off inversely as speed; i.e. a constant-power law. Determine the percentage of rated flux required to increase the speed to 2400 rev/min, and the corresponding value of armature current.

T3.6. Derive expressions for armature current and flux for operation of a d.c. motor at any specified voltage, speed, electromagnetic torque and armature-circuit resistance.

A 500-V d.c. motor at rated load runs at 1500 rev/min and takes an armature current of 50 A. The armature resistance is 0.5 Ω and the shunt-field resistance is 200 Ω The mechanical loss torque is 5 Nm. Determine:

(a) The electromagnetic torque, the mechanical output and the efficiency.
(b) The electromagnetic torque when the flux is weakened to give a speed of 2000 rev/min with armature current allowed to increase by 25%.
(c) The armature current and *per-unit* flux ($k_\phi/k_{\phi(\text{rated})}$) required to sustain operation at 1000 rev/min with terminal voltage halved and armature-circuit resistance doubled. Assume the total mechanical torque is proportional to (speed)^2.

T3.7. A d.c. generator has $R_a = 1\ \Omega$ and $R_f = 480\ \Omega$. Its o.c. characteristic is as follows:

Field current	0.1	0.2	0.4	0.6 A
Generated e.m.f.	85	160	233	264 V

The machine is to be run as a motor from 230 V with an armature current of 30 A. Find:

(a) The additional resistance in the field circuit to give a speed of 750 rev/min.
(b) The additional resistance in the armature circuit to give a speed of 300 rev/min when the field winding is connected directly across the 230-V supply.
(c) The resistance in parallel with the armature terminals, but otherwise as in condition (b), to give a speed of 200 rev/min. What will then be the supply current?

N.B. This is the potential-divider connection.

T3.8. A d.c. series motor was run at constant speed of 1500 rev/min, with varying voltage and load to give the following points:

Terminal voltage	61	71	81	93	102.5	110 V
Current	1.39	1.68	1.98	2.42	2.87	3.22 A

If the motor is supplied at 110 V, calculate the speed/torque curve:

(a) with the natural armature-circuit resistance of 7.4 Ω;
(b) with an additional series resistor of value 30 Ω.

What is the torque at 1500 rev/min in each case and what is the operating mode when the speed goes negative in case (b)?

T3.9. A d.c. series motor was run at 500 rev/min when the following magnetisation characteristic was taken:

Field current	5	10	15	20	25	30 A
Generated e.m.f.	192	276	322	356	384	406 V

The armature and field resistances are each 1 Ω. The motor is connected in series with a resistor of 20 Ω across a 420-V d.c. supply. Across the armature terminals is a further diverter resistor of 21 Ω. Calculate the speed/torque curve with this connection and show the currents and voltages at the various points in the circuit when the field current is 20 A. For what purpose would this characteristic be useful?

T3.10. A d.c. series motor of resistance 0.12 Ω has a magnetisation curve at 750 rev/min as follows:

Field current	50	100	150	200	250	300 A
Generated e.m.f.	110	220	300	360	400	420 V

The motor is controlling a hoist in the dynamic-braking connection. Determine the resistance to be connected across the machine terminals so that when the overhauling torque is 1017 Nm, the speed is limited to

500 rev/min. How much extra resistance will be required to maintain this speed when the total torque falls to 271 Nm?

If the drive is now changed to hoisting with this last resistor in circuit and a supply voltage of 400 V, what will be the operating speed for a motoring torque of 271 Nm?

T3.11. The d.c. series motor of T3.8 is to be braked from the full-load motoring condition (110 V, 1500 rev/min). If the initial braking current is to be limited to twice the full-load value of 3.22 A, what resistor would have to be inserted in series with the armature:

(a) for rheostatic (dynamic) braking;
(b) for plugging (reverse-current) braking?

At what speed, in each case, would the braking torque be equal to the full-load value?

T3.12. Illustrate on a 4-quadrant diagram, the various methods of electrical braking using the following data for a separately excited d.c. machine.

Rating: 500 V, 50 A, 500 rev/min
Armature resistance 0.5 Ω
Rated flux maintained throughout.

Calculate the speed/torque equations for:

(a) regenerative braking into rated voltage supply;
(b) reverse-current braking with the same supply and an additional 10 Ω limiting resistor;
(c) rheostatic braking with a 5-Ω external limiting resistor.

At what speed in each case will full-load torque be developed? Comment on the special features of each mode, pointing out the advantages and limitations. See also the second part of Tutorial Example T6.6.

T3.13. A 500-V, separately excited d.c. motor takes an armature current of 100 A when driving its mechanical load at rated speed of 600 rev/min. The total mechanical torque as a function of speed is given by the following expression:

$$T_m = k[1 + (\omega_m/20\pi) + (\omega_m/20\pi)^2] \text{ Nm},$$

where ω_m is the speed in rad/s. The armature-circuit resistance is 0.4 Ω. Calculate the torque at rated speed and the value of k.

If the speed is increased to 700 rev/min by field weakening, what percentage change of flux and of armature current would be required? Comment on the suitability of this method of speed control for this particular load.

T3.14. A d.c. separately excited motor has a *per-unit* resistance of 0.05 based on rated armature voltage and current. For the 3rd base quantity in the *per-unit* system, take rated flux, which gives rated torque at rated armature current. In the following, *per-unit* quantities are being referred to throughout.

At rated flux and rated torque, what is the rated speed, i.e. when rated voltage is applied?

The motor drives a load which requires a torque of 0.3 at starting and thereafter increases linearly with speed reaching 1.0 at rated speed. What is the expression for this mechanical torque as a function of speed?

Assuming the magnetisation characteristic can be expressed by the following relationship:

$$I_f = \frac{0.6\phi}{1 - 0.4\phi}$$

where ϕ is the flux, calculate flux, field current and armature current for the following conditions:

(a) Voltage = 0.5, speed = 0.5.
(b) Voltage = 1.0, speed = 1.2.
(c) Voltage = 0.5, speed = 0.4 and total armature-circuit resistance = 0.15.

T3.15. A separately excited d.c. motor is permanently coupled to a mechanical load with a total characteristic given by $T_m = 0.24 + 0.8\omega_m$ *per unit*. T_m is unity at rated speed where rated flux and armature current are required to produce the corresponding T_e at rated voltage. *Per-unit* R_a based on the rated terminal voltage and armature current is 0.05. *Per- unit* speeds are based on the speed at rated voltage and flux with zero torque. The required portion of the magnetisation curve of *per-unit* flux against *per-unit* field current may be taken as a straight line passing through $\phi = 0.6$, $I_f = 0$ and $\phi = 1.0$, $I_f = 1.0$. Calculate the *per-unit* field and armature currents to sustain operation at:

(a) half rated voltage and speed;
(b) rated voltage and speed of:
   (i) 0.75 *per unit*;
   (ii) 1.25 *per unit*.

(If *per-unit* ideas are insecure, take rated V as 100 V, rated I_a as 100 A (i.e. R_a = 0.05 Ω), rated speed as 100 rad/s (955 rev/min) and hence $k_{\phi(\text{rated})}$ = (100 − 100 × 0.05)/100 = 0.95 Nm/A.)

Chapter 4. Induction machines

For this chapter, the approximate equivalent circuit will normally be used unless specifically stated otherwise. The magnetising branch may be omitted if no data is given. Also, mechanical loss will be neglected unless the data include this.

T4.1. The primary leakage-impedance per phase of a 3-phase, 440-V, 50-Hz, 4-pole, star-connected induction machine is identical with the referred seondary impedance and is equivalent to a 1-Ω resistor in series with a 10 mH inductor. The magnetising impedance may be considered as connected across the primary terminals and consists of a 300-Ω resistor in parallel with a 200 mH inductor. Calculate, for a slip of 0.05:

(a) the input current and power factor;
(b) the electromagnetic torque;
(c) the mechanical output-power if the mechanical loss-torque is 1 Nm;
(d) the efficiency.

If the mechanical load-torque were to be increased from this figure, at what speed would the motor begin to stall?

T4.2. A 3-phase, 440-V, 50-Hz, 4-pole, star-connected induction machine gave the following line-input readings during parameter tests:

Locked rotor (short circuit)	120 V	25 A	2.0 kW
No-load (running uncoupled)	440 V	8 A	1.5 kW

Separate tests determined the friction and windage loss as 600 W.

Using the approximation that the magnetising branch is connected across the terminals, deduce the value of the equivalent-circuit parameters, dividing the leakage impedance equally to give identical stator and (referred) rotor values. Note that the mechanical loss must be deducted from the no-load input before calculating R_m and X_m.

(a) At the full-load slip of 4% determine the input current and power factor, the rotor (referred) current, electromagnetic torque, output power and efficiency.
(b) What is the starting torque at full voltage and the ratio of starting current to full-load current?

T4.3. Using the data for the machine of the last question, calculate at a speed of 1560 rev/min:

(a) the stator current and power factor;
(b) the rotor current;
(c) the electromagnetic torque;
(d) the mechanical coupling-power;
(e) the efficiency.

Repeat the calculations using the same parameters in the 'exact' circuit and draw a power-flow diagram similar to that of Example 4.8 to illustrate this operating mode.

From the 'exact' circuit calculate the self-inductance of the stator and rotor and the mutual inductance, all referred to the stator winding, for the condition with all three phases excited with balanced currents.

T4.4. A 50-Hz, 6-pole, wound-rotor induction motor has a star-connected stator and a delta-connected rotor. The effective stator/rotor turns-ratio per phase is 2/1. The rotor leakage impedance at standstill is 0.36 + j1.5 Ω, the corresponding figure for the stator being 1.4 + j7 Ω per phase.

If the impressed stator voltage is 220 V/phase, calculate:

(a) the actual rotor current at starting;
(b) the actual rotor current when running at 960 rev/min;
(c) the initial rotor current when, from condition (b), the stator supply sequence (and hence the rotating field) is suddenly reversed;
(d) the electrical input power for condition (c) and the mode of operation;
(e) the required resistance in series with the slip rings, to give maximum torque at starting.

Note: for convenience, refer all quantities to the rotor winding.

T4.5. Determine the value of the starting torque and the maximum torque (in terms of the full-load torque) for an induction motor with a full-load slip of 4%. The primary and secondary leakage reactances are identical when referred to the same winding and the leakage reactance is five times the resistance. Note: use $I_2^2R_2/s$ ratios so that any constants will cancel.

T4.6. In a certain 3-phase induction motor, the stator and referred-rotor impedances are identical, the leakage reactance being four times the resistance. Determine the effect on the starting torque, the speed at which maximum torque occurs and the maximum torque itself of doubling:

(a) the rotor resistance;
(b) the stator resistance.

Hence sketch the shape of the three speed/torque curves corresponding to normal operation, condition (a) and condition (b). Note: let $R_1 = R_2' = R$ and $x_1 = x_2' = 4R$, then use ratios of the appropriate expressions, taking normal T_{max} as 1 *per-unit* torque.

T4.7. On locked-rotor test, an induction motor takes three times full-load current (3 *per-unit*), at half rated-voltage. The motor has a full-load slip of 4% and is to be started against a load requiring 1/3 of full-load torque. An auto-transformer is used to reduce the motor terminal voltage at starting so that it is just sufficient to meet this requirement. What percentage tapping will be necessary and what will then be the supply current expressed in *per unit*?

Note: if the ratio of starting torque to rated torque at full voltage is expressed in terms of the appropriate $I_2{}^2R_2/s$ ratios, the value of the *per-unit* starting torque will be given in terms of *per-unit* current and full-load slip. Hence the required starting current in *per-unit* for the reduced starting torque and the transformer tapping will follow.

T4.8. The speed/electromagnetic-torque curve for a 50-Hz, 4-pole induction motor at rated voltage is given by the following points:

Speed	1470	1440	1410	1300	1100	900	750	350	0 rev/min
Torque	6	12	18	26	30	26	22	14	10 Nm

The mechanical load it drives requires a torque of 14 Nm and the mechanical loss torque can be taken as 1 Nm. What is the speed at normal voltage?

It is required to reduce the speed to 1200 rev/min. Assuming that the load and loss torques do not change with speed:

(a) What voltage reduction would be required?
(b) What percentage increase in rotor resistance would be required if the voltage was left unchanged?

Comment on these two methods of achieving this speed reduction.

T4.9. A 3-phase, 440-V, 50-Hz, 6-pole, delta-connected induction motor drives a fan at 920 rev/min when supplied at rated voltage. The motor has a high-resistance speed/torque characteristic and the following equivalent-circuit parameters:

$$R_1 = 8\,\Omega, \quad R_2' = 16\,\Omega, \quad x_1 - x_2' = 12\,\Omega.$$

Assuming the total mechanical torque varies in proportion to the square of the speed, what supply voltage would be required to run the fan at 460 rev/min and what will then be the rotor current and copper loss?

As an alternative, calculate the required extra rotor-resistance to get half speed with full voltage; the unknown in eqn (4.5) is now R_2'. Calculate the rotor current and copper loss to compare with the previous method. Refer also to T4.22.

T4.10. A certain 3-phase induction motor on locked-rotor test takes full-load current at a power-factor of 0.4 lagging, from a normal-frequency supply. If the motor is operated with a 30% reduction in both voltage and frequency, estimate the new starting torque and maximum torque in terms of normal values. Sketch the two speed/torque curves for comparison. Assume the impedance is divided equally between stator and (referred) rotor.

Note: take $R_1 + R_2' = 0.4$ and $x_1 = x_2'$ as $\sin(\cos^{-1} 0.4)$ and use ratios from appropriate expressions.

T4.11. A 3-phase, 440-V, 50-Hz, 6-pole, star-connected induction motor has the following equivalent-circuit parameters at normal supply frequency:

$$R_1 = 0.2\,\Omega, \quad R_2' = 0.18\,\Omega, \quad x_1 = x_2' = 0.58\,\Omega \text{ per phase.}$$

The machine is run up and controlled from a variable-frequency supply, the voltage of which is directly proportional to the frequency up to 440 V at 50 Hz. Find the supply frequency which gives maximum-starting torque and compare the value of this torque with the starting torque at rated voltage and frequency.

Note: set voltage, synchronous speed and reactances at kV, kω_s and kx respectively in the approximate-circuit expressions, then differentiate with respect to k.

T4.12. For a certain 3-phase induction motor, $R_1 + jx_1 = R_2' + jx_2'$ and $x_1/R_1 = 3$ at normal frequency. Calculate speed/torque coordinates at zero torque, maximum motoring torque and at zero speed for the following conditions:

(a) rated voltage	V	and rated frequency	f;
(b) voltage	$V/3$	and frequency	$f/3$;
(c) voltage	$V/3$	and frequency	f;
(d) voltage	V	and frequency	$1.5f$.

Express speeds as fractions of rated synchronous speed and torques as fractions of the maximum torque at rated voltage and frequency.

Note: use equations (4.5), (4.12) and (4.13) with $R_1 = R_2' = R$ and $x_1 = x_2' = 3R$ at normal frequency. Using torque ratios will permit common terms to cancel. $\hat{s}$ values will be with respect to the *actual* supply frequency.

T4.13. Sketch the speed/torque characteristics for an induction-motor drive supplied from a variable-frequency source, explaining why it is desirable to maintain a particular relationship between supply voltage and frequency.

A 3-phase, 400-V, 50-Hz, 6-pole, star-connected induction motor has leakage impedances $\mathbf{z}_1 = \mathbf{z}_2' = 0.15 + j0.75\,\Omega$ per phase at rated frequency. Calculate the torque at rated voltage and frequency for the rated slip of 3%.

If the same torque is required at starting and also at 750 rev/min, to what values must the supply frequency and voltage be adjusted if the machine flux-per-pole is to be the same as at the rated condition?

T4.14. A 3-phase, 50-Hz, 4-pole induction motor at rated voltage and frequency has the speed/torque characteristic given in Example 4.10. The motor is controlled to maintain the flux-per-pole-constant at any particular torque. Estimate the frequency required to produce:

(a) maximum torque at starting;
(b) a speed of 750 rev/min with a torque of 9 Nm.

T4.15. Show that if rotor leakage reactance is neglected, maximum torque for a constant-current induction-motor drive occurs when the rotor current and magnetising current are equal in value at $I_1/\sqrt{2}$. Show also that for maximum torque during a constant-current acceleration, the rotor frequency must be constant.

Using the data from T4.1, but neglecting the magnetising resistance, calculate the maximum torque:

(a) with rated voltage and frequency applied – use the approximate circuit;
(b) with a constant-current drive at the same rotor current as in (a). Include x_2' and calculate the required I_1 as well as the maximum torques:
    (i) neglecting saturation of X_m and;
    (ii) assuming saturation reduces X_m to 62.8/3 ohms per phase.

Check whether condition (b) (ii) is feasible by calculating the value of the required line voltage to supply the primary current at starting and at 300 rev/min. The 'exact' circuit will have to be used for this purpose.

T4.16. A 3-phase, 660-V, 50-Hz, 4-pole, delta-connected induction motor has $\mathbf{z}_1 = \mathbf{z}_2' = 0.15 + j0.75\,\Omega$ per phase at standstill. The full-load slip is 3%. Compare the torque developed at full load with that developed immediately after making the following alternative changes from the full-load condition:

(a) reversal of two primary-supply leads;
(b) disconnection of a.c. supply from primary and replacement by a d.c. voltage across two lines, previously adjusted so that the same air-gap flux will exist as before the changeover.

What is the initial *per-unit* current in each case, based on full-load current?

T4.17. A 3-phase, 1100-V, 50-Hz, 6-pole, star-connected induction machine when operating at full load as a motor running at 980 rev/min takes a total current of 113.5 – j76.3 A per phase, the current through the magnetising branch being –3.8 – j59.3 A per phase, with the terminal voltage as the reference phasor. The primary and (referred) secondary impedances are of equal value at standstill at 0.1 + j0.4 Ω per phase. Calculate X_m from E_1/I_m (neglecting R_m) where $E_1 = |\mathbf{V}_1 - \mathbf{I}_1\mathbf{z}_1|$ and obtain the equivalent circuit for a d.c. dynamic braking condition, with an extra rotor (referred) resistance of 3.5 Ω per phase. What is then the electromagnetic torque when switched over from motoring, assuming the air-gap flux is unchanged?

To maintain this air-gap flux what is the required stator current: (a) in a.c. terms and (b) in d.c. terms; two of the phases being connected in series for d.c. excitation purposes? What excitation voltage would be required?

Calculate also the speed at which maximum torque occurs when dynamic braking and the ratio of this torque to that which occurs immediately after changing over the connections from rated motoring load.

T4.18. A 3-phase, 400-V, 50-Hz, 4-pole, star-connected, double-cage induction motor has the following equivalent circuit parameters per phase at standstill:

Primary	0.0625 + j0.25	*per unit*, (0.5 + j2 Ω);
Outer cage	0.25 + j0.075	*per unit*, (2 + j0.6 Ω);
Inner cage	0.0375 + j0.3125	*per unit*, (0.3 + j2.5 Ω).

The *per-unit* impedances are based on rated voltage and any convenient output, say 20 kW, for which the ohmic impedances are shown (base impedance = 8 Ω). Calculate the *per-unit* starting torque and also the mechanical power when the speed is 1410 rev/min. Use *per-unit* values in the calculation but check the answers using real values.

T4.19. For the motor of Example 4.19, p. 117 calculate the three input line-currents and electromagnetic torque for a positive-sequence slip of 3%, if the line voltages are only slightly unbalanced at 440 V, 430 V and 430 V. There will be no zero-sequence currents and calculation of the other currents, though somewhat tedious, is facilitated if polar coordinates are used to get $\mathbf{V}_+/\mathbf{Z}_+$ and $\mathbf{V}_-/\mathbf{Z}_-$ for each phase. These must be added using Cartesian coordinates of course. The positive- and negative-sequence phase voltages must be obtained for each phase, noting that the former lag 30° behind the line voltages, Figure 1.6c, whereas the latter lead the line voltages by 30°.

T4.20. Solve the single-phase induction-motor problem of Example 4.21, p. 122, for a slip of 0.03, using the exact equivalent circuit though neglecting the very small effect of R_m; i.e.

$$X_m/2 = 20\,\Omega,\ R_1/2 = R_2'/2 = 0.1\,\Omega,\ x_1/2 = x_2'/2 = 0.5\,\Omega.$$

Note: combine the primary, forward-circuit and backwards-circuit impedances in series to determine the total current. Hence follow the e.m.f.s $\mathbf{E}_f$ and $\mathbf{E}_b$ and the currents, $\mathbf{I}_f$, $\mathbf{I}_b$, $\mathbf{I}_{mf}$ and $\mathbf{I}_{mb}$. Compare answers with those from the approximate circuit in Example 4.21. The calculation of the torque pulsation will be deferred to Tutorial Example T7.7, which could be referred to with advantage.

T4.21. Using the answers of Tutorial Problem T4.20 and assuming that the iron loss and mechanical loss are unchanged from the operating condition of Example 4.2(b) (p. 85) determine the output, power factor and efficiency as a single-phase motor running at 3% slip. What would these values be if the motor was 2-phase, having two windings in quadrature, each phase having an identical equivalent circuit to that of Example 4.2(b)? Make a table comparing the answers.

T4.22. Consideration is to be given to controlling the fan-motor drive of Tutorial Example T4.9 by means of a slip-power recovery scheme to compare the performance. The rotor power-factor at 460 rev/min is to remain the same as at 920 rev/min but the impedance drop of the equivalent circuit, which previously was supplied entirely by the stator voltage (121.8 V), will now be provided by rated voltage $\mathbf{V}_1$ $(440 + j0) - \mathbf{V}_3'/s$; eqn (4.19). $\mathbf{V}_3'$ will have to be in phase with $\mathbf{V}_1$ if there is to be no change of power factor. The value of the current will be less however because the torque expression is different, eqn (4.20). Determine $\mathbf{V}_3'$ and this new current, the VA, the corresponding rotor copper loss and the slip power recovered, $3(V_3' I_2' \cos \varphi_3)$. Check that the total rotor *circuit power* is the same as for T4.9. Also determine the maximum VA rating of the injected-power source to give speed control over the whole range to zero speed. Compare, on a phasor diagram, the three methods of speed reduction.

T4.23. Show that the linear speed of a rotating field at frequency f, number of pole pairs p, pole pitch $\tau = \pi d/2p$ metres, is $v = 2\tau f$ metres/s. Hence determine the pole pitch of a linear induction motor required to have a speed of 200 miles per hour with a slip of 0.5, from a 50-Hz supply.

Chapter 5. Synchronous machines

Unless specifically stated otherwise, the following problems will assume that the air gap is uniform, and that resistance and mechanical losses are neglected.

T5.1. A 3-phase, 3.3-kV, 50-Hz, 6-pole, star-connected synchronous motor gave the following test points when run at synchronous speed as a generator, on open circuit and then on short circuit:

Open-circuit test	Line voltage	2080	3100	3730	4060	4310 V
	Field current	25	40	55	70	90 A
Short-circuit test	Armature current 100 A with field current = 40 A.					

The armature leakage impedance and all power losses may be neglected.

Calculate the excitation current required at rated voltage and frequency when operating with a mechanical output of 500 kW and an input power factor of 0.8 leading. Allow for the saturation of the magnetising reactance X_{mu} (= X_{su} since $x_{a1} = 0$).

What overload torque, gradually applied, would pull the machine out of synchronism with this calculated field-current maintained? Note that saturation affects both E_f and X_s to the same degree. What then is the permissible overload in *per unit* based on full-load torque? (Overload = Max./Rated.)

T5.2. Recalculate the answers for Examples 5.5 and 5.6 where appropriate, but neglecting the resistance, to check the errors in the approximation.

T5.3. A 3-phase, 500-kVA, 6600-V, 50-Hz, 6-pole, star-connected synchronous motor has a synchronous impedance per phase which can be taken as j70 Ω. At its normal rating, the motor is excited to give unity power-factor at the input terminals. Find:

(a) the rated current;
(b) the e.m.f. behind synchronous impedance (E_f);
(c) the rated electromagnetic torque;
(d) the pull-out torque with excitation as in (b);
(e) the required increase in excitation (E_f) which will just permit an overload margin of 100% before pulling out of synchronism;

(f) the load angle, armature current and power factor if this excitation is maintained at rated load.

T5.4. A 3-phase, 6600-V, star-connected synchronous motor has a synchronous impedance of $(0 + j30)\ \Omega$ per phase. When driving its normal load, the input current is 100 A at a power factor of 0.9 lagging.

The excitation is now increased by 50% above the value required to sustain the above condition, the mechanical load being unaltered. What changes in machine behaviour will take place?

With the new excitation, to what value and power factor will the armature current settle down if the mechanical load is removed altogether?

T5.5. A 3-phase, 11-kV, 50-Hz, 6-pole, star-connected synchronous motor is rated at 1 MVA and power factor 0.9 leading. It has a synchronous impedance which may be taken as entirely inductive of value $j120\ \Omega$ per phase. Calculate the rated current, the e.m.f. behind synchronous impedance, the electromagnetic torque and the output power.

Calculate the new values of armature current and excitation required to give operation at rated power but at 0.8 p.f. lagging. Express these values as fractions of rated values.

With the same excitation and armature current, at what power and power factor would the machine operate if running as a generator?

Note: use the cosine rule on the voltage triangle.

T5.6. Make a concise comparison of synchronous- and induction-machine performance features.

A 3-phase, 1000-kVA, 6.6-kV synchronous motor operates at unity power-factor when at its rated load condition. Its synchronous impedance can be taken as $j40\ \Omega$ per phase. Determine the e.m.f. behind synchronous impedance.

For a 50% change of this excitation voltage (both increaed and decreased values), what changes would take place in the motor performance?

T5.7. A 3-phase, 440-V, 50-Hz, 6-pole, star-connected synchronous motor has a synchronous impedance of $(0 + j10)\ \Omega$ per phase. At its normal rating, the armature current is 20 A and the power factor is 0.9 leading. The load torque is slowly increased from this condition to 300 Nm. By what percentage must the excitation be increased to avoid pulling out of step with this load torque?

T5.8. A round rotor machine has 3-phase windings on stator and rotor, the stator being connected to the 3-phase mains. Explain briefly why the machine will work in the induction mode at all speeds other than one particular value, with the rotor short circuited, whereas, in the synchronous mode, this speed is the only value at which it *will* operate when the rotor is d.c. supplied.

A 3-phase, 6.6-kV, star-connected synchronous motor has a synchronous impedance of (0 + j48) Ω per phase, and is running at its rated output of 650 kW. What will be the input current and power factor when the excitation is so adjusted that the e.m.f. behind synchronous impedance is 1.5 times the applied voltage?

If the supply voltage was to suffer a fall of 50%, would you expect the machine to continue supplying the load at synchronous speed? Would the reaction to this voltage fall have been generally similar if the machine had instead been motoring in the induction mode? Give reasons for your answers.

T5.9. Solve the numerical examples of T5.2 to T5.8 using the phasor loci and operating charts as in the solution of Example 5.16. Note that the Examples 5.5 to 5.7, for which the resistance was included, can also be solved by this method but with $V/(R_a + jX_s)$ defining the circle centre, the excitation term drawn from this being $E_f/(R_a + jX_s)$.

T5.10. A 3-phase, 3.3-kV industrial plant has the following induction-motor drives:

	IM 1	IM 2	IM3
Rated output	50	100	150 kW
Full-load efficiency	93%	94%	94.5%
Full-load power factor	0.89	0.91	0.93

A star-connected synchronous motor rated at 150 kVA is to be installed and is to be overexcited to improve the overall plant power-factor to unity when <u>all</u> machines are operating at their full ratings.

What is the required excitation in terms of the e.m.f. (E_f) if $\mathbf{Z}_s = 0 + j50\ \Omega$ per phase? What power output will the synchronous motor be delivering if its efficiency is 95% when it is operating at rated kVA?

T5.11. A 3.3-kV, 3-phase industrial installation has an overall power factor of 0.88 lagging. A 200-kVA synchronous motor is added to the system and is run at zero power factor and slightly reduced rating to provide power-factor correction only. The total load is then 350 kVA at unity power factor.

If it was decided to run the synchronous machine at its rated kVA and use it as a source of mechanical power, what gross mechanical power would it produce with the overall power factor of the installation at:

(a) unity;
(b) reduced to 0.96 lagging?

Note: in part (b) I_{PS} and I_{QS} are both unknowns but are the quadrature components of the rated synchronous machine current I_S,

i.e. $I_{PS} = \sqrt{I_s^2 - I_{QS}^2}$.

T5.12. Convert eqn (5.12) to an expression giving the total torque in *per unit* for any frequency k times the base frequency at which E_f and X_s are specified. It will be the same form as eqn (5.20) except for a factor $-\sin(\delta - \alpha)$ multiplying the first term, see also Tutorial Example T7.6.

Refer to the above equation and the preamble before Example 5.15, using the same values of R_a and X_s as in this example, to determine the required excitation (E_f) in *per unit* when operating at rated frequency, rated voltage and current and a power factor of 0.8 leading. If the load is increased slowly, what maximum value of electromagnetic torque with the corresponding current will be reached before pulling out of synchronism?

What value of terminal voltage would be required, with the excitation maintained, to sustain this maximum torque when (a) k = 0.3, and (b) k = 0.1?

T5.13. Solve Tutorial Examples T5.3 to T5.8 using the current-source equivalent circuit of Figure 5.3c and the sine or cosine rules, where appropriate.

T5.14. A salient-pole synchronous motor has $V = 1$ *per unit*, $X_d = 0.9$ *per unit* and $X_q = 0.6$ *per unit*. Neglect R_a. The current is 1 *per unit* at power factor 0.8 leading. Calculate the required excitation, power and torque in *per unit* and also the components I_d and I_q. What is the maximum torque?

Note: Refer to the phasor diagram of Figure 5.4a, the relationships $I_q = I_a \cos(\delta - \varphi)$; $I_d = I_a \sin(\delta - \varphi)$ and eqn (5.22) to solve for δ; eqn (5.23) to solve for E_f.

T5.15. The same synchronous motor as in the previous question has its e.m.f. E_f reduced to 1 *per unit*, the power remaining the same. What will now be the load angle, power factor, I_a, I_d and I_q? What will be the maximum torque and its reduction from the previous maximum value with the higher excitation?

Note: with the data given, for a specified power of 0.8, an explicit solution is not possible but using eqn (5.24) a simple iteration will quickly produce the value of δ. Eqns (5.22) and (5.23) then solve for the currents.

T5.16. Use eqns (5.15), (5.16) or (5.17) as appropriate to check any, or all, of the various torque calculations for uniform air-gap induction and synchronous machines.

T5.17. A permanent-magnet synchronous machine has an open-circuit e.m.f. equal, at rated speed, to the terminal voltage. The value of X_d is 0.8 *per* *unit* and $X_q = 1$ *per unit.* Note that the direct-axis flux has to negotiate the low-permeability permanent-magnet material whereas for the quadrature axis, the flux has a soft-iron path in the rotor structure, hence $X_q > X_d$. Refer to Example 5.22 and show that maximum torque occurs at an angle greater than 90° and calculate its value in *per unit.* What is the load angle, power factor and current at 1 *per-unit* torque. A short iteration procedure with eqn (5.24) will be necessary to obtain δ and eqns (5.22) and (5.23) will yield the current components etc.

Chapter 6. Transient behaviour; closed loop control

T6.1. Consider an idealised thermal system and a small change of temperature rise $\Delta\theta$ taking place in time Δt. It has a heat source of P watts, a heat storage capacity of $M \cdot S$ joules per °C where M = mass in kg and S = specific heat in joules per kg per °C, where M = mass in kg and S = specific heat in joules per kg per °C change, and radiates heat at the rate of $K.\theta$ *watts per second, where* K has the units of watts/°C and θ is the temperature rise above the surroundings. Balance the heat generated in time Δt against the heat stored and radiated and hence show that the temperature rises exponentially in accordance with eqn (6.2b).

Assuming that an electrical machine can be so represented, calculate the maximum and minimum temperature rises occurring eventually when it has been subjected repeatedly to the following duty cycle for an appropriate period:

(a) full-load ON for a time equal to the thermal time constant, τ;
(b) load reduced to zero for a time equal to one-half of the thermal time constant.

Express the temperature rises in terms of the final temperature rise θ_m which would occur if the full load were to be left on indefinitely.

Starting from cold, what would be the temperature rises at times τ, 1.5τ and 2.5τ? What is the r.m.s. value of the duty cycle in terms of the full-load power P?

T6.2. A 230-V, 50-hp, 935-rev/min, separately excited d.c. motor has a rated armature current of 176 A. The armature-circuit resistance is 0.065 Ω. If a starting resistance of 0.75 Ω is connected in series, what will be the initial starting current and torque exerted with rated field current maintained?

With a Coulomb-friction load of 271.3 Nm and a total coupled inertia of 3 58 kgm^2, what will be the final balancing speed and the time taken to reach 98% of this speed?

If the motor were to be supplied from a constant-current source instead, set at a value corresponding to the initial starting current above, what would then be the time to full speed from rest? Explain fully the difference between the two time periods calculated.

T6.3. The machine having the same data as in Examples 3.2 and 3.3 is to be started from a 220-V d.c. supply which is first connected directly across the field winding. The series winding is not in circuit. An external armature resistor is used to limit the maximum current to 80 A and is left in circuit while the motor runs against a constant torque T_m corresponding to this field flux and rated armature current, 40 A.

Calculate the expression for electromagnetic torque T_e as the speed changes. Develop the differential equation for speed when the total coupled inertia is 13.5 kgm^2.

What are the electromechanical time-constant and the final steady-state speed in rev/min? Consider the next step of the starting period when it can be assumed that on current falling to 60 A, the circuit resistance is reduced so that the current again increases to 80 A. What will be the e.m.f. at this instant and the value of the total circuit resistance required to produce this result?

Set up the new transient speed equation and find:

(a) the new electromechanical time-constant;
(b) the next balancing speed if the resistance is left in circuit;
(c) the time for the speed to rise to 400 rev/min if, on reaching the speed in (b), the mechanical torque is suddenly reduced to zero.

Neglect the circuit inductance.

T6.4. The machine of the last Tutorial Example has an armature rating of 220 V, 40 A at 500 rev/min. The armature-circuit resistance is 0.25 Ω and the rotational inertia is 13.5 kgm^2. Calculate the rated flux in Nm/A and the stored-energy constant, [stored energy $\frac{1}{2}J\omega^2_{m\,(base)}$/rated power, see eqn (8.9)]. $\omega_{m\,(base)}$ is 1 *per-unit* machine speed in radians/sec at full voltage and full flux.

An estimate of the armature-circuit inductance for a d.c. machine in *per unit*, eqn (8.10), is given by:

$$L\ per\ unit = \frac{K}{2\pi \times f \text{ at base speed}}$$

where K = 0.6 for a non-compensated machine and K = 0.25 for a fully compensated machine. f at base speed is the conductor frequency (pn_{base}) at base speed. Make an estimate of the inductance for this 4-pole compensated machine and hence calculate the electrical and electromechanical time-constants to see whether the speed response is likely to be oscillatory. Check the time-constants using actual and *per-unit* expressions.

T6.5. A 250-V, 500-rev/min d.c. separately excited motor has an armature resistance of 0.13 Ω and takes an armature current of 60 A when delivering full-load power at rated flux, which is maintained constant throughout. Calculate the speed at which a braking torque equal to the full-load torque will be developed when:

(a) regeneratively braking at normal terminal voltage;
(b) plugging braking but with an extra resistor to limit the initial torque on changeover to 3 *per unit*;
(c) dynamically braking with an extra resistor to limit the peak current to 2 *per unit.*

For cases (b) and (c), write down the torque balance-equation and hence find the maximum total inertia in each case which could be reduced to zero speed in 2 sec, the full-load friction torque being coupled throughout.

For case (b), what is the total time to reverse to 95% of the final balancing speed with the extra resistance in circuit? Note the reversal of friction torque with rotation reversal.

T6.6. For a 3-phase induction motor, express the rotor copper loss as a function of electromagnetic torque and slip. Nothing that $T_e = J\,d\omega_m/dt = -J\omega_s\,ds/dt$ (see Example 6.16), integrate the expression for copper loss over a range of slip and show that the energy dissipated in rotor heat $= \frac{1}{2}J\omega_s^2\,(s_1^2 - s_2^2)$. Hence show that for acceleration from zero to synchronous speed with zero load and for deceleration from synchronous speed to zero under dynamic braking conditions, the energy loss in rotor heating is equal to the stored mechanical energy in the rotating mass at synchronous speed. Show also that when plugging from synchronous speed to zero speed, the rotor heat energy is equal to three times this kinetic energy at synchronous speed.

Now consider the d.c. separately excited machine under similar conditions. ω_s is replaced by the no-load speed $\omega_0 = V/k_\phi$, and $\omega_m = \omega_0(1 - s)$ where $s = I_a R/(k_\phi \omega_0)$. Noting also that I_a = armature copper loss/$I_a R$ where $I_a R = s k_\phi \omega_0$, use the d.c. machine equations where appropriate to show that the rotor energy loss is the same in terms of the kinetic energy, as for the induction machine for the same transient conditions.

T6.7. A 300-V, d.c. series motor driven at 500 rev/min as a separately excited generator, with the armature loaded to give the same current as in the field winding, gave the following characteristic:

Terminal voltage	142	224	273	305	318 V
Field current	15	25	35	45	55 A

The field and armature resistances are each 0.15 Ω.

The motor is to be braked from normal motoring speed where it is developing an electromagnetic torque of 300 Nm when supplied from 300 V. An external resistor is to be inserted to limit the initial current to 55 A. It can be assumed that on changeover, the response of flux following the increase of current is completed before the speed changes significantly so that the e.m.f. rises also.

(a) Determine the speed when motoring at 300 Nm, from 300 V, and the corresponding values of current and flux (k_ϕ).
(b) Calculate the required resistor values for plugging and for dynamic braking.
(c) Estimate the times in (b) for the speed to fall to such a level that the braking torque is equal in magnitude to the full-load torque, the load torque being still coupled and the inertia is 10 kgm^2.

Note: the motor curves must be calculated in order to solve part (a) but on braking only two specific points are required and the average braking torque can be used. Nevertheless, it may be a good idea to sketch the braking curves to clarify the method.

T6.8. Using the data of Example 4.10 and the associated figure, calculate the time to accelerate from zero speed to 1400 rev/min at full voltage and with the natural rotor resistance. Take the total drive inertia as 0.05 kgm^2 as in Example 6.7, p. 207 and take points on the curves at speed intervals of 200 rev/min.

T6.9. A 10-pole, 50-Hz induction motor drives a d.c. pulse generator through a flywheel coupled between the two machines. The pulse requires a generator input torque of 2713 Nm for 4 sec and the combined inertia of machines and flywheel is 1686 kgm^2. The no-load speed of the motor is 597 rev/min and the speed at rated torque of 977 Nm is 576 rev/min. Assuming the fall of speed with torque is linear, what is the peak motor-torque at the end of the load pulse?

T6.10. A mine winder requires the following duty cycle for its d.c. motor:

Time period	Condition	Torque required
0–20 sec	Constant acceleration up to 45 rev/min	2.712×10^5 Nm
20–50 sec	Constant speed of 45 rev/min	1.356×10^5 Nm
50–70 sec	Regenerative braking at constant torque	-0.678×10^5 Nm
70–90 sec	Speed now zero; rest period	

Plot the torque and power duty-cycles, find the average power throughout the cycle and draw a line at the appropriate height representing this average power. The area above (and below) this line represents the magnitude of the energy pulsation in watt-seconds. If this is provided by the stored energy of the flywheel and inertia of the motor-generator set supplying the winder motor, then the m.g. set motor will be shielded from the peak. The m.g. set speed will have to fall, under control, from ω_1 to ω_2 to release this energy of magnitude $\frac{1}{2}J\omega_1^2 - \frac{1}{2}J\omega_2^2$ watt-seconds. By equating, the value of J follows. Determine this inertia if the m.g. set motor speed falls from 740 to 650 rev/min.

What is the peak power when motoring and when regenerating? Neglect machine losses.

T6.11. Referring to Figure 6.8, p. 224 showing the duty cycle for a mine hoist, the numerical values are as follows:

Torque (Nm × 10^6)	$T_1 = 0.6$, $T_2 = 0.55$, $T_3 = 0.3$, $T_4 = -0.1$, $T_5 = -0.35$, $T_6 = -0.4$
Time (seconds)	$t_1 = 10$, $t_2 = 80$, $t_3 = 10$, $t_4 = 20$.

The motor speed reaches a maximum of 40 rev/min. Determine the average and the r.m.s. torque – refer to eqn (6.16). Calculate the various power ordinates and from the power/time curve obtain the average and the r.m.s. power:

(a) using the actual motor speed,
(b) assuming the motor speed is constant at the maximum value.

T6.12. A speed control system for an electrical drive which drives a roll for paper uptake in a paper-manufacturing plant has an analogue input voltage v_i and a feedback voltage v_f. The error voltage $v_e = (v_i - v_f)$ is the input to a current source amplifier of gain K. The amplifier's current output i is the input to a motor which has a transfer function

$$\frac{T_e(p)}{i(p)} = \frac{1}{(1 + 0.1p)}$$

where T_e is the motor torque. The load has a transfer function

$$\frac{\omega_m(p)}{T_e(p)} = \frac{1}{10 + Jp}$$

where ω_m is the shaft speed. The speed sensor output is $v_f = 20\,\omega_m$.

For an empty roll, inertia $J = 50\,\text{kgm}^2$. Show that the open-loop transfer function in this case is:

$$\frac{v_f(p)}{v_i(p)} = \frac{4K}{(p+10)(p+0.2)}$$

(a) Find the value of K required to give the system a damping factor of 1.0 and estimate the settling time to within 2% of the final speed when a step change is applied to v_i.

(b) When the roll is full, $J = 500\,\text{kgm}^2$. Find the damping factor and the settling time to 2% in this case.

Chapter 7. Power-electronic/electrical machine drives

T7.1. A d.c. permanent-magnet motor is supplied from a 50-V source through a fixed-frequency chopper circuit. At normal motor rating the armature current is 30 A and the speed is 1000 rev/min. The armature resistance is 0.2 Ω. If the current pulsations can be taken as relatively small so that the mean current can be used in calculations, what is the required duty-cycle ratio of the chopper if the motor is to operate at a mean torque corresponding to the full rating and at a speed of 400 rev/min?

T7.2. A battery-driven vehicle is powered by a d.c. series motor. The time-constant of armature and field together is 0.2 sec, the resistance being 0.1 Ω. At a speed of 1000 rev/min, the mean generated volts/field A over the operating range of current is 0.9. A fixed frequency, 200-Hz chopper is used to control the speed and when this is 1000 rev/min, the mark: space ratio is 3:2. The battery voltage is 200 V. Find the maximum and minimum values of the current pulsation and hence determine the mean torque and power output if the mechanical losses are 1000 W.

T7.3. Referring to the derivation of eqn (B.3) in Appendix B, show that the voltage loss due to overlap is also equal to $E_{do} \sin^2 \mu/2$ when the alternative method is used for averaging

$$\int_0^{\mu} \frac{(e_b - e_a)}{2}\, d(\omega t)$$

over the pulse period $2\pi/m$. The instantaneous expressions for the e.m.f.s are: $e_a = E_p \cos(\omega t - \pi/m)$ and $e_b = E_p \cos(\omega t + \pi/m)$.

A 3-phase diode bridge is to provide an output of 220 V, 50 A d.c. from a 3-phase, 440-V supply and a suitable transformer. Allow for a 30° overlap angle, a 1-V drop for each diode and 4 V for the transformer resistance losses, referred to the d.c. terminals. If the transformer flux can be expressed as 3 r.m.s. volts/turn find a suitable number of primary and secondary turns for a delta/star transformer and its kVA rating. Note that all voltage drops must be added to the required terminal voltage to obtain the necessary value of E_{do}.

T7.4. On a thyristor converter/d.c. machine system, the converter mean voltage falls from 500 V on no load to 460 V when delivering 100 A, there being no gate firing delay. The d.c. motor has an armature resistance of 0.3 Ω. Determine the required firing-delay angle α under the following conditions:

(a) As a motor taking 50 A and excited to produce an e.m.f. E_m = 400 V.
(b) As a motor at the normal rating, 460 V, 100 A, 1000 rev/min. Calculate $k_{\phi R}$.
(c) Regenerating at rated terminal-voltage and current.
(d) Motoring at half-speed, the total torque being proportional to (speed)^2 and the flux being set at rated value.

If speed increase is required by field weakening, what permissible torque can the motor deliver at 1250 rev/min without exceeding rated current and what would then be the flux in *per unit*? The firing-delay angle is 0°.

T7.5. Referring to Figures 8.4 and 8.5, the induction machine equivalent circuit parameters at 50 Hz are as follows:

$$R_1 = 6.7\,\Omega, \quad R_2' = 7.7\,\Omega, \quad L_{11} = L_{22}' = 0.436\,\text{H}, \quad M' = 0.41\,\text{H},$$

$$x_1 = x_2' = \omega(L_{11} - M') = 8.168\,\Omega.$$

Values are per phase, with all phases excited. The motor is 4-pole, 50-Hz and star connected, the rated line voltage being 220 V.

For reference purposes and to relate the normal values to Figures 8.4 and 8.5, first use the exact equivalent circuit to work out the rated input current, power factor, rotor current and electromagnetic torque at a slip of 0.06. Also estimate the maximum torque from the approximate circuit and express this in *per-unit* based on the rated torque. This will help to explain the behaviour on Figure 8.5 when the motor pulls out of synchronism.

The motor is now to be supplied from a quasi-square-wave, voltage-source inverter as shown on Figure 7.13, with a d.c. link voltage so adjusted that the r.m.s. value of the fundamental of phase-voltage waveform is the same as its rated sinusoidal voltage. Determine the d.c. link voltage required.

Using the method described in Section 7.4 culminating in eqn (7.8), estimate the value of the 6th-harmonic torque pulsation as a function of time. Compare this with the value measured from Figure 8.4 which is based on 1 per-unit torque = 1.9 Nm and 1 per-unit current = 1.3 A. (The computed peak value of pulsation is ±0.327 Nm.)

T7.6. The cylindrical-rotor synchronous machine of Example 5.15 (and Tutorial Example T5.12) is provided with load-angle (δ) control through a position detector. By differentiating the torque expression of eqn (5.13) determine the angle at which δ must be set to produce maximum torque:

(a) at rated voltage and frequency;
(b) at 0.3 × rated frequency;
(c) at 0.1 × rated frequency.

Use eqn (5.13) modified to *per-unit* terms and expressed for any frequency k × the base frequency at which E_f and X_s are specified.

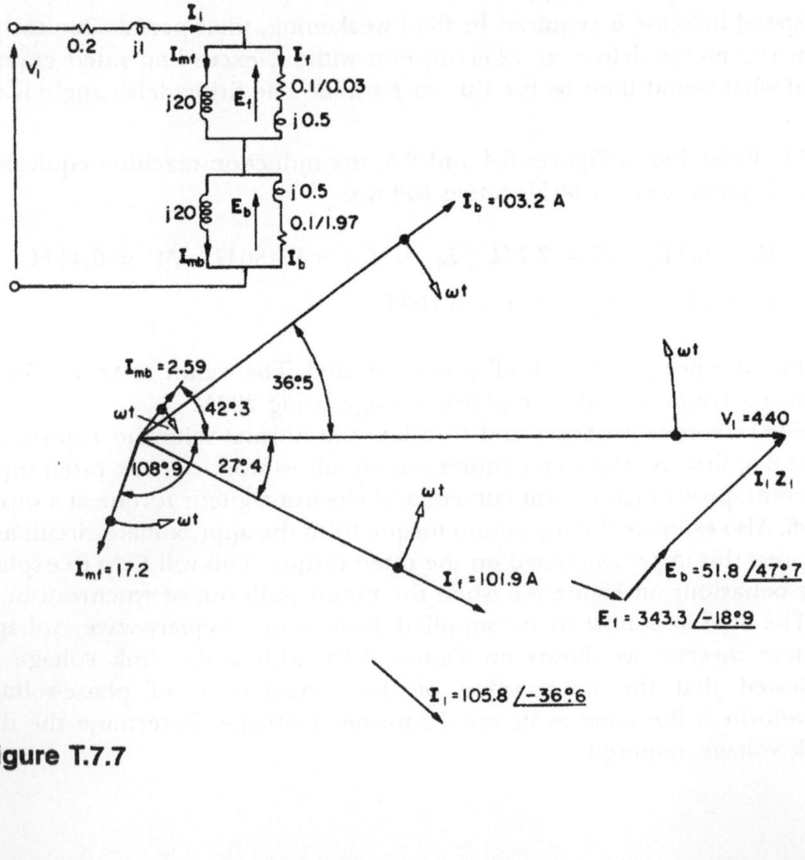

Figure T.7.7

Check the value of $\alpha = \tan^{-1} R_a/kX_s$ at each frequency and show that $\delta - \alpha$ is $-90°$ at maximum torque for parts (a), (b) and (c).

Check also, using the answers to T5.12 for V and E_f, that the maximum torque for (b) and (c) is the same as for (a) if the excitation is maintained at the part (a) setting.

T7.7. For this problem, reference back will be necessary to Example 4.21, Tutorial Example T4.20, eqn (7.7) in Section 7.4 and the general approach of Tutorial Example T7.5. Use the answers obtained in Tutorial Example T4.20 shown on Figure T.7.7 on p. 358, to calculate the average torque and the pulsating torque for the single-phase induction-motor operation. The average torque is due to the reaction of $\mathbf{I}_f$ with $\mathbf{I}_{mf}$, minus the reaction of $\mathbf{I}_b$ and $\mathbf{I}_{mb}$. The pulsating torque is due to the reaction of $\mathbf{I}_f$ with $\mathbf{I}_{mb}$ plus the reaction of $\mathbf{I}_b$ with $\mathbf{I}_{mf}$. Compare the average-torque answer with that obtained in T4.20 and express the pulsating torque as a fraction of the average torque.

T7.8. A permanent-magnet d.c. motor has an armature resistance of 0.5 Ω. It operates from a 200-V d.c. supply via a bipolar transistor chopper and is required to drive a fan which has a load torque proportional to (speed)^2. The fan requires a torque of 9 Nm at 1000 rev/min.

Find the range of modulation factor 'δ' necessary for speed control over the range from 1000 to 2000 rev/min. $k_\phi = 0.6$ Nm/A.

Chapter 8. Mathematical and computer simulation of machine drives

This chapter, concerned with machine modelling and simulation really needs a computer to give tutorial practice. Equations are given in some cases for fairly substantial programs so that to check Figures 8.4 and 8.5 for example would require much time for program development. The data are available for the purpose however and once the program is working successfully relatively small changes are needed to explore a variety of problems. Note that for Figures 8.4 and 8.5, Tutorial Example T7.5 gives the machine data, the inertia being artificially lowered to $J = 0.009$ kgm^2 so that acceleration could be speeded up and the various transients observed on the graph plot over a few cycles. A d–q simulation of these two figures should give almost identical results using rather longer time increments. For the inverter waveform this could mean an increase from $h = 10\,\mu$s to $100\,\mu$s say, and for a sinusoidal waveform, from $h = 100\,\mu$s to 0.5 ms. Apart from such major computational exercises, the following suggested problems give an opportunity to think around the ideas presented in this chapter.

T8.1. The steps in the development of eqn (8.6) have been given in broad outline but it would be a useful exercise to make the substitutions suggested and work through the matrix multiplications in detail to obtain $\mathbf{Z}' = \mathbf{C}_t \cdot \mathbf{Z} \cdot \mathbf{C}$.

T8.2. A single-phase a.c. load is connected through a bilateral thyristor-type semiconductor switch. It conducts if the voltage across it is 1 V or more, in either direction, and the time angle ($\theta = \omega t$) of the sine-wave supply voltage $e = E \sin \omega t$ is greater than the firing-delay angle α measured from voltage zero for the positive half wave, and greater than ($\pi + \alpha$) for the negative half wave. Draw up a logic-check diagram, similar to that of Example 8.3, which will check the conduction condition.

Consider how this could be applied to a star-connected 3-phase source supplying a 3-phase star-connected induction motor which is voltage controlled for speed variation.

T8.3. From the expressions for direct- and quadrature axis-flux linkages, prove that the final equations for voltage and torque given in Appendix C in terms of flux linkages are the same as those derived from the impedance matrix for the rotor reference frame.

T8.4. Calculate *vdss* and *vqss* at time $t = 0$ for a 3-phase stator voltage when $v_A = 100 \sin \omega t$.

T8.5. For the same voltage as in T8.4 and at the same instant, calculate *vdsr* and *vqsr* assuming that at $t = 0$, (a) $\theta = 0$ and (b) $\theta = \pi/6$.

Answers to tutorial examples

T2.1. Primary 400 turns; Secondary 14 turns per section.

T2.2. (a) $N_1 = 1800$, $N_2 = 312$, $N_3 = 12$. (b) $I_1 = 9\,\text{A}$, $\cos\varphi = 0.845$, 1030 kVA, 870 kW.

T2.3. $\cos\varphi = 0.986$ lead; 0.101 *per unit* at 80°.5 lag.

T2.4. (a) 96 V, 25.1 V, –52.5 V. (b) 98.2% and 98.559% at u.p.f. (c) 98.56%. (d) $I_0 = 0.106 - j0.28\,\text{A}$. (e) 14.78 A at 0.79 p.f., $I_2 = 145.2\,\text{A}$.

T2.5. Answers are on attached figure.

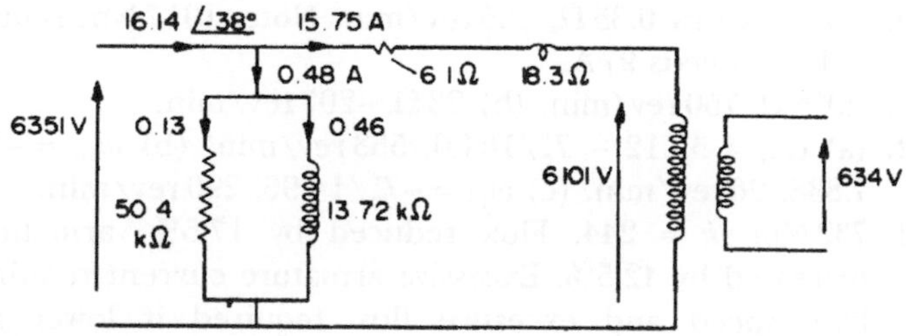

Figure T.2.5

T2.6. 95.71%; 96.53%.

T2.7. (a) $\mathbf{I}_A = \mathbf{I}_B \times 0.827\ \underline{/112°.7}$, transformer A is working with reverse power flow. (b) $\mathbf{I}_A = \mathbf{I}_B \times 0.7\ \underline{/0°}$.

T2.8. $\mathbf{I}_{total} = 1640\ \underline{/-38°}$; circulating current = $178\ \underline{/-79°}$. A terminal voltage increased from 375 line V to 392.5 line V. Note: $\mathbf{Z}_{ASC} = 0.0032 + j0.0181$, $\mathbf{Z}_{BSC} = 0.0016 + j0.00716$ ohms per phase.

T2.9. 95.3%.

T3.1. $k_{\phi R} = 9.068\,\text{Nm/A}$, $T_{loss} = 24.7\,\text{Nm}$. (a) $I_f = 0.9$, 55.6 Ω. (b) 525 rev/min. (c) 553 rev/min. (d) ≏ 160 Ω. (e) 9.64 Ω, 5.2 kW (7 hp).

T3.2. 296 rev/min; flux and torque reduced to 29.6% of rated value.

T3.3. (a) 87.64 A. (b) 802.4 Nm. (c) 38.5 Nm. ≏ 990 Ω. 37.1 kW.

T3.4. 5.946 Nm/A, 1308 Nm, 102.7 kW; or 0.588, 0.646, and 0.973 *per unit.*

T3.5. 5.84 Nm; 32.7%; 51.4 A.

T3.6. (a) 151.2 Nm, 23 kW, 87.5%. (b) 139.9 Nm. (c) 32.35 A, 0.687 *per unit.*

T3.7. (a) ≏ 670 Ω. (b) 2.52 Ω. (c) 6.85 Ω, 46.5 A.

T3.8. (a) 1.77 Nm. (b) 0.53 Nm. The mode is plugging.

T3.9. ω_m (rad/s)/T_e (Nm) points −88.5/−36.6, −32.3/309. Remaining answers on attached figure. Mode gives limited speed on zero load and has some regenerative capability; see Figure E.3.18(b).

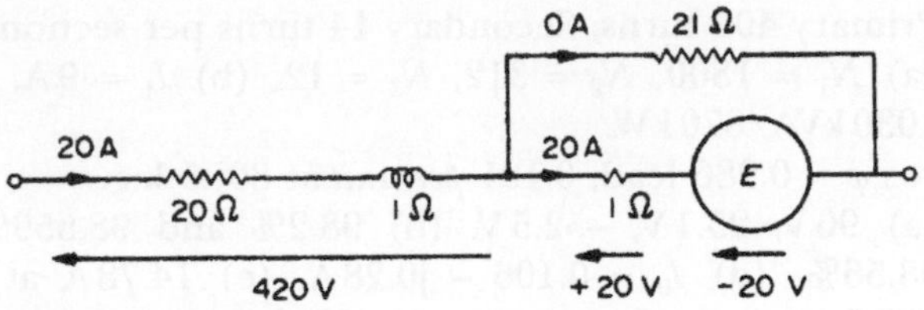

Figure T.3.9

T3.10. 1.04 Ω; extra 0.35 Ω; 865 rev/min. Note: 1017 Nm requires 214 A; 271 Nm needs 97 A.

T3.11. (a) 6 Ω, 750 rev/min. (b) 23 Ω, −207 rev/min.

T3.12. (a) $\omega_m = 55.12 - T_e/164.6$, 553 rev/min. (b) $\omega_m = -55.12 - T_e/7.838$, 26 rev/min. (c) $\omega_m = -T_e/14.96$, 290 rev/min.

T3.13. 732 Nm, k = 244. Flux reduced by 17.5%, armature current increased by 42.5%. Excessive armature current required beyond base speed and excessive flux required if lower speeds are attempted by strengthening the flux (see also T3.15).

T3.14. 0.95 *per unit*; $T_m = 0.3 + 0.737\omega_m$. (a) 0.928, 0.885, 0.72. (b) 0.769, 0.666, 1.54. (c) 1.034, 1.058, 0.575.

T3.15. (a) 0.965, 0.628. (b) (i) 1.725, 0.65. (b) (ii) 0.33, 1.694.

T4.1. (a) 14 A at 0.85 p.f. lag. (b) 51.5 Nm. (c) 7.54 kW. (d) 82.6%· 1265 rev/min.

T4.2. $R_1 = R_2' = 0.53\,\Omega$; $x_1 = x_2' = 1.28\,\Omega$; $R_m = 215\,\Omega$; $X_m = 32.1\,\Omega$. (a) 22 A at 0.86 p.f. lag, 18.1 A, 84.5 Nm, 12.1 kW, 84%. (b) 85.1 Nm, 4.5/1.

T4.3. (a) 21.5 A at 0.837 p.f. lead ('exact' = 19.9 A at 0.821 p.f. lead). (b) 19.6 A (18.7 A). (c) 96.9 Nm (88.9 Nm). (d) 16.4 kW (15.1 kW). (e) 83.5% (82.4%). $L_{11} = L_{22}' = 106$ mH, $M' = 102$ mH. See flow diagram on Figure T.4.3. Actual directions shown, values in kW.

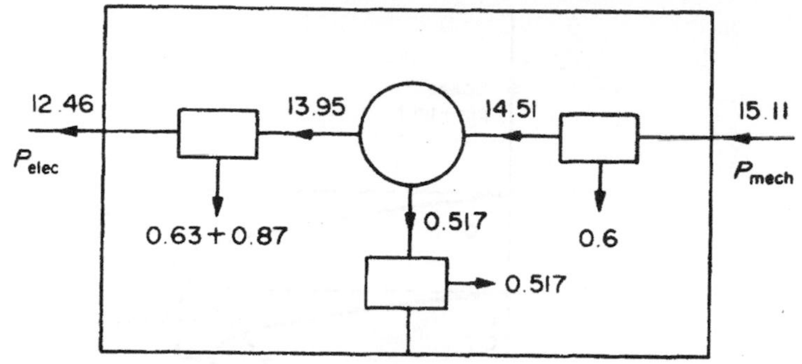

FIG. T4.3.

Figure T.4.3

T4.4. (a) 33 A. (b) 11.1 A. (c) 33.4 A. (d) 1.79 kW; plugging mode. (e) 0.97 Ω.

T4.5. $T_{start} = 0.298\ T_{fl}$; $T_{max} = 1.405\ T_{fl}$.

T4.6. Answers on Figure T.4.6. Torques are expressed in terms of normal peak torque.

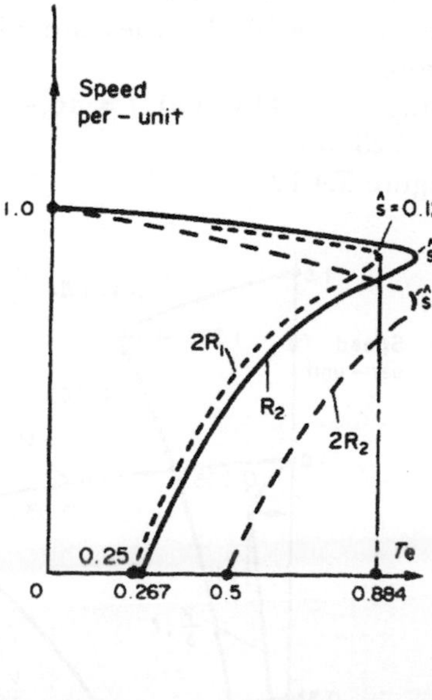

Figure T.4.6

T4.7. $T_s = I^2 \times s_{fl} \times T_{fl}$ (all in *per unit*). Tapping at 0.481. Motor current = 2.89 *p.u.* and transformer input current = 1.39 *p.u.*

T4.8. 1425 rev/min. (a) Reduction to 71.7% voltage (can be estimated by drawing curves). (b) Extra $3R_2$. (See also T4.9 and T4.22.)

T4.9. 6.32 Nm at 460 rev/min requires 121.8 V and 2.73 A giving 357 W rotor copper loss; or, extra 465 Ω (referred) and 0.51 A giving same

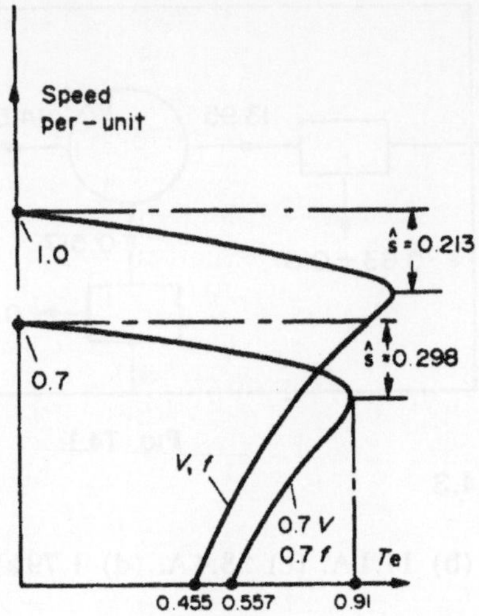

Figure T.4.10

rotor loss (same torque), but much reduced stator copper loss at the lower current. See also T4.22.

T4.10. Answers on Figure T.4.10. Torques are expressed in terms of normal peak torque.

T4.11. k = 0.328 giving V = 144 V and f = 16.4 Hz. Torque = 377 Nm compared with 223 Nm.

T4.12. Answers on Figure T.4.12.

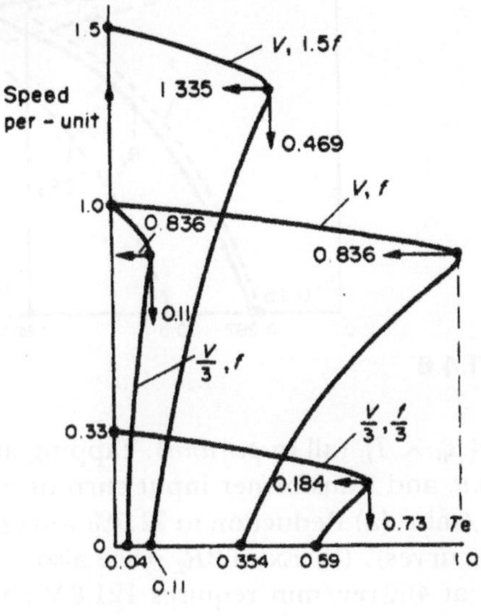

Figure T.4.12

T4.13. 266 Nm; 1.5 Hz, 22.6 line V; 39 Hz, 314 line V.
T4.14. (a) 13.3 Hz. (b) 28 Hz (f_2 = 3 Hz).
T4.15. (a) I_2' = 26.24 A, T_{max} = 83.7 Nm. (b) (i) f = 0.7579 Hz, I_1 = 39.08 A, T_{max} = 867.4 Nm. (b) (ii) $\hat{f}$ = 2.076 Hz, I_1 = 42.67 A, T_{max} = 316.6 Nm. V_1 = 111.7 line V at starting. At 300 rev/min, $\hat{f}$ = 12.076 Hz and V_1 = 357 line V, so less than available supply V.
T4.16. Rated torque = 1446 Nm; rated current = 123 A. (a) T_e = 0.19 *p.u.* I_2' = 3.54 *p.u.* (b) T_e = 1.35 *p.u.* I_2' = 6.6 *p.u.* Note E_1 at synchronous speed = 621.9 V.
T4.17. X_m = 10 Ω, T_e = 2722 Nm, I_2' = 161 A. (a) I_1 = 177 A. (b) I_{dc} = 217 A, V_{dc} = 43.5 V. Max. torque = 4340 Nm (1.59 ×) at 346 rev/min.
T4.18. T_s = 0.802 *p.u.* or 102 Nm. Power = 0.842 *p.u.* or 16.9 kW for T_e = 0.896 *p.u.*
T4.19. $\mathbf{V}_{A+} = (1/\sqrt{3})433.3073\underline{/-30°}$, $\mathbf{V}_{A-} = (1/\sqrt{3})\,6.6927\underline{/+30°}$,
$\mathbf{Z}_+ = 3.576\underline{/16°.258}$
$\mathbf{Z}_- = 1.011\underline{/81°.427}$, I_A = 73.8 A, I_B = 67.8 A, I_C = 68.4 A.
T_e = 622.4 Nm, negative-sequence torque negligible.
T4.20. Z_f = 3.091 + j0.9903, Z_b = 0.048 + j0.488, Z_{input} = 3.339 + j2.478 = $4.16\underline{/36°.6}$.
T_e = 434.1 Nm. Remaining answers shown on Figure T.7.7 for Tutorial Example T7.7.

T4.21.

	Single phase	Two phase	Three phase
Output kW	32.1	31.6	47.9
Power factor	0.8	0.908	0.908
Efficiency %	81.6	87.9	88.8
Stator current A	105.8	77.9	77.9

Iron loss 1.94 kW. Mechanical loss 1 kW. Note that if the stator had been fully wound for 2-phase operation, the output would have been nearer to that for 3-phase operation.
T4.22. V_3' = 171.8 V, I_2' = 0.758 A, 390 VA, Rotor loss = 27.6. W. Slip-power recovered = 329.4 W. VA rating from rated I_2' (2.1 A) and max. V_3' = 440 V is 2770 VA. Phasor diagrams for comparison on Figure T.4.22.
T4.23. τ = 1.789 m.

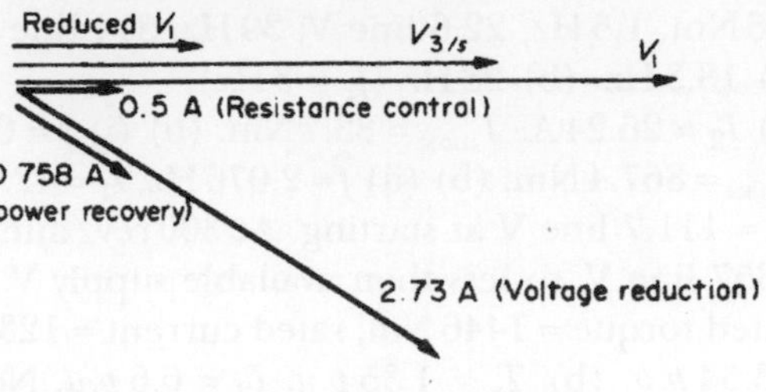

Figure T.4.22

T5.1. X_{su} = 19.1 Ω. k_{fs} = 76.6 line V/A, I_f = 77.4 A. Max. T_e = 10 620 Nm = 2.22 *p.u.*

T5.2. (5.5) – 18°.7, 31 830 Nm, 229 A at 0.95 lead, 6910 line V, 2010 kW. (5.6) 3750 line V, 95.7 A at 0.98 lead, 550 kW.

T5.3. (a) I_{aR} = 43.74 A. (b) $\mathbf{E}_f = 4888\underline{/-38°.9}$ (phase V). (c) 4775 Nm. (d) 7620 Nm. (e) For 9550 Nm, E_f = 10.61 kV (line). (f) δ = –30°, I_a = 48.65 A at 0.898 lead.

T5.4. –δ reduced to 29°.3 from 47°.7. I_a reduced to 96 A and power-factor to 0.937 lead. On no load the current becomes 57 A at zero leading-power-factor (sync. capacitor).

T5.5. 52.49 A, 10.72 kV/phase, 8594 Nm, 900 kW. I_a becomes 59.1 A and E_f = 6.05 kV. (1.125 *p.u.* and 0.56 × $E_{f(rated)}$.) 900 kW at 0.8 p.f. lead.

T5.6. $\mathbf{E}_f = 5173\underline{/-42°.6}$, I_{aR} = 87.48 A. 1.5E_f gives δ = –26°.8 and I_a = 117.1 A at 0.747 p.f. lead. 0.5 E_f requires –sin δ > 1 to sustain load, so pulls out of step.

T5.7. $\mathbf{E}_f = 385.8\underline{/-27°.8}$ Normal pull-out torque = 280.8 Nm so E_f must be increased by 6.8%.

T5.8. 61.5 A at 0.91 p.f. lead –sin δ = 0.4775 so would still be <1 if V (or E_f) was halved. For induction motor $T \alpha V^2$ so much more sensitive to V reduction.

T5.10. I_{QS} = 24.3 A, I_{PS} = 9.88 A, Power = 53.65 kW, E_f = 1.66 *p.u.* (3160 V/phase).

T5.11. At zero load, kVAr = 188.6. (a) 66.7 kW. (b) I_{QS} = 5.05 A, I_{PS} = 34.63 A. Power = 198 kW.

T5.12. $\mathbf{E}_f = 1.766\underline{/-28°}$; and at T_{max} = 1.61, δ = –87°.13, and $\mathbf{I}_a = 1.985\underline{/-24°.47}$ A. (a) 0.363. (b) 0.18. See also T7.6.

T5.14. E_f = 1.692, Power = Torque in *per unit* = 0.8. I_d = –0.832, I_q = 0.554, T_{max} = 1.955 *p.u.* = 2.44 × T_{fl} at δ = –75°.1.

T5.15. δ = –30°.1797, I_a = 0.851 A at 0.94 p.f. lag, I_d = –0.1506, I_q = 0.8379, ψ = –10°.19. T_{max} = 1.223 *p.u.*, reduction of 37.5%; at δ = –68°.5.

T5.17. δ = –100°.73, T_{max} = 1.274. δ = –62°, I_d = –0.663, I_q = 0.883, I_a = 1.104 at 0.906 lagging power factor. Ψ = –36°.9.

T6.1. $P\Delta t = MS\Delta\theta + K\theta\Delta t$, $\tau = MS/K$, $\theta_m = P/K$. $0.813\theta_m$, $0.493\theta_m$. $0.632\theta_m$, $0.383\theta_m$, $0.773\theta_m$. $\sqrt{2/3} \cdot P$.

T6.2. 282 A, 630 Nm, 561 rev/min, 2.35 sec. 0.98 sec to full speed – constant torque maintained till maximum speed reached, so acceleration constant at maximum.

T6.3. Required (total) armature-circuit resistance to limit current to 80 A = 2.75 Ω. k_ϕ from mag. curve = 4.43 Nm/A; T_e = 177.2 Nm; τ_m = 1.89 s; 236.8 rev/min; E = 55 V; 2.063 Ω. (a) 1.418 s. (b) 296 rev/min. (c) 1.24 sec. final speed = 474.2 rev/min.

T6.4. 4.011 Nm/A, 2.31 seconds. 0.0125 H (0.00228 *per unit*). τ_e = 0.05 sec, τ_m = 0.21 sec. τ_m/τ_e = 4.18, greater than 4 so just not oscillatory, but would be if uncompensated.

T6.5. (a) 531.7 rev/min. (b) 2.6 Ω extra, –177.3 rev/min, J = 34 kgm^2, 15 sec total. (c) 1.89 Ω extra, 249.7 rev/min, J = 19.3 kgm^2.

T6.7. (a) 445 rev/min, I_a = 49.5 A, k_ϕ = 6.1 Nm/A. (b) 10.45 Ω and 5 Ω extra. (c) 0.27 sec and 0.11 sec, to speeds of 363 rev/min and 411 rev/min.

T6.8. 2.58 sec.

T6.9. 1680 Nm.

T6.10. Average power = 319.6 kW. Energy area = 16 770 kW sec, J = 24 450 kgm^2. 1278 kW, –318.5 kW.

T6.11. 0.083 × 10^6 Nm, 0.234 × 10^6 Nm. (a) 314.1 kW, 692.9 kW, (b) 349 kW, 981.1 kW.

T6.12. (a) K = 6; t = 0.78 s; (b) ζ = 3.1; t = 15 s.

T7.1. δ = 0.472. (E reduced to 17.6 V from 44 V at rated speed)

T7.2. 125.9 A to 113.9 A. Mean torque = 123.9 Nm and power output = 11.97 kW.

T7.3. E_{do} = 242.3 V, E_p = 253.7 V, E = 103.6 V. N_2 = 35 turns per phase, N_1 = 149 turns per phase. R.M.S. secondary current = 40.8 A, kVA = 12.68.

T7.4. (a) 29°.5. (b) 0°, $k_{\phi R}$ = 4.106. (c) 147°. (d) 62°.3. 328 Nm at k_ϕ = 0.8 *p.u.*

T7.5. I_1 = 1.286 A, at 47°.2 lag; I'_2 = 0.882 A; T_e = 1.907 Nm. Max. T_e = 6.09 Nm = 3.2 *p.u.* D.C. link voltage = 282.2 V. I'_5 = 0.3095 A, I'_7 = 0.158 A, φ_5 = 84°.4, φ_7 = 86°. Torque pulsation = $-0.368 \sin(6\omega t$ + 7.°.3). Note: Approximate circuit overestimates I_{m1}, mostly explaining discrepancy from computed pulsation of 0.327 Nm.

T7.6. (a) δ = –87.°13. (b) δ = –80°.53. (c) δ = –63°.41. Eqn (5.13) in *per unit* = $(-E_f/Z_s^2)\,(VkX_s \sin\delta - VR_a \cos\delta + kE_fR_a)$.

T7.7. M′ = 0.06366 H, p = 4. Forwards torque = $4M'I_fI_{mf} \sin[-27°.4 - (-108°.9)]$. Backwards torque = $4M'I_bI_{mb} \sin[36°.5 - 42°.3]$ Pulsating torque = $4M'[I_fI_{mb} \sin(\varphi_f - \varphi_{mb}) + I_b I_{mf} \sin(\varphi_b - \varphi_{mf})]$

where $\varphi_f - \varphi_{mb} = [-69°.7 + 2\omega t]$ and $\varphi_b - \varphi_{mf} = [145°.4 - 2\omega t]$. Average torque = 440.96 – 6.88 = 434.1. Pulsation = 439.9 sin(2ωt + 26°). Ratio = 1.013, so pulsation slightly bigger than average value.

T7.8. δ varies from 0.35 to 0.78.

T8.2. Answer on Figure T.8.2. For 3-phase motor, each phase would have to be checked in this way, making appropriate corrections for time-

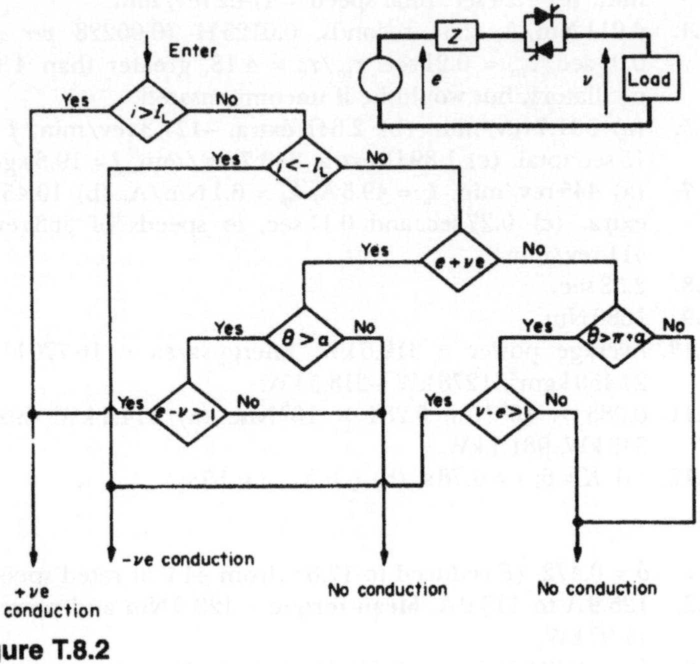

Figure T.8.2

phase shift of waveforms. At least two thyristor-units must be free to conduct for current to flow at all. Voltage equations must then be set up in accordance with the overall conducting pattern.

T8.3. Necessary answers given with problem.

T8.4. $v_A = 0$, $v_B = -86.6$ V, $v_C = 86.6$ V. $vdss = 0$, $vqss = -122.47$ V.

T8.5. (a) $vdsr = 0$ V, $vqsr = 122.47$ V. (b) $vdsr = -106.7$ V, $vqsr = 61.24$ V.

References

1. HINDMARSH, JOHN, *Electrical Machines and Their Applications*, 4th edition, Pergamon Press, 1984.
2. HANCOCK, N. N. *Matrix Analysis of Electrical Machinery*, 2nd edition, Pergamon Press, 1974.
3. HANCOCK, N. N. *Electric Power Utilisation*, Pitman, 1967.
4. HINDMARSH, JOHN. Electrical Machines Teaching: An Integrated Theoretical/Practical Course. *International Journal of Applied Engineering Education*. Now published under the title: *International Journal of Engineering Education* at Tempus Publications, Berliner Tor, 20099 Hamburg. Part I, The Lecture course. Vol. 2, No. 4, 1986, Part II. The Laboratory course, Vol. 3, No. 1, 1987.
5. BROWN, J. E. and JHA, C. S. The starting of a 3-phase Induction Motor connected to a single-phase supply. *Proceedings I.E.E.*, Paper 2860 U; April 1959.
6. ALLEN, T. P. and STEVEN, R. E. *Selected Calculations in Electric Power*, Hodder & Stoughton, 1979.
7. CHALMERS, B. J. and ONBILGIN, G. Analysis of highly saturated synchronous motor with cylindrical excited rotor, *International Journal of Electrical Engineering Education* (I.J.E.E.E. UMIST) **16**(1), January 1979.
8. I E.E. Conference Proceedings on 'Power Electronics and Variable-speed Drives' and 'Electrical Machines and Drives'; both generally published biennially (in alternate years) from 1983 onwards.
9. HUGHES, A. *Electric Motors and Drives*, Butterworth-Heinemann 1993.
10. MURPHY, J. and TURNBULL, F. G., *Power Electronic Control of AC Motors*, Pergamon Press, 1988.
11. HUGHES, A., CORDA, J. and ANDRADE, D. A. An inside look at cage motors with vector control. I.E.E. Int. Conf. on Electrical Machines and Drives, 1993, I.E.E. Conf. Publ. 376, pp. 258–64.
12. NASAR, S. A. and SCOTT, J. B., Time-domain formulation of the dynamics of induction motors, *I.J.E.E.E.*, **11** (1), January 1974.
13. WESTI J. G. W., DC, induction, reluctance and PM motors for electric vehicles, *I.E.E. Power Engineering Journal*, **8** (2), April 1994, pp. 77–88.
14. WILLIAMS, S. and SMITH, I. R., Fast Digital Computation of 3-phase Thyristor Bridge Circuits, *Proceedings I.E.E.*, **120** (7), July 1973

Printed and bound by CPI Group (UK) Ltd, Croydon, CR0 4YY
03/10/2024
01040432-0018